Lial Video Library Workbook with Integrated Review

Christine Verity

Beginning Algebra

Twelfth Edition

Margaret L. Lial
American River College
John Hornsby
University of New Orleans
Terry McGinnis

PEARSON

Boston Columbus Indianapolis New York San Francisco
Amsterdam Cape Town Dubai London Madrid Milan Munich Paris Montreal Toronto
Delhi Mexico City São Paulo Sydney Hong Kong Seoul Singapore Taipei Tokyo

Reproduced by Pearson from electronic files supplied by the author.

ISBN-13: 978-0-13-419708-1
ISBN-10: 0-13-419708-9

1 2 3 4 5 6 RRD-H 19 18 17 16 15

www.pearsonhighered.com

PEARSON

CONTENTS for INTEGRATED REVIEW WORKSHEETS

CONTENTS for LIAL VIDEO LIBRARY WORKBOOK

Chapter 1 The Real Number System

Learning Objectives

Identify the place value of a digit.
Write decimals as fractions or mixed numbers.
Add and subtract negative decimals.
Multiply positive and negative decimals.
Divide a number by a decimal.
Write fractions as equivalent decimals.

Key Terms

Use the vocabulary terms listed below to complete each statement in exercises 1–14.

decimal places	**factor**	**product**	**repeating decimal**
decimals	**decimal point**	**place value**	
quotient	**dividend**	**divisor**	**equivalent**
numerator	**denominator**	**mixed number**	

1. We use ____________________ to show parts of a whole.

2. A ____________________ is assigned to each place to the left or right of the decimal point.

3. The dot that separates the whole number part from the fractional part of a decimal number is called the ____________________.

4. Each number in a multiplication problem is called a ____________________.

5. When multiplying decimal numbers, first multiply the numbers, then find the total number of ____________________ in both factors.

6. The answer to a multiplication problem is called the ____________________.

7. In a division problem, the number being divided is called the ____________________.

8. The number $0.8\overline{3}$ is an example of a ____________________.

9. The answer to a division problem is called the ____________________.

10. In the problem $6.39 \div 0.9$, 0.9 is called the ____________________.

11. A fraction and a decimal that represent the same portion of a whole are ____________________.

12. The ____________________ of a fraction is the dividend.

13. The ____________________ of a fraction shows the number of equal parts in a whole.

14. A ____________________ consists of a whole number part and a fractional or decimal part.

Objective **Identify the place value of a digit.**

Video Examples

Review these examples:

Identify the place value of each digit.

486.92

4 hundreds
8 tens
6 ones
.
9 tenths
2 hundredths

Now Try:

Identify the place value of each digit.

862.93

____

Write the decimal number that has the specified place values.

7 thousandths, 2 hundredths, 5 ones,
3 ten-thousandths, 4 tenths

5 ones
.
4 tenths
2 hundredths
7 thousandths
3 ten-thousandths

The number is 5.9273

Write the decimal number that has the specified place values.

8 hundredths, 5 tenths, 7 ones
2 thousandths, 9 ten-thousandths

____

Practice Exercises

Identify the digit that has the given place value.

1. 43.507 tenths **1.** ____________

hundredths ____________

Write the decimal number that has the specified place values.

2. 1 ten-thousandth, 8 tenths, 2 ones, 7 thousandths, 3 hundredths **2.** ____________

Identify the place value of each digit in these decimals.

3. 37.082 3 **3.** ____________

7 ____________

0 ____________

8 ____________

2 ____________

Objective Write decimals as fractions or mixed numbers.

Video Examples

Review these examples.

Write each decimal as a fraction or mixed number in lowest terms.

17.216

$$17.216 = 17\frac{216}{1000} = 17\frac{216 \div 8}{1000 \div 8} = 17\frac{27}{125}$$

0.88

$$0.88 = \frac{88}{100} = \frac{88 \div 4}{100 \div 4} = \frac{22}{25}$$

Now Try:

Write each decimal as a fraction or mixed number in lowest terms.

6.04

____

0.35

____

Practice Exercises

Write each decimal as a fraction or mixed number in lowest terms.

4. 0.001 **4.** ______________

5. 3.6 **5.** ______________

6. 0.95 **6.** ______________

Objective **Add and subtract negative decimals.**

Video Examples

Review these examples:

Add or subtract as indicated.

$-5.8+(-21)$

Both addends are negative, so the sum will be negative. To begin, $|-5.8|$ is 5.8, and $|-21|$ is 21. Then add the absolute values.

$$\begin{array}{r} 5.8 \\ +\,21.0 \\ \hline 26.8 \end{array} \leftarrow \text{Write in a decimal point and one 0}$$

$-5.8+(-21)=-26.8$

$-2.74-(-5.9)$

Rewrite subtractions as adding the opposite.

$-2.74-(-5.9)$

$-2.74+5.9$

5.9 has the greater absolute value and is positive, so the answer will be positive.

$-2.74-(-5.9)=3.16$

$$\begin{array}{r} 5.90 \\ -\;2.74 \\ \hline 3.16 \end{array}$$

Now Try:

Add or subtract as indicated.

$-6.4+(-32)$

____

$-5.29-(-8.6)$

____

Practice Exercises

Find each difference.

7. 18.1 – 84.6 **7.** __________

8. 87.6 – (–90.4) **8.** __________

9. 1.71 – 12.68 **9.** __________

Objective **Multiply positive and negative decimals.**

Video Examples

Review these examples:

Find the product of 6.23 and –5.4.

Step 1 Multiply the numbers as if they were whole numbers.

$$\begin{array}{r} 6.23 \\ \times \quad 5.4 \\ \hline 2492 \\ 3115 \\ \hline 33642 \end{array}$$

Step 2 Count the total number of decimal places in both factors.

$$\begin{array}{rl} 6.23 & \leftarrow 2 \text{ decimal places} \\ \times \quad 5.4 & \leftarrow 1 \text{ decimal place} \\ \cline{2-2} 2492 & 3 \text{ total decimal places} \\ 3115 & \\ \hline 33642 & \end{array}$$

Step 3 Count over 3 places in the product and write the decimal point. Count from right to left.

$$\begin{array}{rl} 6.23 & \leftarrow 2 \text{ decimal places} \\ \times \quad 5.4 & \leftarrow 1 \text{ decimal place} \\ \cline{2-2} 2492 & 3 \text{ total decimal places} \\ 3115 & \\ \hline 33.642 & \end{array}$$

Step 4 The factors have different signs, so the product is negative. 6.23 times –5.4 is –33.642.

Find the product: $(-0.035)(-0.07)$.

Start by multiply, then count decimal places.

$$\begin{array}{rl} 0.035 & \leftarrow 3 \text{ decimal places} \\ \times \quad 0.07 & \leftarrow 2 \text{ decimal places} \\ \hline 245 & 5 \text{ total decimal places} \end{array}$$

After multiplying, the answer has only three decimal places, but five are needed, so write two zeros on the left side of the answer. Then count over 5 places and write in the decimal point.

$$\begin{array}{rl} 0.035 & \leftarrow 3 \text{ decimal places} \\ \times \quad 0.07 & \leftarrow 2 \text{ decimal places} \\ \hline .00245 & 5 \text{ total decimal places} \end{array}$$

The final product is 0.00245, which has five decimal places. The product is positive because the factors have the same sign.

Now Try:

Find the product of 2.51 and –4.3.

____

Find the product $(-0.062)(-0.03)$.

____

Practice Exercises

Find each product.

10. $\begin{array}{r} -19.3 \\ \times\ 4.7 \\ \hline \end{array}$ **10.** ______________

11. $\begin{array}{r} 0.682 \\ \times\ 3.9 \\ \hline \end{array}$ **11.** ______________

12. $(-0.074)(-0.05)$ **12.** ______________

Objective **Divide a number by a decimal.**

Video Examples

Review this example.

Divide.

$$\frac{41.2}{0.005}$$

Move the decimal point in the divisor three places to the right so 0.005 becomes the whole number 5. Move the decimal point in the dividend the same number of places and write in two extra 0s.

$$5\overline{)41200.}\quad 8240.$$

Now Try:

Divide.

$$0.0024\overline{)48.984}$$

____

Practice Exercises

Find each quotient. Round answers to the nearest thousandth, if necessary.

13. $0.9\overline{)3.4166}$

13. ____________

14. $3.4\overline{)436.05}$

14. ____________

15. $-0.07 \div (-0.00043)$

15. ____________

Objective Write fractions as equivalent decimals.

Video Examples

Review these examples:

Write the fraction as a decimal.

$\frac{7}{8}$

$\frac{7}{8}$ means $7 \div 8$. Write it as $8\overline{)7}$. Write extra zeros in the dividend so you can continue dividing until the remainder is zero.

$$\begin{array}{r} 0.875 \\ 8\overline{)7.000} \\ \underline{6\,4} \\ 60 \\ \underline{56} \\ 40 \\ \underline{40} \\ 0 \end{array}$$

Therefore, $\frac{7}{8} = 0.875$

$3\frac{3}{16}$

One method is to divide 3 by 16 to get 0.1875 for the fraction part. Then add the whole number part to 0.1875.

$$\frac{3}{16} \rightarrow \begin{array}{r} 0.1875 \\ 16\overline{)3.0000} \\ \underline{1\,6} \\ 140 \\ \underline{128} \\ 120 \\ \underline{112} \\ 80 \\ \underline{80} \\ 0 \end{array} \rightarrow \begin{array}{r} 3.0000 \\ +\ 0.1875 \\ \hline 3.1875 \end{array}$$

So $3\frac{3}{16} = 3.1875$.

Now Try:

Write the fraction as a decimal.

$\frac{1}{16}$

____

$4\frac{3}{8}$

____

A second method is to first write $3\frac{3}{16}$ as an improper fraction and then divide numerator by denominator.

$$3\frac{3}{16}=\frac{51}{16}$$

$$\frac{51}{16}\rightarrow 51\div 16\rightarrow 16\overline{)51}\rightarrow \begin{array}{r} 3.1875 \\ 16\overline{)51.0000} \\ \underline{48} \\ 30 \\ \underline{16} \\ 140 \\ \underline{128} \\ 120 \\ \underline{112} \\ 80 \\ \underline{80} \\ 0 \end{array}$$

So $3\frac{3}{16}=3.1875$.

Write $\frac{5}{9}$ as a decimal and round to the nearest thousandth.

$\frac{5}{9}$ means $5\div 9$. To round to thousandths, divide out one more place, to ten-thousandths.

$$\frac{5}{9}\rightarrow 5\div 9\rightarrow 9\overline{)5}\rightarrow \begin{array}{r} 0.5555 \\ 9\overline{)5.0000} \\ \underline{45} \\ 50 \\ \underline{45} \\ 50 \\ \underline{45} \\ 50 \\ \underline{45} \\ 5 \end{array}$$

Written as a repeating decimal, $\frac{5}{9}=0.\overline{5}$.

Rounded to the nearest thousandth, $\frac{5}{9}=0.556$.

Write $\frac{5}{12}$ as a decimal and round to the nearest thousandth.

____

Practice Exercises

Write each fraction or mixed number as a decimal. Round to the nearest thousandth, if necessary.

16. $\frac{1}{8}$ **16.** ______________

17. $4\frac{1}{9}$ **17.** ______________

18. $19\frac{17}{24}$ **18.** ______________

Chapter 2 Linear Equations and Inequalities in One Variable

Learning Objectives

Identify variables, constants, and expressions.
Use the distributive property.
Translate word phrases into algebraic expressions.
Write percents as decimals.
Write decimals as percents.
Write percents as fractions.
Write fractions as percents.
Solve applications of percent.
Classify numbers and graph them on number lines.
Use the < and > symbols to compare integers.

Key Terms

Use the vocabulary terms listed below to complete each statement in exercises 1−12.

variable	**constant**	**expression**	**number line**
sum	**difference**	**product**	**quotient**
increased by	**less than**	**double**	**per**
percent	**ratio**	**decimals**	**integers**

1. An ________________________ tells the rule for doing something.

2. A ____________________ is a letter that represents a number that varies or changes, depending on the situation.

3. A ______________________ is a number that is added or subtracted in an expression.

4. __________________ and __________________ are words that mean addition.

5. _________________ and ________________ are words that mean multiplication.

6. __________________ and __________________ are words that mean division.

7. __________________ and _________________ are words that mean subtraction.

8. To compare two quantities that have the same type of units, use a ____________.

9. ____________________ means per one hundred.

10. ____________________ represent parts of a whole.

11. A _________________ is used to show how numbers relate to each other.

12. The whole numbers together with their opposites and 0 are called ____________.

Objective **Identify variables, constants, and expressions.**

Video Examples

Review this example:

Write an expression for this rule. Identify the variable and the constant.

Order the class limit minus 8 lunches because some students will brown bag.

Let c represent the variable and 8 is the constant.

$c - 8$

Now Try:

Write an expression for this rule. Identify the variable and the constant.

Order the class limit plus 4 extra lunches because some students are football players.

____

Practice Exercises

Identify the parts of each expression. Choose from **variable**, **constant**, *and* **coefficient**.

1. $-7+h$

1. ____________

____

2. $-2w$

2. ____________

____

3. $9k + 1$

3. ____________

____

Objective **Use the distributive property.**

Video Examples

Review these examples.
Use the distributive property to rewrite each expression.

$7(p-6)$

$$7(p-6)=7[p+(-6)]$$
$$=7p+7(-6)$$
$$=7p-42$$

$-3(5x-2)$

$$-3(5x-2)=-3[5x+(-2)]$$
$$=-3(5x)+(-3)(-2)$$
$$=(-3\cdot 5)x+(-3)(-2)$$
$$=-15x+6$$

$4\cdot 8+4\cdot 7$

$$4\cdot 8+4\cdot 7=4(8+7)$$

$5\cdot 3+5x+5m$

$$5\cdot 3+5x+5m=5(3+x+m)$$

Now Try:
Use the distributive property to rewrite each expression.

$17(x-6)$

____

$-4(2x-5)$

____

$3\cdot 11+3\cdot 7$

____

$12y+12\cdot 6+12x$

____

Rewrite each expression.

$-(5x+7)$

$$-(5x+7)=-1\cdot(5x+7)$$
$$=-1\cdot 5x+(-1)\cdot 7$$
$$=-5x-7$$

$-(-p-5r+9x)$

$$-(-p-5r+9x)$$
$$=-1\cdot(-1p-5r+9x)$$
$$=-1\cdot(-1p)-1\cdot(-5r)-1\cdot(9x)$$
$$=p+5r-9x$$

$6a+6b+6$

$$6a+6b+6=6a+6b+6\cdot 1$$
$$=6(a+b+1)$$

Rewrite each expression.

$-(3x+4)$

____

$-(-4x-5y+z)$

____

$3x+3y+3$

____

Practice Exercises

Use the distributive property to rewrite each expression. Simplify if possible.

4. $n(2a-4b+6c)$ **4.** ________________

5. $-2(5y-9z)$ **5.** ________________

6. $-(-2k+7)$ **6.** ________________

Objective **Translate word phrases into algebraic expressions.**

Video Examples

Review these examples:

Write each phrase as an algebraic expression. Use x as the variable.

-66 added to a number

Algebraic expression: $-66 + x$ or $x + (-66)$

39 minus a number

Algebraic expression: $39 - x$

Now Try:

Write each phrase as an algebraic expression. Use x as the variable.

-20 added to a number

____

48 minus a number

____

Write the phrase as an algebraic expression. Use x as the variable.

17 subtracted from 8 times a number

Algebraic expression: $8x - 17$

Write the phrase as an algebraic expression. Use x as the variable.

81 subtracted from 6 times a number

____

Practice Exercises

Write an algebraic expression using x as the variable.

7. The product of -6 and a number **7.** ____________

8. The quotient of a number and 10 **8.** ____________

9. One more than three times a number **9.** ____________

Objective **Write percents as decimals.**

Video Examples

Review these examples.

Write each percent as a decimal.

39%

$39\% = 39 \div 100 = 0.39$

7%

$7\% = 7 \div 100 = 0.07$

89.9%

$89.9\% = 89.9 \div 100 = 0.899$

169%

$169\% = 169 \div 100 = 1.69$

Now Try:

Write each percent as a decimal.

23% __________

8% __________

15.3% __________

302% __________

Write each percent as a decimal by dropping the percent symbol and moving the decimal point two places to the left.

56%

Drop the percent sign and move the decimal point two places to the left.

$56\% = 56.\% = 0.56$

0.2%

Two zeros are attached so the decimal point can be moved two places to the left.

$0.2\% = 0.002$

Write each percent as a decimal by dropping the percent symbol and moving the decimal point two places to the left.

48% __________

0.8% __________

Practice Exercises

Write each percent as a decimal.

10. 42% **10.** __________

11. 310% **11.** __________

12. 18.9% **12.** __________

Objective **Write decimals as percents.**

Video Examples

Review these examples.

Write each decimal as a percent by moving the decimal point two places to the right.

0.31

Decimal point is moved two places to the right and percent symbol is attached.
$0.31 = 31\%$

1.6

0 is attached so the decimal point can be moved two places to the right.
$1.6 = 1.60 = 160\%$

0.904

$0.904 = 90.4\%$

Now Try:

Write each decimal as a percent by moving the decimal point two places to the right.

0.43

____

2.3

____

0.751

____

Practice Exercises

Write each decimal as a percent.

13. 0.2

14. 0.564

15. 4.93

13. ____________

14. ____________

15. ____________

Objective **Write percents as fractions.**

Video Examples

Review these examples.

Write the percent as a fraction in lowest terms.

62.5%

Write 62.5 over 100.

$$62.5\% = \frac{62.5}{100}$$

To get a whole number in the numerator, multiply the numerator and denominator by 10.

$$\frac{62.5}{100} = \frac{62.5(10)}{100(10)} = \frac{625}{1000}$$

Now write the fraction in lowest terms.

$$\frac{625}{1000} = \frac{625 \div 125}{1000 \div 125} = \frac{5}{8}$$

Now Try:

Write the percent as a fraction in lowest terms.

43.6%

____

Write the percent as a fraction or mixed number in lowest terms.

350%

$$350\% = \frac{350}{100} = \frac{350 \div 50}{100 \div 50} = \frac{7}{2} = 3\frac{1}{2}$$

Write the percent as a fraction or mixed number in lowest terms.

175%

____

Practice Exercises

Write each percent as a fraction or mixed number in lowest terms.

16. 140% **16.** __________

17. $18\frac{1}{3}\%$ **17.** __________

18. 55.6% **18.** __________

Objective **Write fractions as percents.**

Video Examples

Review this example.

Write the fraction as a percent.

$$\frac{5}{16}$$

$$\frac{5}{16}=\left(\frac{5}{16}\right)(100\%)=\left(\frac{5}{16}\right)\left(\frac{100}{1}\%\right)$$
$$=\left(\frac{5}{4\cdot 4}\right)\left(\frac{4\cdot 25}{1}\%\right)$$
$$=\frac{125}{4}\%=31\frac{1}{4}\%$$

Now Try:

Write the fraction as a percent.

$$\frac{15}{16}$$

____

Practice Exercises

Write each fraction or mixed number as a percent. If you're using a calculator, first work each one by hand. Then use your calculator and round to the nearest tenth of a percent, if necessary.

19. $\frac{47}{50}$ **19.** __________

20. $\frac{11}{40}$ **20.** __________

21. $\frac{64}{75}$ **21.** __________

Objective Solve applications of percent.

Video Examples

Review these examples.

Janelle had budgeted \$250 for new school clothes but ended up spending \$390. The amount she spent was what percent of her budget?

Step 1 Read the problem. It is about comparing her budget to the amount spent.
Unknown: The percent of her budget
Known: \$250 budgeted, \$390 spent.

Step 2 Assign a variable. Let p be the unknown percent.

Step 3 Write an equation.

$$\underbrace{\text{Amount spent}}_{390} \quad \underset{=}{\text{is}} \quad \underbrace{\text{what percent}}_{p} \quad \underset{\cdot}{\text{of}} \quad \underbrace{\text{amount budgeted?}}_{250}$$

Step 4 Solve the equation.

$$390 = p \cdot 250$$

$$\frac{390}{250} = \frac{p \cdot \cancel{250}}{\cancel{250}}$$

$$1.56 = p$$

$$1.56 = 156\%$$

Step 5 State the answer. Janelle spent 156% of her budget.

Step 6 Check the solution. 100% is \$250, and 50% is \$125. So \$250 + \$125 = \$350, or 150%, which is close to 156%.

Now Try:

Total Fitness Club predicted that 240 new members would join after Christmas. It actually had 396 new members join. The actual number joining is what percent of the predicted number?

____

David has 4.5% of his earnings deposited into a money market. If this amounts to \$146.25 per month, find his monthly earnings.

Step 1 Read the problem. It is about earnings.
Unknown: monthly salary
Known: \$146.25 is 4.5% of monthly earnings.

Step 2 Assign a variable. Let n = monthly salary.

A multivitamin contains 10 micrograms of vitamin K. If this is 13% of the recommended daily dosage, what is the recommended daily dosage of vitamin K? Round the answer to the nearest whole number.

____

Step 3 Write an equation.

4.5%	of	how much	is	$146.25
↓	↓	↓	↓	↓
0.045	·	n	=	146.25

Step 4 Solve the equation.

$$0.045 \cdot n = 146.25$$

$$\frac{\cancel{0.045} \cdot n}{\cancel{0.045}} = \frac{146.25}{0.045}$$

$$n = 3250$$

Step 5 State the answer. David's monthly earnings is $3250.

Step 6 Check the solution. Round 4.5% to 5%. 10% of $3250 is $325, then 5% is half of $325 or $162.50, which is close to the number given.

Practice Exercises

Use the six problem-solving steps to answer each question. Round percent answers to the nearest tenth of a percent.

22. Members who are between 25 and 45 years of age make up 92% of the total membership of an organization. If there are 850 total members in the organization, find the number of members in the 25 to 45 age group.

22. ________________

23. Payroll deductions are 35% of Jason's gross pay. If his deductions total $350, what is his gross pay?

23. ________________

24. Vera's Antique Shoppe says that of its 5100 items in stock, 4233 are just plain junk, while the rest are antiques. What percent of the number of items in stock is antiques?

24. ________________

Objective **Classify numbers and graph them on number lines.**

Video Examples

Review these examples.

Use an integer to express the boldface italic number in the application.

In August, 2012, the National Debt was approximately $***16*** trillion.

Use –$16 trillion because "debt" indicates a negative number.

Now Try:

Use an integer to express the boldface italic number in the application.

Death Valley is ***282*** feet below sea level.

____

Graph each number on a number line.

$-3\frac{1}{2}, -\frac{3}{2}, 0, \frac{7}{2}, 1$

To locate the improper fractions on the number line, write them as mixed numbers or decimals.

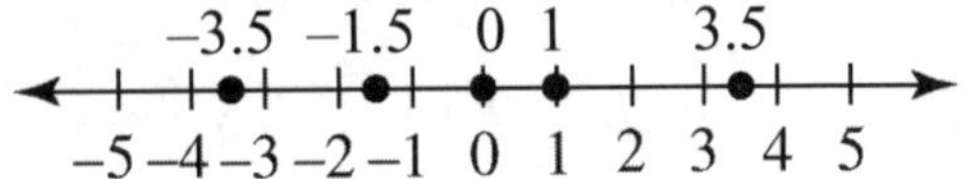

Graph each number on a number line.

$\frac{1}{2}, 0, -3, -\frac{5}{2}$

–5 –4 –3 –2 –1 0 1 2 3 4 5

List the numbers in the following set that belong to each set of numbers.

$\left\{-6, -\frac{5}{6}, 0, 0.\overline{3}, \sqrt{3}, 4\frac{1}{5}, 6, 6.7\right\}$

Whole numbers

Answer: 0 and 6

Integers

Answer: –6, 0, and 6

Rational numbers

Answer: $-6, -\frac{5}{6}, 0, 0.\overline{3}, 4\frac{1}{5}, 6, 6.7$

Irrational numbers

Answer: $\sqrt{3}$

List the numbers in the following set that belong to each set of numbers.

$\left\{-10, -\frac{5}{8}, 0, 0.\overline{4}, \sqrt{5}, 5\frac{1}{2}, 7, 9.9\right\}$

Whole numbers

____

Integers

____

Rational numbers

____

Irrational numbers

____

Practice Exercises

Use a real number to express each number in the following applications.

25. Last year Nina lost 75 pounds. **25.** ______________

26. Between 1970 and 1982, the population of Norway increased by 279,867. **26.** ______________

Graph the group of rational numbers on a number line.

27. –4.5, –2.3, 1.7, 4.2 **27.**

–5 –4 –3 –2 –1 0 1 2 3 4 5

Objective **Use the < and > symbols to compare integers.**

Video Examples

Review these examples.

Write < or > between each pair of integers to make a true statement.

0 _____ 7

0 is to the left of 7 on the number line, so 0 is less than 7. Write $0 < 7$.

–2 _____ –5

–2 is to the right of –5, so –2 is greater than –5. Write $-2 > -5$.

–6 _____ 7

–6 is to the left of 7, so –6 is less than 7. Write $-6 < 7$.

Now Try:

Write < or > between each pair of integers to make a true statement.

0 _____ –8

____

–3 _____ –9

____

11 _____ –4

____

Practice Exercises

Write < or > in each blank to make a true statement.

28. –23 ____ –32

29. –6 ____ 0

30. –5 ____ –3

28. __________

29. __________

30. __________

Chapter 3 Linear Equations and Inequalities in Two Variables; Functions

Learning Objectives

Classify numbers and graph them on number lines.
Graph intervals on a number line.
Evaluate algebraic expressions, given values for the variables.

Key Terms

Use the vocabulary terms listed below to complete each statement in exercises 1−19.

natural numbers	**whole numbers**	**number line**	**integers**
integers	**negative number**	**positive number**	
rational number	**set-builder notation**		**coordinate**
irrational number	**real numbers**	**signed numbers**	**variable**
inequalities	**interval**	**interval notation**	**constant**
linear inequality	**algebraic expression**		

1. The set {0, 1, 2, 3, …} is called the set of ______________________________.

2. A ______________________________ is a symbol, usually a letter, used to represent an unknown number.

3. The whole numbers together with their opposites and 0 are called ____________________.

4. The set { 1, 2, 3, …} is called the set of ______________________________.

5. A collection of numbers, variables, operation symbols, and grouping symbols is an ______________________________.

6. A ______________________________ shows the ordering of the real numbers on a line.

7. A real number that is not a rational number is called a(n) ____________________.

8. The number that corresponds to a point on the number line is the ____________________ of that point.

9. A number located to the left of 0 on a number line is a ____________________.

10. A number located to the right of 0 on a number line is a ____________________.

11. Numbers that can be represented by points on the number line are ____________________.

12. ______________________ uses a variable and a description to describe a set.

13. A number that can be written as the quotient of two integers is a ______________________.

14. Positive numbers and negative numbers are ______________________.

15. A ______________________ is a fixed, unchanging number.

16. A portion of a number line is called a(n) ______________________.

17. A(n) ______________________ can be written in the form $Ax + B < C$, $Ax + B \leq C$, $Ax + B > C$, or $Ax + B \geq C$, where A, B, and C are real numbers with $A \neq 0$.

18. Algebraic expressions related by $<$, $\leq$, $>$, or $\geq$ are called ______________________.

19. The ______________________ for $a \leq x < b$ is $[a, b)$.

Objective **Classify numbers and graph them on number lines.**

Video Examples

<table>
<tr><th>Review these examples:</th><th>Now Try:</th></tr>
<tr><td>Use an integer to express the boldface italic number in the application.

In August, 2012, the National Debt was approximately $16 trillion.

Use –$16 trillion because “debt” indicates a negative number.</td><td>Use an integer to express the boldface italic number in the application.
Death Valley is 282 feet below sea level.

__________</td></tr>
<tr><td>Graph each number on a number line.
$-3\frac{1}{2},\ -\frac{3}{2},\ 0,\ \frac{7}{2},\ 1$

To locate the improper fractions on the number line, write them as mixed numbers or decimals.
–3.5 –1.5 0 1 3.5
–5 –4 –3 –2 –1 0 1 2 3 4 5</td><td>Graph each number on a number line.
$\frac{1}{2},\ 0,\ -3,\ -\frac{5}{2}$

–5 –4 –3 –2 –1 0 1 2 3 4 5</td></tr>
<tr><td>List the numbers in the following set that belong to each set of numbers.
$\left\{-6,\ -\frac{5}{6},\ 0,\ 0.\bar{3},\ \sqrt{3},\ 4\frac{1}{5},\ 6,\ 6.7\right\}$

Whole numbers

Answer: 0 and 6

Integers

Answer: –6, 0, and 6

Rational numbers

Answer: $-6,\ -\frac{5}{6},\ 0,\ 0.\bar{3},\ 4\frac{1}{5},\ 6,\ 6.7$

Irrational numbers

Answer: $\sqrt{3}$</td><td>List the numbers in the following set that belong to each set of numbers.
$\left\{-10, -\frac{5}{8},\ 0,\ 0.\bar{4},\ \sqrt{5},\ 5\frac{1}{2},\ 7,\ 9.9\right\}$
Whole numbers

____

Integers

____

Rational numbers

____

Irrational numbers

__________</td></tr>
</table>

Practice Exercises

Use a real number to express each number in the following applications.

1. Last year Nina lost 75 pounds. 1. ______________

2. Between 1970 and 1982, the population of Norway increased by 279,867. 2. ______________

Graph the group of rational numbers on a number line.

3. $-4.5, -2.3, 1.7, 4.2$ 3.

−5 −4 −3 −2 −1 0 1 2 3 4 5

Objective **Graph intervals on a number line.**

Video Examples

Review this example:

Write the inequality in interval notation, and graph the interval.

$x > -3$

The statement $x > -3$ says that x can represent any value greater than –3, but cannot equal –3, written $(-3, \infty)$. We graph this interval by placing a parenthesis at –3 and drawing an arrow to the right. The parenthesis indicates that –3 is not part of the graph.

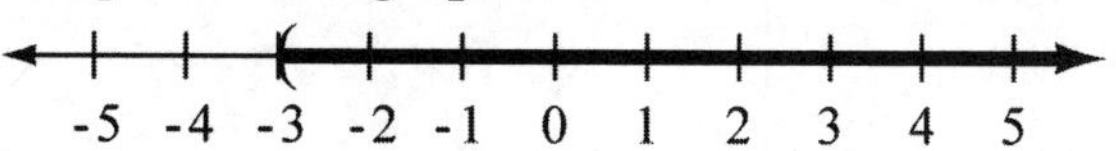

Now Try:

Write the inequality in interval notation, and graph the interval.

$x > -1$

-5 -4 -3 -2 -1 0 1 2 3 4 5

Practice Exercises

Write each inequality in interval notation and graph the interval.

4. $3 < a$

4. ________________

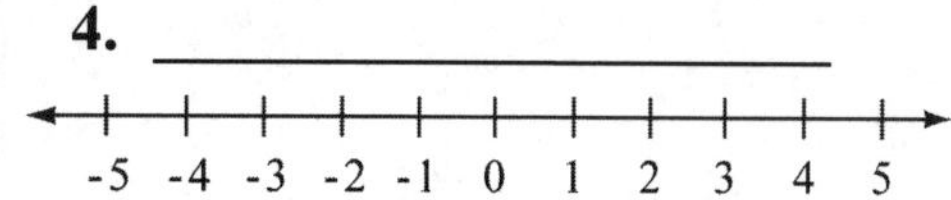

5. $y \geq -2$

5. ________________

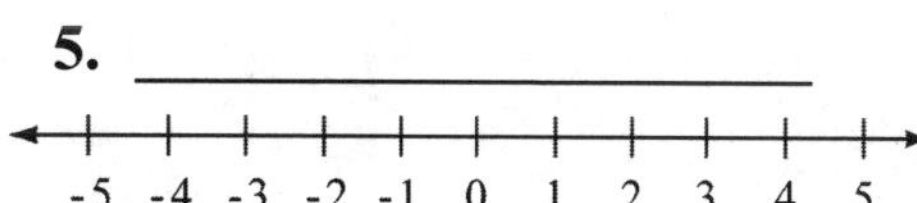

6. $x < -4$

6. ________________

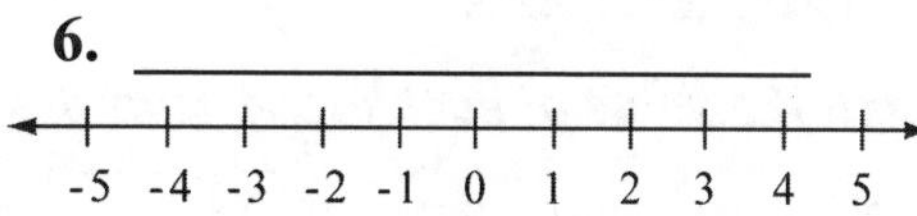

Objective **Evaluate algebraic expressions, given values for the variables.**

Video Examples

Review these examples:

Find the value of the algebraic expression for $x = 4$ and then $x = 7$.

$5x^2$

For $x = 4$,

$5x^2 = 5 \cdot 4^2$ Let $x = 4$.

$= 5 \cdot 16$ Square 4.

$= 80$ Multiply.

For $x = 7$,

$5x^2 = 5 \cdot 7^2$ Let $x = 7$.

$= 5 \cdot 49$ Square 7.

$= 245$ Multiply.

Now Try:

Find the value of the algebraic expression for $x = 6$ and then $x = 9$.

$7x^2$

____

Find the value of the expression for $x = 7$ and $y = 6$.

$3x + 4y + 2$

Replace x with 7 and y with 6.

$3x + 4y + 2 = 3 \cdot 7 + 4 \cdot 6 + 2$

$= 21 + 24 + 2$ Multiply.

$= 47$ Add.

Find the value of the expression for $x = 8$ and $y = 4$.

$5x + 6y + 1$

____

Practice Exercises

Find the value of each expression if $x = 2$ *and* $y = 4$.

7. $9x - 3y + 2$

7. __________

8. $\dfrac{2x + 3y}{3x - y + 2}$

8. __________

9. $\dfrac{3y^2 + 2x^2}{5x + y^2}$

9. __________

Chapter 4 Systems of Linear Equations and Inequalities

Learning Objectives

Evaluate algebraic expressions, given values for the variables.
Determine whether a number is a solution of an equation.
Solve equations with fractions or decimals as coefficients.
Graph linear inequalities in two variables.
Use linear equations to solve application problems.

Key Terms

Use the vocabulary terms listed below to complete each statement in exercises 1−6.

variable **constant** **algebraic expression**

linear inequality in two variables **equation** **solution**

1. A(n) ____________________________ is a symbol, usually a letter, used to represent an unknown number.

2. A collection of numbers, variables, operation symbols, and grouping symbols is an____________________________________.

3. An inequality that can be written in the form $Ax + By < C$, $Ax + By > C$, $Ax + By \leq C$, or $Ax + By \geq C$ is called a(n) ______________________________.

4. The _____________________ of an equation is a number that can replace the variable so that the equation is true.

5. A(n) ____________________________ is a fixed, unchanging number.

6. A(n) _____________________ is a statement that says two expressions are equal.

Objective **Evaluate algebraic expressions, given values for the variables.**

Video Examples

Review these examples:

Find the value of each algebraic expression for $x = 4$ and then $x = 7$.

$5x^2$

For $x = 4$,

$$\begin{aligned} 5x^2 &= 5\cdot 4^2 && \text{Let } x = 4. \\ &= 5\cdot 16 && \text{Square 4.} \\ &= 80 && \text{Multiply.} \end{aligned}$$

For $x = 7$,

$$\begin{aligned} 5x^2 &= 5\cdot 7^2 && \text{Let } x = 7. \\ &= 5\cdot 49 && \text{Square 7.} \\ &= 245 && \text{Multiply.} \end{aligned}$$

Now Try:

Find the value of each algebraic expression for $x = 6$ and then $x = 9$.

$7x^2$

____

Find the value of each expression for $x = 7$ and $y = 6$.

$3x+4y+2$

Replace x with 7 and y with 6.

$$\begin{aligned} 3x+4y+2 &= 3\cdot 7+4\cdot 6+2 \\ &= 21+24+2 && \text{Multiply.} \\ &= 47 && \text{Add.} \end{aligned}$$

Find the value of each expression for $x = 8$ and $y = 4$.

$5x+6y+1$

____

Practice Exercises

Find the value of each expression if $x = 2$ *and* $y = 4$.

1. $9x-3y+2$ **1.** ____________

2. $\dfrac{2x+3y}{3x-y+2}$ **2.** ____________

3. $\dfrac{3y^2+2x^2}{5x+y^2}$ **3.** ____________

Objective **Determine whether a number is a solution of an equation.**

Video Examples

Review these examples:

Is 8 a solution of this equation?

$$3y+4=35$$

Replace y with 8.

$$3y+4=35$$
$$3(8)+4=35$$
$$24+4=35$$
$$28\neq 35 \quad \text{False}$$

The false statement shows that 8 is not a solution of $3y+4=35$.

Now Try:

Is 6 a solution of this equation?

$$35=5r$$

____

Practice Exercises

Decide whether the given number is a solution of the equation.

4. $b-5=18$; 13

4. ____________

5. $5+8m=3$; -1

5. ____________

6. $-5y+1=6$; -1

6. ____________

Objective **Solve equations with fractions or decimals as coefficients.**

Video Examples

Review these examples:

Solve $\frac{1}{4}(x+3)-\frac{2}{5}(x+1)=2$.

To clear fractions, multiply by 20, the LCD.

$$\frac{1}{4}(x+3)-\frac{2}{5}(x+1)=2$$

Step 1 $$20\left[\frac{1}{4}(x+3)-\frac{2}{5}(x+1)\right]=20(2)$$

$$20\left[\frac{1}{4}(x+3)\right]+20\left[-\frac{2}{5}(x+1)\right]=20(2)$$

$$5(x+3)-8(x+1)=40$$

$$5x+15-8x-8=40$$

$$-3x+7=40$$

Step 2 $$-3x+7-7=40-7$$

$$-3x=33$$

Step 3 $$\frac{-3x}{-3}=\frac{33}{-3}$$

$$x=-11$$

Step 4 Check to confirm that {–11} is the solution set.

Solve $0.2x+0.04(10-x)=0.06(4)$.

To clear decimals, multiply by 100.

$$0.2x+0.04(10-x)=0.06(4)$$

Step 1 $$100[0.2x+0.04(10-x)]=100[0.06(4)]$$

$$100(0.2x)+100[0.04(10-x)]=100[0.06(4)]$$

$$20x+4(10)+4(-x)=24$$

$$20x+40-4x=24$$

$$16x+40=24$$

Step 2 $$16x+40-40=24-40$$

$$16x=-16$$

Step 3 $$\frac{16x}{16}=\frac{-16}{16}$$

$$x=-1$$

Step 4 Check to confirm that {–1} is the solution set.

Now Try:

Solve $\frac{1}{7}(x+5)-\frac{1}{2}(x+4)=-2$.

____

Solve
$0.5x+0.04(5-8x)=0.07(8)$.

____

Practice Exercises

Solve each equation and check your solution.

7. $\frac{3}{8}x-\frac{1}{3}x=\frac{1}{12}$ **7.** ________________

8. $\frac{1}{3}(2m-1)-\frac{3}{4}m=\frac{5}{6}$ **8.** ________________

9. $0.45a-0.35(20-a)=0.02(50)$ **9.** ________________

Objective **Graph linear inequalities in two variables.**

Video Examples

Review these examples:

Graph $3x-2y\leq 6$.

The inequality $3x-2y\leq 6$ means that

$3x-2y<6$ or $3x-2y=6$.

We begin by graphing the line $3x-2y=6$ with intercepts (0, –3) and (2, 0). This boundary line divides the plane into two regions, one of which satisfies the inequality. We use the test point (0, 0) to see whether the resulting statement is true or false, thereby determining whether the point is in the shaded region or not.

$$3x-2y\leq 6$$
$$3(0)-2(0)\overset{?}{\leq} 6$$
$$0-0\overset{?}{\leq} 6$$
$$0\leq 6 \text{ True}$$

Since the last statement is true, we shade the region that includes the test point (0, 0). The shaded region, along with the boundary line, is the desired graph.

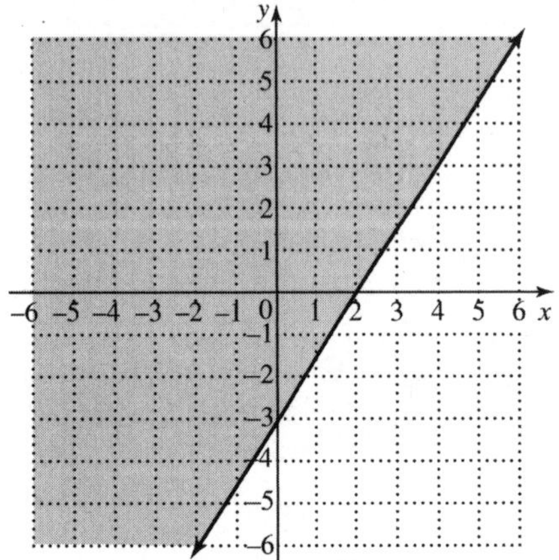

Now Try:

Graph $2x+5y\leq -8$.

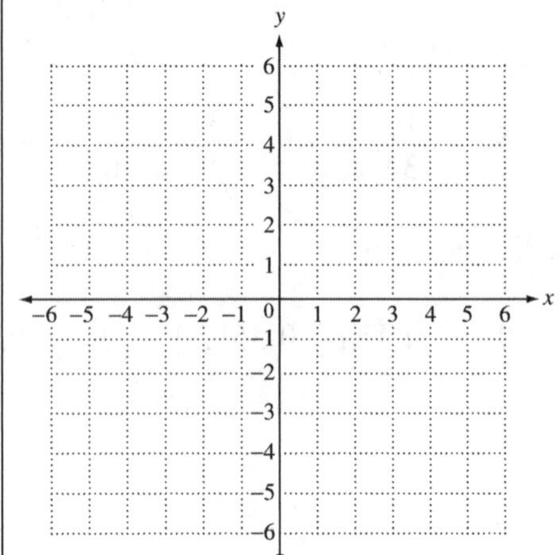

Graph $x - 4 \leq -1$.

First, solve the inequality for x.

$x \leq 3$

Now graph the line $x = 3$, a vertical line through the point (3, 0). Use a solid line, and choose (0, 0) as a test point.

$0 \leq 3$ True

Because $0 \leq 3$ is true, we shade the region containing (0, 0).

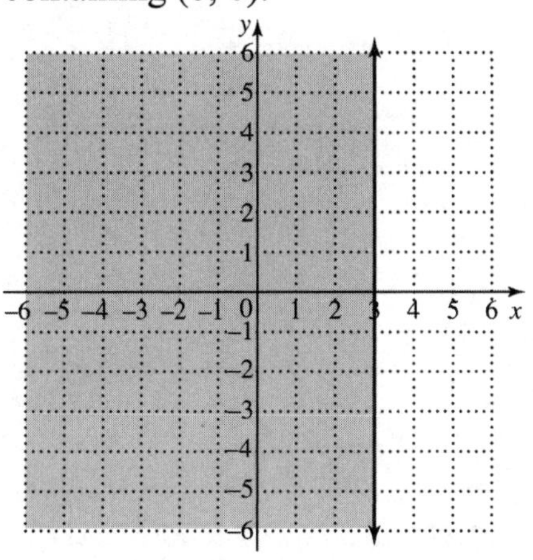

Graph $y \geq -1$.

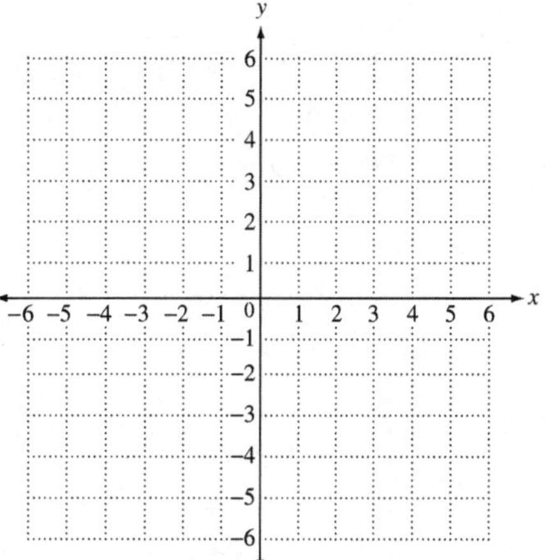

Practice Exercises

Graph each linear inequality.

9. $y \geq x - 1$

9.

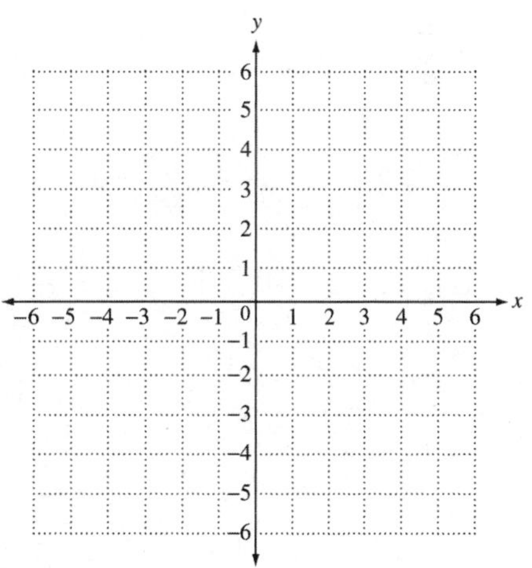

10. $y > -x + 2$

10.

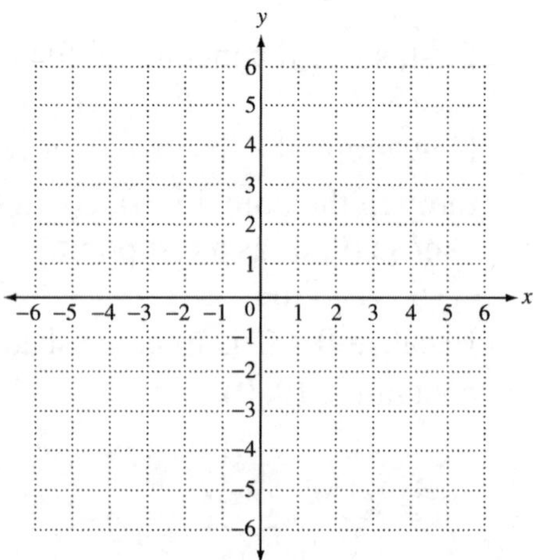

11. $3x - 4y - 12 > 0$

11.

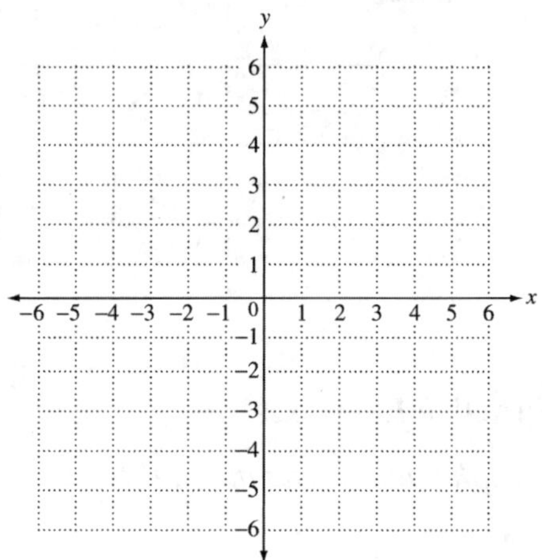

Objective **Use linear equations to solve application problems.**

Video Examples

Review this example:

How many gallons of a 20% alcohol solution must be mixed with 15 gallons of a 12% alcohol solution to obtain a 14% alcohol solution?

Step 1 Read the problem. Find the amount of the 20% alcohol solution.

Step 2 Assign a variable. Let x = the amount of the 20% alcohol solution.

Liters of Solution	Rate (as a decimal)	Liters of Pure Acid
x	0.20	$0.2x$
15	0.12	0.12(15)
$x+15$	0.14	$0.14(x+15)$

Step 3 Write an equation.

$$0.20x+0.12(15)=0.14(x+15)$$

Step 4 Solve.

$$0.20x+1.8=0.14x+2.1$$

$$0.06x=0.3$$

$$x=5$$

Step 5 State the answer. 5 gallons of the 20% alcohol solution must be added.

Step 6 Check.

$$0.20(5)+0.12(15)\stackrel{?}{=}0.14(5+15)$$

$$2.8=2.8$$

The answer checks.

Now Try:

How many ounces of a 35% solution of acid must be mixed with a 60% solution to get 20 ounces of a 50% solution?

____

Practice Exercises

Solve each problem.

13. How many pounds of peanuts worth \$3 per pound must be mixed with mixed nuts worth \$5.50 per pound to make 40 pounds of a mixture worth \$5 per pound? **13.** ________________

14. How many pounds of candy worth \$7 per pound must be mixed with candy worth \$4.50 per pound to make 100 pounds of candy worth \$6 per pound? **14.** ________________

Chapter 5 Exponents and Polynomials

Learning Objectives

Review the use of exponents.
Decide whether an expression gives a positive or negative result.
Combine like terms.

Key Terms

Use the vocabulary terms listed below to complete each statement in exercises 1–6.

exponential expression	**base**	**power**
numerical coefficient	**term**	**like terms**

1. In the term $4x^2$, "4" is the__________________________________.

2. A number, a variable, or a product or quotient of a number and one or more variables raised to powers is called a __________________________________.

3. Terms with exactly the same variables, including the same exponents, are called ______________________________.

4. 2^5 is read "2 to the fifth ______________________".

5. A number written with an exponent is called a(n) ____________________________.

6. The ________________ is the number being multiplied repeatedly.

Objective **Review the use of exponents.**

Video Examples

Review these examples.

Name the base and exponent of each expression. Then evaluate.

3^4

Base: 3
Exponent: 4
Value: $3^4 = 3 \cdot 3 \cdot 3 \cdot 3 = 81$

-3^4

Base: 3
Exponent: 4
Value: $-3^4 = -1 \cdot (3 \cdot 3 \cdot 3 \cdot 3) = -81$

$(-3)^4$

Base: –3
Exponent: 4
Value: $(-3)^4 = (-3)(-3)(-3)(-3) = 81$

Name the base and exponent of each expression. Then evaluate.

2^6

____

-2^6

____

$(-2)^6$

____

Practice Exercises

Write the expression in exponential form and evaluate, if possible.

1. $\left(\frac{1}{3}\right)\left(\frac{1}{3}\right)\left(\frac{1}{3}\right)\left(\frac{1}{3}\right)\left(\frac{1}{3}\right)$

1. ____________

Evaluate each exponential expression. Name the base and the exponent.

2. $(-4)^4$

2. ____________
base____________
exponent____________

3. -3^8

3. ____________
base____________
exponent____________

Objective Decide whether an expression gives a positive or negative result.

Practice Exercises

Decide whether each expression gives a positive or negative result.

4. $(-5)^5$ **4.** ____________

5. -5^4 **5.** ____________

6. $(-5)^6$ **6.** ____________

Objective **Combine like terms.**

Video Examples

Review these examples:

Combine like terms in each expression.

$8r+5r+4r$

$$8r+5r+4r=(8+5+4)r$$
$$=17r$$

$9x+x$

$$9x+x=9x+1x$$
$$=(9+1)x$$
$$=10x$$

$15x^2-8x^2$

$$15x^2-8x^2=(15-8)x^2$$
$$=7x^2$$

Now Try:

Combine like terms in each expression.

$14r+7r+2r$

____

$18x+x$

____

$17x^2-9x^2$

____

Simplify each expression.

$8k-5-4(7-3k)$

$$8k-5-4(7-3k)=8k-5-4(7)-4(-3k)$$
$$=8k-5-28+12k$$
$$=20k-33$$

$-\frac{3}{5}(x-10)-\frac{1}{10}x$

$$-\frac{3}{5}(x-10)-\frac{1}{10}x=-\frac{3}{5}x-\frac{3}{5}(-10)-\frac{1}{10}x$$
$$=-\frac{3}{5}x+6-\frac{1}{10}x$$
$$=-\frac{6}{10}x+6-\frac{1}{10}x$$
$$=-\frac{7}{10}x+6$$

Simplify each expression.

$7k-9-5(3-6k)$

____

$-\frac{3}{4}(x-8)-\frac{1}{2}x$

____

Practice Exercises

Simplify.

7. $12y-7y^2+4y-3y^2$ **7.** ______________

8. $-4(x+4)+2(3x+1)$ **8.** ______________

9. $2.5(3y+1)-4.5(2y-3)$ **9.** ______________

Chapter 6 Factoring and Applications

Learning Objectives

Find prime factorizations.
Multiply a monomial and a polynomial.
Multiply two polynomials.
Solve application problems involving linear equations.

Key Terms

Use the vocabulary terms listed below to complete each statement in exercises 1−5.

factors	**composite number**	**prime number**
factorizations		**prime factorization**

1. The numbers that can be multiplied to give a specific number (product) are ____________________ of that number.

2. A ________________________ has at least one factor other than itself and 1.

3. In a _________________________________ every factor is a prime number.

4. Two different factors of a _______________________ are itself and 1.

5. Numbers that are multiplied to give a product are ______________________.

Objective **Find prime factorizations**

Video Examples

Review these examples:

Find the prime factorization of 18.

Try to divide 18 by the first prime, 2.

$18 \div 2 = 9.$

so

$18 = 2 \cdot 9.$

Try to divide 9 by the prime, 3.

$9 \div 3 = 3.$

so

$18 = 2 \cdot 3 \cdot 3.$

Because all factors are prime, the prime factorization of 18 is $2 \cdot 3 \cdot 3$.

Find the prime factorization of 72.

$3\overline{)3}$ quotient 1 — Continue to divide until the quotient is 1. Divide 3 by 3.

$3\overline{)9}$ Divide 9 by 3.

$2\overline{)18}$ Divide 18 by 2.

$2\overline{)36}$ Divide 36 by 2.

$2\overline{)72}$ Divide 72 by 2.

Because all factors (divisors) are prime, the prime factorization of 72 is

$72 = 2 \cdot 2 \cdot 2 \cdot 3 \cdot 3$ or $2^3 \cdot 3^2$.

Find the prime factorization of 375.

$5\overline{)5}$ quotient 1 — Continue to divide until the quotient is 1. Divide 5 by 5.

$5\overline{)25}$ Divide 25 by 5.

$5\overline{)125}$ Divide 125 by 5.

$3\overline{)375}$ Divide 375 by 3.

Because all factors (divisors) are prime, the prime factorization of 375 is

$375 = 3 \cdot 5 \cdot 5 \cdot 5$ or $3 \cdot 5^3$.

Now Try:

Find the prime factorization of 24.

____

Find the prime factorization of 56.

____

Find the prime factorization of 150.

____

Practice Exercises

Find the prime factorization of each number. Write the answer with exponents when repeated factors appear.

1. 28 **1.** ________________

2. 72 **2.** ________________

3. 450 **3.** ________________

Lial Video Library Workbook with Integrated Review
for Beginning Algebra, 12th edition.

Objective **Multiply a monomial and a polynomial.**

Video Examples

Review this example:

Find the product.

$5x^2(7x+3)$

Use the distributive property.

$$5x^2(7x+3) = 5x^2(7x) + 5x^2(3)$$
$$= 35x^3 + 15x^2$$

Now Try:

Find the product.

$8x^3(4x+8)$

____

Practice Exercises

Find each product.

4. $7z(5z^3+2)$ **4.** ______________

5. $2m(3+7m^2+3m^3)$ **5.** ______________

6. $-3y^2(2y^3+3y^2-4y+11)$ **6.** ______________

Objective **Multiply two polynomials.**

Video Examples

Review these examples:

Multiply $(x^2+6)(5x^3-4x^2+3x)$.

Multiply each term of the second polynomial by each term of the first.

$$(x^2+6)(5x^3-4x^2+3x)$$
$$=x^2(5x^3)+x^2(-4x^2)+x^2(3x)$$
$$+6(5x^3)+6(-4x^2)+6(3x)$$
$$=5x^5-4x^4+3x^3+30x^3-24x^2+18x$$
$$=5x^5-4x^4+33x^3-24x^2+18x$$

Multiply $(2x^3+7x^2+5x-1)(4x+6)$ vertically.

Write the polynomials vertically.

$$\begin{array}{r} 2x^3+7x^2+5x-1 \\ \underline{4x+6} \end{array}$$

Begin by multiplying each term in the top row by 6.

$$\begin{array}{r} 2x^3\ +7x^2\ +5x\ -1 \\ 4x\ +6 \\ \hline 12x^3+42x^2+30x-6 \end{array}$$

Now multiply each term in the top row by $4x$. Then add like terms.

$$\begin{array}{r} 2x^3\ +7x^2\ +5x\ -1 \\ 4x\ +6 \\ \hline 12x^3\ +42x^2+30x-6 \\ 8x^4+28x^3+20x^2\ -4x \\ \hline 8x^4+40x^3+62x^2+26x-6 \end{array}$$

The product is $8x^4+40x^3+62x^2+26x-6$.

Now Try:

Multiply $(x^3+9)(4x^4-2x^2+x)$

____

Multiply $(4x^3-3x^2+6x+5)(7x-3)$ vertically.

____

Find the product of $-16m^3+12m^2+4$ and $\frac{1}{4}m^2+\frac{3}{4}$.

Multiply each term of the second polynomial by each term of the first.

$$\left(-16m^3+12m^2+4\right)\left(\frac{1}{4}m^2+\frac{3}{4}\right)$$

$$=-16m^3\left(\frac{1}{4}m^2\right)-16m^3\left(\frac{3}{4}\right)+12m^2\left(\frac{1}{4}m^2\right)$$

$$+12m^2\left(\frac{3}{4}\right)+4\left(\frac{1}{4}m^2\right)+4\left(\frac{3}{4}\right)$$

$$=-12m^3+9m^2+3-5m^5+3m^4+m^2$$

$$=-4m^5+3m^4-12m^3+10m^2+3$$

The product is $-4m^5+3m^4-12m^3+10m^2+3$.

Find the product of $12x^3-36x^2+6$ and $\frac{1}{6}x^2+\frac{5}{6}$.

____

Practice Exercises

Find each product.

7. $(x+3)(x^2-3x+9)$ **7.** ____________

8. $(2m^2+1)(3m^3+2m^2-4m)$ **8.** ____________

9. $(3x^2+x)(2x^2+3x-4)$ **9.** ____________

Objective **Solve application problems involving linear equations.**

Video Examples

Review this example:

Find two consecutive odd integers such that if three times the smaller is added to twice the larger, the sum is 69.

Step 1 Read the problem. We must find two consecutive odd integers.

Step 2 Assign a variable.

Let x = the lesser consecutive odd integer.
Then $x + 2$ = the greater consecutive odd integer.

Step 3 Write an equation.

Three times the smaller	is added to	twice the larger	is	69.
↓	↓	↓	↓	↓
$3x$	$+$	$2(x+2)$	$=$	69

Step 4 Solve the equation.

$$3x+2x+4=69$$
$$5x+4=69$$
$$5x=65$$
$$x=13$$

Step 5 State the answer. The lesser integer is 13. The greater is $13 + 2 = 15$.

Step 6 Check. Three times the smaller is 39, added to twice the larger, 30, is a sum of 69. The answers check.

Now Try:

The sum of four consecutive even integers is 4. Find the integers.

____

Practice Exercises

Solve each problem.

10. Find two consecutive even integers such that the smaller, added to twice the larger, is 292.

10. ____________

11. Find two consecutive integers such that the larger, added to three times the smaller, is 109.

11. ____________

12. Find three consecutive odd integers whose sum is 363.

12. ____________

Chapter 7 Rational Expressions and Applications

Learning Objectives

- Write a fraction in lowest terms using a common factor.
- Multiply signed fractions.
- Divide signed fractions.
- Add and subtract unlike fractions.
- Rewrite mixed numbers as improper fractions, or the reverse.
- Solve application problems containing mixed numbers.
- Solve application problems using linear equations.

Key Terms

Use the vocabulary terms listed below to complete each statement in exercises 1−8.

equivalent fractions	**common factor**	**lowest terms**
like fractions	**unlike fractions**	**least common denominator**
mixed number	**improper fraction**	

1. A fraction is written in ______________________________ when its numerator and denominator have no common factor other than 1.

2. A ________________________ is a number that can be divided into two or more whole numbers.

3. Two fractions are ______________________________ when they represent the same portion of a whole.

4. A(n) ________________________ includes a fraction and a whole number written together.

5. A mixed number can be rewritten as a(n) ________________________.

6. Fractions with different denominators are called ____________________.

7. Fractions with the same denominator are called ________________________.

8. The ______________________________ of two whole numbers is the smallest whole number divisible by both of the numbers.

Objective **Write a fraction in lowest terms using a common factor.**

Video Examples

Review these examples:

Write each fraction in lowest terms.

$\frac{32}{48}$

Divide both numerator and denominator by 16.

$$\frac{32}{48}=\frac{32\div 16}{48\div 16}=\frac{2}{3}$$

$\frac{28}{56}$

Suppose we thought that 4 was the greatest common factor of 28 and 56. Dividing by 4 would give

$$\frac{28}{56}=\frac{28\div 4}{56\div 4}=\frac{7}{14}$$

But $\frac{7}{14}$ is not in lowest terms, because 7 and 14 have a common factor of 7. So we divide by 7.

$$\frac{7}{14}=\frac{7\div 7}{14\div 7}=\frac{1}{2}$$

The fraction $\frac{28}{56}$ could have been written in lowest terms in one step by dividing by 28, the greatest common factor of 28 and 56.

$$\frac{28}{56}=\frac{28\div 28}{56\div 28}=\frac{1}{2}$$

Now Try:

Write each fraction in lowest terms.

$\frac{18}{27}$ ____________

$\frac{27}{45}$ ____________

Practice Exercises

Write each fraction in lowest terms.

1. $\frac{14}{49}$ **1.** ____________

2. $\frac{8}{36}$ **2.** ____________

3. $\frac{30}{42}$ **3.** ____________

Objective **Multiply signed fractions.**

Video Examples

Review these examples:

Multiply. Write the product in lowest terms.

$-\frac{7}{9}\cdot-\frac{5}{11}$

Multiply the numerators and multiply the denominators.

$$-\frac{7}{9}\cdot-\frac{5}{11}=\frac{7\cdot5}{9\cdot11}=\frac{35}{99}$$

The answer is in lowest terms because 35 and 99 have no common factor other than 1.

$-\frac{9}{7}\left(\frac{7}{15}\right)$

Multiplying a negative number times a positive number gives a negative product.

$$-\frac{9}{7}\left(\frac{7}{15}\right)=-\frac{3\cdot3\cdot7}{7\cdot3\cdot5}=-\frac{\overset{1}{\cancel{3}}\cdot3\cdot\overset{1}{\cancel{7}}}{\underset{1}{\cancel{7}}\cdot\underset{1}{\cancel{3}}\cdot5}=-\frac{3}{5}$$

Now Try:

Multiply. Write the product in lowest terms.

$-\frac{10}{11}\cdot-\frac{4}{13}$ ____________

$-\frac{11}{7}\left(\frac{14}{33}\right)$ ____________

Find $\frac{3}{8}$ of $\frac{4}{9}$.

Recall that "of" indicates multiplication.

$$\frac{3}{8}\cdot\frac{4}{9}=\frac{3\cdot2\cdot2}{2\cdot2\cdot2\cdot3\cdot3}=\frac{\overset{1}{\cancel{3}}\cdot\overset{1}{\cancel{2}}\cdot\overset{1}{\cancel{2}}}{\underset{1}{\cancel{2}}\cdot\underset{1}{\cancel{2}}\cdot2\cdot\underset{1}{\cancel{3}}\cdot3}=\frac{1}{6}$$

Find $\frac{2}{7}$ of $\frac{21}{40}$. ____________

Practice Exercises

Multiply. Write the products in lowest terms.

4. $-\frac{10}{42}\cdot\frac{3}{5}$ **4.** ____________

5. $\frac{6}{18}\cdot\frac{9}{2}$ **5.** ____________

6. $\frac{5}{9}$ of 81 **6.** ____________

Objective **Divide signed fractions.**

Video Examples

Review these examples:

Divide. Write each quotient in lowest terms.

$$\frac{5}{7} \div \frac{10}{3}$$

$$\frac{5}{7} \div \frac{10}{3} = \frac{5}{7} \cdot \frac{3}{10} = \frac{\overset{1}{\cancel{5}} \cdot 3}{7 \cdot 2 \cdot \underset{1}{\cancel{5}}} = \frac{3}{14}$$

$$6 \div \left(-\frac{1}{5}\right)$$

$$6 \div \left(-\frac{1}{5}\right) = \frac{6}{1} \cdot \left(-\frac{5}{1}\right) = -\frac{6 \cdot 5}{1 \cdot 1} = -\frac{30}{1} = -30$$

$$-\frac{7}{9} \div (-6)$$

$$-\frac{7}{9} \div (-6) = -\frac{7}{9} \cdot \left(-\frac{1}{6}\right) = \frac{7 \cdot 1}{3 \cdot 3 \cdot 3 \cdot 2} = \frac{7}{54}$$

Now Try:

Divide. Write each quotient in lowest terms.

$$\frac{5}{9} \div \frac{20}{3}$$ ____________

$$10 \div \left(-\frac{1}{8}\right)$$ ____________

$$-\frac{6}{13} \div \left(-\frac{3}{26}\right)$$ ____________

Practice Exercises

Divide. Write the quotients in lowest terms.

7. $\frac{7}{8} \div (-21)$

7. ____________

8. $-\frac{5}{12} \div \frac{15}{8}$

8. ____________

9. $-\frac{2}{3} \div \left(-\frac{7}{9}\right)$

9. ____________

Objective **Add and subtract unlike fractions.**

Video Examples

Review these examples:

Find each sum or difference.

$\frac{1}{2}+\frac{1}{6}$

Step 1 The larger denominator (6) is the LCD.

Step 2 $\frac{1}{2}=\frac{1\cdot 3}{2\cdot 3}=\frac{3}{6}$ and $\frac{1}{6}$ already has the LCD.

Step 3 Add the numerators.

$$\frac{1}{2}+\frac{1}{6}=\frac{3}{6}+\frac{1}{6}=\frac{3+1}{6}=\frac{4}{6}$$

Step 4 Write $\frac{4}{6}$ in lowest terms.

$$\frac{4}{6}=\frac{2\cdot \overset{1}{\cancel{2}}}{\underset{1}{\cancel{2}}\cdot 3}=\frac{2}{3}$$

$\frac{3}{8}-\frac{7}{12}$

Step 1 The LCD is 24.

Step 2 $\frac{3}{8}=\frac{3\cdot 3}{8\cdot 3}=\frac{9}{24}$ and $\frac{7}{12}=\frac{7\cdot 2}{12\cdot 2}=\frac{14}{24}$

Step 3 Subtract the numerators.

$$\frac{3}{8}-\frac{7}{12}=\frac{9}{24}-\frac{14}{24}=\frac{9-14}{24}=\frac{-5}{24},\text{ or }-\frac{5}{24}$$

Step 4 $-\frac{5}{24}$ is in lowest terms.

$-\frac{7}{18}+\frac{5}{12}$

Step 1 Use prime factorization to find the LCD.

$18=2\cdot 3\cdot 3$

$12=2\cdot 2\cdot 3$

$\text{LCD}=2\cdot 2\cdot 3\cdot 3=36$

Step 2

$-\frac{7}{18}=-\frac{7\cdot 2}{18\cdot 2}=-\frac{14}{36}$ and $\frac{5}{12}=\frac{5\cdot 3}{12\cdot 3}=\frac{15}{36}$

Step 3 Add the numerators.

$$-\frac{7}{18}+\frac{5}{12}=-\frac{14}{36}+\frac{15}{36}=\frac{-14+15}{36}=\frac{1}{36}$$

Step 4 $\frac{1}{36}$ is in lowest terms.

Now Try:

Find each sum or difference.

$\frac{5}{9}+\frac{7}{18}$

____

$\frac{8}{15}-\frac{7}{10}$

____

$-\frac{5}{24}+\frac{7}{9}$

____

Practice Exercises

Find each sum or difference. Write all answers in lowest terms.

10. $\frac{1}{6}+\frac{2}{15}$ **10.** ________________

11. $-\frac{1}{2}+\frac{7}{12}$ **11.** ________________

12. $\frac{33}{40}-\frac{7}{24}$ **12.** ________________

Objective Rewrite mixed numbers as improper fractions, or the reverse.

Video Examples

Review these examples:

Write $9\frac{3}{4}$ as an improper fraction.

Step 1 $9\frac{3}{4}$ $4 \cdot 9 = 36$ Then $36 + 3 = 39$

Step 2 $9\frac{3}{4} = \frac{39}{4}$

Write the improper fraction as an equivalent mixed number in simplest form.

$\frac{18}{7}$

Divide 18 by 7.

$$\begin{array}{r} 2 \\ 7\overline{)18} \\ \underline{14} \\ 4 \end{array}$$

The quotient 2 is the whole number part. The remainder 4 is the numerator of the fraction, and the denominator stays as 7.

$\frac{18}{7} = 2\frac{4}{7}$

Now Try:

Write $5\frac{5}{6}$ as an improper fraction.

____

Write the improper fraction as an equivalent mixed number in simplest form.

$\frac{27}{8}$

____

Practice Exercises

Write each mixed number as an improper fraction.

13. $-8\frac{2}{7}$

13. ____________

14. $-1\frac{7}{9}$

14. ____________

Write the improper fraction as a mixed number in simplest form.

15. $\frac{26}{3}$

15. ____________

Objective **Solve application problems containing mixed numbers.**

Video Examples

Review this example:

First, estimate the answer to the application problem. Then find the exact answer.

George's daughter grew $1\frac{1}{3}$ inches last year and $2\frac{1}{5}$ inches this year. How much has her height increased over the two years?

First, round each mixed number to the nearest whole number.

$1\frac{1}{3}$ rounds to 1 and $2\frac{1}{5}$ rounds to 2

Using the rounded numbers, we add.

$1+2=3 \leftarrow$ Estimate

To find the exact answer, use the original mixed numbers and add.

$$1\frac{1}{3}+2\frac{1}{5}=\frac{4}{3}+\frac{11}{5}=\frac{20}{15}+\frac{33}{15}=\frac{20+33}{15}$$
$$=\frac{53}{15}=3\frac{8}{15}$$

Her height increased $3\frac{8}{15}$ in. over the two years.

This result is close to the estimate of 3 in.

Now Try:

First, estimate the answer to the application problem. Then find the exact answer.

A plumber has three pieces of pipe measuring $2\frac{1}{5}$ ft, $3\frac{3}{4}$ ft, and $4\frac{1}{8}$ ft. What is the total length of pipe?

Estimate ____________

Exact ____________

Practice Exercises

First, estimate the answer to each application problem. Then find the exact answer.

16. A living room has dimensions $3\frac{3}{4}$ meters by $3\frac{1}{3}$ meters. What is the area of the room?

16.
Estimate ____________

Exact ____________

17. Suppose that a pair of pants requires $3\frac{1}{8}$ yd of material. How much material would be needed for 6 pairs of pants?

17.
Estimate ____________

Exact ____________

Objective **Solve application problems using linear equations.**

Video Examples

Review these examples:

A driver averaged 58 mph and took 10 hours to drive from Little Rock to Indianapolis. What is the distance between Little Rock and Indianapolis?

We must find the distance, given the rate and time using $rt = d$.

$58 \cdot 10 = 580$ miles

The distance is 580 miles.

Now Try:

A driver averaged 54 mph and took 5 hours to drive from Los Angeles to Las Vegas. What is the distance between Los Angeles and Las Vegas?

____

A car and a truck leave Oklahoma City at the same time and travel west on the same route. The car travels at a constant rate of 62 mph. The truck travels at a constant rate of 68 mph. In how many hours will the distance between them be 30 miles?

Step 1 Read the problem.

Step 2 Assign a variable. We are looking for time.
Let t = the number of hours until the distance between them is 30 miles.

	Rate	Time	Distance
Truck	68	t	$68t$
Car	62	t	$62t$

Step 3 Write an equation.

$68t - 62t = 30$

Step 4 Solve.

$6t = 30$

$t = 5$

Step 5 State the answer. It will take 5 hours for the truck and car to be 30 miles apart.

Step 6 Check. After 5 hours, the truck travels $68 \cdot 5 = 340$ miles and the car travels $62 \cdot 5 = 310$ miles. The difference is $340 - 310 = 30$, as required.

A car and a truck leave Dallas at the same time and travel north on the same route. The car travels at a constant rate of 67 mph. The truck travels at a constant rate of 72 mph. In how many hours will the distance between them be 25 miles?

____

Practice Exercises

Solve each problem.

18. A driver averages 52 mph and took 10 hours to drive from Charlotte, North Carolina to Orlando, Florida. What is the distance between Charlotte and Orlando? **18.** ________________

19. A driver averages 50 mph and took 9 hours to drive from Denver, Colorado to Albuquerque, New Mexico. What is the distance between Denver and Albuquerque? **19.** ________________

20. A car and a truck leave Billings, Montana at the same time and travel east on the same route. The truck travels at a constant rate of 78 mph. The car travels at a constant rate of 67 mph. In how many hours will the distance between them be 33 miles? **20.** ________________

Chapter 8 Roots and Radicals

Learning Objectives

Use exponents to write repeated factors.
Simplify expressions containing exponents.
Combine like terms.
Use the four steps for solving a linear equation.

Key Terms

Use the vocabulary terms listed below to complete each statement in exercises 1−3.

exponent **term** **like terms**

1. A number, a variable, or a product or quotient of a number and one or more variables raised to powers is called a(n) ______________________________.

2. Terms with exactly the same variables, including the same exponents, are called ______________________________.

3. A(n) ______________________ tells how many times a number is used as a factor in repeated multiplication.

Objective **Use exponents to write repeated factors.**

Video Examples

Review these examples:

Given the factored form, give the exponential form, simplified form, and how it is read.

$4 \cdot 4 \cdot 4$

Exponential form: 4^3
Simplified: 64
Read as: 4 cubed, or 4 to the third power

$(3)(3)$

Exponential form: 3^2
Simplified: 9
Read as: 3 squared, or 3 to the second power

11

Exponential form: 11^1
Simplified: 11
Read as: 11 to the first power

Now Try:

Given the factored form, give the exponential form, simplified form, and how it is read.

$5 \cdot 5 \cdot 5 \cdot 5$

____

$(6)(6)$

____

23

____

Practice Exercises

Rewrite each number in factored form as a number in exponential form; then state how the exponential form is read.

1. $12 \cdot 12 \cdot 12$

1. ____________

____

2. $6 \cdot 6 \cdot 6 \cdot 6 \cdot 6$

2. ____________

____

3. $7 \cdot 7 \cdot 7 \cdot 7 \cdot 7 \cdot 7 \cdot 7$

3. ____________

____

Objective **Simplify expressions containing exponents.**

Video Examples

Review these examples:
Simplify.

$(-4)^2$

$(-4)^2 = (-4)(-4) = 16$

$(-3)^3$

$$\begin{aligned}(-3)^3 &= (-3)(-3)(-3)\\ &= 9(-3)\\ &= -27\end{aligned}$$

$(-4)^4$

$(-4)^4 = (-4)(-4)(-4)(-4) = 256$

Now Try:
Simplify.

$(-7)^2$ __________

$(-10)^3$ __________

$(-3)^4$ __________

Practice Exercises

Simplify.

4. $(-1)^{91}$

5. $2^4 \cdot (-3)^2$

6. $(-2)^5 \cdot (-3)^2$

4. __________

5. __________

6. __________

Objective **Combine like terms.**

Video Examples

Review these examples:

Combine like terms in each expression.

$8r + 5r + 4r$

$$8r + 5r + 4r = (8 + 5 + 4)r$$
$$= 17r$$

$9x + x$

$$9x + x = 9x + 1x$$
$$= (9 + 1)x$$
$$= 10x$$

$15x^2 - 8x^2$

$$15x^2 - 8x^2 = (15 - 8)x^2$$
$$= 7x^2$$

Now Try:

Combine like terms in each expression.

$14r + 7r + 2r$

____

$18x + x$

____

$17x^2 - 9x^2$

____

Simplify each expression.

$8k - 5 - 4(7 - 3k)$

$$8k - 5 - 4(7 - 3k) = 8k - 5 - 4(7) - 4(-3k)$$
$$= 8k - 5 - 28 + 12k$$
$$= 20k - 33$$

$-\frac{3}{5}(x - 10) - \frac{1}{10}x$

$$-\frac{3}{5}(x - 10) - \frac{1}{10}x = -\frac{3}{5}x - \frac{3}{5}(-10) - \frac{1}{10}x$$
$$= -\frac{3}{5}x + 6 - \frac{1}{10}x$$
$$= -\frac{6}{10}x + 6 - \frac{1}{10}x$$
$$= -\frac{7}{10}x + 6$$

Simplify each expression.

$7k - 9 - 5(3 - 6k)$

____

$-\frac{3}{4}(x - 8) - \frac{1}{2}x$

____

Practice Exercises

Simplify.

7. $12y-7y^2+4y-3y^2$ **7.** ________________

8. $-4(x+4)+2(3x+1)$ **8.** ________________

9. $2.5(3y+1)-4.5(2y-3)$ **9.** ________________

Objective **Use the four steps for solving a linear equation.**

Video Examples

Review these examples:

Solve $-5x+8=23$.

Step 1 There are no parentheses, fractions, or decimals in this equation, so this step is not necessary.

$$-5x+8=23$$

Step 2 $$-5x+8-8=23-8$$

$$-5x=15$$

Step 3 $$\frac{-5x}{-5}=\frac{15}{-5}$$

$$x=-3$$

Step 4 Check by substituting –3 for x in the original equation.

$$-5x+8=23$$

$$-5(-3)+8\stackrel{?}{=}23$$

$$15+8\stackrel{?}{=}23$$

$$23=23 \quad \text{True}$$

The solution, –3, checks, so the solution set is $\{-3\}$.

Now Try:

Solve $-8x+11=59$.

____

Solve $4x+3=6x-11$.

Step 1 There are no parentheses, fractions, or decimals in this equation, so begin with Step 2.

$$4x+3=6x-11$$

Step 2 $$4x+3-4x=6x-11-4x$$

$$3=2x-11$$

$$3+11=2x-11+11$$

$$14=2x$$

Step 3 $$\frac{14}{2}=\frac{2x}{2}$$

$$7=x$$

Step 4 Check by substituting 7 for x in the original equation.

$$4x+3=6x-11$$

$$4(7)+3\stackrel{?}{=}6(7)-11$$

$$28+3\stackrel{?}{=}42-11$$

$$31=31 \quad \text{True}$$

The solution, 7, checks, so the solution set is $\{7\}$.

Solve $5x+4=8x-20$.

____

Solve $9a-(4+3a)=2a+5$.

$$9a-(4+3a)=2a+5$$

Step 1 $\quad 9a-4-3a=2a+5$

$$6a-4=2a+5$$

Step 2 $\quad 6a-4-2a=2a+5-2a$

$$4a-4=5$$

$$4a-4+4=5+4$$

$$4a=9$$

Step 3 $\quad \dfrac{4a}{4}=\dfrac{9}{4}$

$$a=\frac{9}{4}$$

Step 4 Check that the solution set is $\left\{\frac{9}{4}\right\}$.

Solve $10a-(11+3a)=5a+4$.

____

Practice Exercises

Solve each equation and check your solution.

10. $7t+6=11t-4$

10. ________________

11. $3a-6a+4(a-4)=-2(a+2)$

11. ________________

12. $3(t+5)=6-2(t-4)$

12. ________________

Chapter 9 Quadratic Equations

Learning Objectives

Simplify, then use the addition property of equality.
Simplify, then use the multiplication property of equality.
Solve equations using the distributive, addition, and division properties.

Key Terms

Use the vocabulary terms listed below to complete each statement in exercises 1–3.

distributive property

division property of equality

addition property of equality

1. The ______________________________ states that if $a = b$, then $a + c = b + c$.

2. The ______________________________ states that $a(b + c) = ab + bc$.

3. The ______________________________ states that if $a = b$, then $\frac{a}{c} = \frac{b}{c}$ for $c \neq 0$.

Objective Simplify, and then use the addition property of equality.

Video Examples

Review these examples:

Solve $5t-16+t+4=9+5t+6$.

$$5t-16+t+4=9+5t+6$$
$$6t-12=15+5t$$
$$6t-12-5t=15+5t-5t$$
$$t-12=15$$
$$t-12+12=15+12$$
$$t=27$$

Check by substituting 27 in the original equation. The solution set is $\{27\}$.

Now Try:

Solve $8t-9+t+7=12+8t+15$.

____

Solve $4(3+6x)-(5+23x)=19$.

$$4(3+6x)-(5+23x)=19$$
$$4(3)+4(6x)-1(5)-1(23x)=19$$
$$12+24x-5-23x=19$$
$$x+7=19$$
$$x+7-7=19-7$$
$$x=12$$

Check by substituting 12 in the original equation. The solution set is $\{12\}$.

Solve $5(7+8x)-(29+39x)=14$.

____

Practice Exercises

Solve each equation. First simplify each side of the equation as much as possible. Check each solution.

1. $3(t+3)-(2t+7)=9$ **1.** _______________

2. $-4(5g-7)+3(8g-3)=15-4+3g$ **2.** _______________

3. $3.6p+4.8+4.0p=8.6p-3.1+0.7$ **3.** _______________

Objective **Simplify, and then use the division property of equality.**

Video Examples

Review this example:
Solve $9m + 4m = 39$.

$$9m + 4m = 39$$
$$13m = 39$$
$$\frac{13m}{13} = \frac{39}{13}$$
$$m = 3$$

Check by substituting 3 in the original equation.
The solution set is $\{3\}$.

Now Try:
Solve $12m + 8m = 80$.

____

Practice Exercises

Solve each equation and check your solution.

4. $-7b + 12b = 125$

4. ______________

5. $3w - 7w = 20$

5. ______________

6. $-11h - 6h + 14h = -21$

6. ______________

Objective **Solve equations using the distributive, addition, and division properties.**

Video Examples

Review these examples:

Solve this equation and check the solution:
$-12 = 6(y-2)$.

We can use the distributive property to simplify the right side of the equation. Then use the steps to solve for y.

$$-12 = 6(y-2)$$
$$-12 = 6 \cdot y - 6 \cdot 2$$
$$-12 = 6y - 12$$
$$-12 = 6y + (-12)$$
$$\underline{12} \qquad \underline{\qquad 12}$$
$$0 = 6y$$
$$\frac{0}{6} = \frac{6y}{6}$$
$$0 = y$$

The solution is 0.

Check Go back to the original equation and replace y with 0.

$$-12 = 6(y-2)$$
$$-12 = 6(0-2)$$
$$-12 = 6[0+(-2)]$$
$$-12 = 6(-2)$$
$$-12 = -12$$

When y is replaced with 0, the equation balances, so 0 is the correct solution.

Now Try:

Solve this equation and check the solution: $8(m-3) = -24$.

____

Solve this equation and check the solution:
$3 + 6(n+5) = 5 + 2n$.

Step 1 Use the distributive property on the left side.

$$3 + 6(n+5) = 5 + 2n$$
$$3 + 6n + 30 = 5 + 2n$$

Step 2 Combine like terms on the left side.

$$6n + 33 = 5 + 2n$$

Solve this equation and check the solution: $5 + 8(m+2) = 5m + 6$.

____

Step 3 Add –2*n* to both sides.

$$\underline{-2n} \qquad \underline{-2n}$$

$$4n + 33 = 5 + 0$$

$$4n + 33 = 5$$

Step 3 To get 4*n* by itself, add –33 to both sides.

$$\underline{\quad -33} \quad \underline{-33}$$

$$4n + 0 = -28$$

Step 4 Divide both sides by 4, the coefficient of the variable term 4*n*.

$$\frac{4n}{4} = \frac{-28}{4}$$

$$n = -7$$

Step 5 Check

$$3 + 6(n + 5) = 5 + 2n$$

$$3 + 6(-7 + 5) = 5 + 2(-7)$$

$$3 + 6(-2) = 5 + (-14)$$

$$3 + (-12) = -9$$

$$-9 = -9$$

When *n* is replaced with –7, the equation balances, so –7 is the correct solution.

Practice Exercises

Solve each equation and check.

7. $40 = 5(b + 8)$

7. ______________

8. $t + 28 - 10 = 4(t + 6) - 24$

8. ______________

9. $-87 + 3g = 9(g - 5) + 8g$

9. ______________

Name: Date:
Instructor: Section:

Chapter R PREALGEBRA REVIEW

R.1 Fractions

Learning Objectives

1. Learn the definition of *factor*.
2. Write fractions in lowest terms.
3. Convert between improper fractions and mixed numbers.
4. Multiply and divide fractions.
5. Add and subtract fractions.
6. Solve applied problems that involve fractions.
7. Interpret data from a circle graph.

Key Terms

Use the vocabulary terms listed below to complete each statement in exercises 1–9.

numerator	**denominator**	**proper fraction**
improper fraction	**equivalent fractions**	**lowest terms**
prime number	**composite number**	**prime factorization**

1. Two fractions are ________________________ when they represent the same portion of a whole.

2. A fraction whose numerator is larger than its denominator is called an ____________________.

3. In the fraction $\frac{2}{9}$, the 2 is the ________________________.

4. A fraction whose denominator is larger than its numerator is called a ____________________.

5. The ________________________ of a fraction shows the number of equal parts in a whole.

6. A ________________________ has at least one factor other than itself and 1.

7. In a ________________________ every factor is a prime number.

8. The factors of a ________________________ are itself and 1.

9. A fraction is written in ________________________ when its numerator and denominator have no common factor other than 1.

Name: Date:
Instructor: Section:

Objective 1 Learn the definition of *factor*.

Video Examples

Review this example for Objective 1:

1. Write the number in prime factored form.

48

We use a factor tree, as shown below. The prime factors are boxed.

Divide by the least prime factor of 54, which is 2. $54 = 2 \cdot 27$

Divide 27 by 3 to find two factors of 27. $54 = 2 \cdot 3 \cdot 9$

Now factor 9 as $3 \cdot 3$. $54 = 2 \cdot 3 \cdot 3 \cdot 3$

$$54 \to \boxed{2} \cdot 27 \to \boxed{3} \cdot 9 \to \boxed{3} \cdot \boxed{3}$$

Now Try:

1. Write the number in prime factored form.
210

____

Objective 1 Practice Exercises

For extra help, see Example 1 on page 2 of your text.

Write each number in prime factored form.

1. 98 **1.** __________

2. 256 **2.** __________

3. 546 **3.** __________

Objective 2 Write fractions in lowest terms.

Video Examples

Review this example for Objective 2:

2. Write the fraction in lowest terms.

$\frac{15}{25}$

$$\frac{15}{25} = \frac{3 \cdot 5}{5 \cdot 5} = \frac{3}{5} \cdot \frac{5}{5} = \frac{3}{5} \cdot 1 = \frac{3}{5}$$

Now Try:

2. Write the fraction in lowest terms.
$\frac{9}{15}$

____

Name: Date:
Instructor: Section:

Objective 2 Practice Exercises

For extra help, see Example 2 on page 3 of your text.

Write each fraction in lowest terms.

4. $\frac{42}{150}$ **4.** ______________

5. $\frac{180}{216}$ **5.** ______________

6. $\frac{132}{292}$ **6.** ______________

Objective 3 Convert between improper fractions and mixed numbers.

Video Examples

Review these examples for Objective 3:

3. Write $\frac{53}{6}$ as a mixed number.

We divide the numerator of the improper fraction by the denominator.

$$\begin{array}{r} 8 \\ 6\overline{)53} \\ \underline{48} \\ 5 \end{array} \qquad \frac{53}{6} = 8\frac{5}{6}$$

4. Write $5\frac{3}{8}$ as an improper fraction.

We multiply the denominator of the fraction by the whole number and add the numerator to get the numerator of the improper fraction.

$8 \cdot 5 + 3 = 40 + 3 = 43$

The denominator of the improper fraction is the same as the denominator in the mixed number, which is 8 here. Thus, $5\frac{3}{8} = \frac{43}{8}$

Now Try:

3. Write $\frac{74}{5}$ as a mixed number.

____

4. Write $12\frac{2}{7}$ as an improper fraction.

____

Name: Date:
Instructor: Section:

Objective 3 Practice Exercises

For extra help, see Examples 3–4 on page 4 of your text.

Write the improper fraction as a mixed number.

7. $\frac{321}{15}$

7. ______________

Write each mixed number as an improper fraction.

8. $13\frac{5}{9}$

8. ______________

9. $22\frac{2}{11}$

9. ______________

Objective 4 Multiply and divide fractions.

Video Examples

Review these examples for Objective 4:

5. Find the product, and write it in lowest terms.

$\frac{5}{12}\cdot\frac{3}{10}$

$$\frac{5}{12}\cdot\frac{3}{10}=\frac{5\cdot 3}{12\cdot 10}$$ Multiply numerators. Multiply denominators.

$$=\frac{5\cdot 3}{4\cdot 3\cdot 2\cdot 5}$$ Factor the denominator.

$$=\frac{1}{4\cdot 2}$$ $\frac{3}{3}=1$ and $\frac{5}{5}=1$

$$=\frac{1}{8}$$ Write in lowest terms.

6. Find the quotient, and write it in lowest terms.

$\frac{2}{5}\div\frac{8}{7}$

$$\frac{2}{5}\div\frac{8}{7}=\frac{2}{5}\cdot\frac{7}{8}$$ Multiply by the reciprocal.

$$=\frac{2\cdot 7}{5\cdot 4\cdot 2}$$ Multiply and factor.

$$=\frac{7}{20}$$

Now Try:

5. Find the product, and write it in lowest terms.

$\frac{7}{15}\cdot\frac{3}{14}$

____

6. Find the quotient, and write it in lowest terms.

$\frac{6}{7}\div\frac{9}{8}$

____

Name: Date:
Instructor: Section:

Objective 4 Practice Exercises

For extra help, see Examples 5–6 on pages 4–6 of your text.

Find each product or quotient, and write it in lowest terms.

10. $\frac{25}{11}\cdot\frac{33}{10}$ **10.** ________

11. $\frac{5}{4}\div\frac{25}{28}$ **11.** ________

12. $4\frac{3}{8}\cdot 2\frac{4}{7}$ **12.** ________

Objective 5 Add and subtract fractions.

Video Examples

Review these examples for Objective 5:

7. Add. Write the sum in lowest terms.

$\frac{5}{24}+\frac{7}{24}$

Add numerators. Keep the same denominator.

$\frac{5}{24}+\frac{7}{24}=\frac{5+7}{24}=\frac{12}{24}$, or $\frac{1}{2}$

Write in lowest terms.

8. Add. Write the sum in lowest terms.

$\frac{5}{21}+\frac{3}{14}$

Step 1 To find the LCD, factor the denominators to prime factored form.

$21 = 3\cdot 7$ and $14 = 2\cdot 7$

7 is a factor of both denominators.

21 14
/\ /\

Step 2 LCD $= 3\cdot 7\cdot 2 = 42$

In this example, the LCD needs one factor of 3, one factor of 7 and one factor of 2.

Now Try:

7. Add. Write the sum in lowest terms.

$\frac{5}{16}+\frac{7}{16}$

____

8. Add. Write the sum in lowest terms.

$\frac{7}{12}+\frac{3}{8}$

____

Step 3 Now we can use the second property of 1 to write each fraction with 42 as the denominator.

$\frac{5}{21}=\frac{5}{21}\cdot\frac{2}{2}=\frac{10}{42}$ and $\frac{3}{14}=\frac{3}{14}\cdot\frac{3}{3}=\frac{9}{42}$

Now add the two equivalent fractions to get the sum.

$$\frac{5}{21}+\frac{3}{14}=\frac{10}{42}+\frac{9}{42}$$
$$=\frac{19}{42}$$

9. Subtract. Write difference in lowest terms.

$\frac{26}{9}-\frac{5}{9}$

Subtract numerators. Keep the same denominator.

$$\frac{26}{9}-\frac{5}{9}=\frac{26-5}{9}$$
$$=\frac{21}{9}$$
$$=\frac{7}{3}, \text{ or } 2\frac{1}{3}$$

9. Subtract. Write difference in lowest terms.

$\frac{11}{18}-\frac{7}{18}$

____

Objective 5 Practice Exercises

For extra help, see Example 7–9 on pages 7–10 of your text.

Find each sum or difference, and write it in lowest terms.

13. $\frac{23}{45}+\frac{47}{75}$ **13.** __________

14. $2\frac{3}{4}+7\frac{2}{3}$ **14.** __________

15. $12\frac{5}{6}-7\frac{7}{8}$ **15.** __________

Name: Date:
Instructor: Section:

Objective 6 Solve applied problems that involve fractions.

Video Examples

Review this example for Objective 6:

10. Pauline and her two children picked cherries. Pauline picked $2\frac{3}{4}$ quarts, Jennie picked $1\frac{2}{3}$ quarts, and Dan picked $1\frac{1}{2}$ quarts. How many quarts of cherries did they pick?

Use Method 2.

$$\begin{array}{r} 2\frac{3}{4}=2\frac{9}{12} \\ 1\frac{2}{3}=1\frac{8}{12} \\ +\ 1\frac{1}{2}=1\frac{6}{12} \\ \hline 4\frac{23}{12} \end{array}$$

Since $\frac{23}{12}=1\frac{11}{12}$, $4\frac{23}{12}=4+1\frac{11}{12}=5\frac{11}{12}$ quarts.

Now Try:

10. A punch is made with $3\frac{1}{3}$ cups of ginger ale, $1\frac{1}{2}$ cups of orange juice, $1\frac{2}{3}$ cups of lemonade, and $2\frac{1}{4}$ cups of pineapple juice.

Find the total number of cups in the punch.

____

Objective 6 Practice Exercises

For extra help, see Example 10 on page 10 of your text.

Solve each applied problem. Write each answer in lowest terms.

16. Arnette worked $24\frac{1}{2}$ hours and earned \$9 per hour. How much did she earn?

16. __________

17. Debbie made a shirt with $3\frac{1}{8}$ yards of material, a dress with $4\frac{7}{8}$ yards, and a jacket with $3\frac{3}{4}$ yards. How many yards of material did she use?

17. __________

18. Three sides of a parking lot are $35\frac{1}{4}$ yards, $42\frac{7}{8}$ yards, and $32\frac{3}{4}$ yards. If the total distance around the lot is $145\frac{1}{2}$ yards, find the length of the fourth side.

18. __________

Name: Date:
Instructor: Section:

Objective 7 Interpret data from a circle graph.

Video Examples

In August 2014, 1300 workers were surveyed on where they eat during their lunch time. The circle graph shows the approximate fractions of locations.

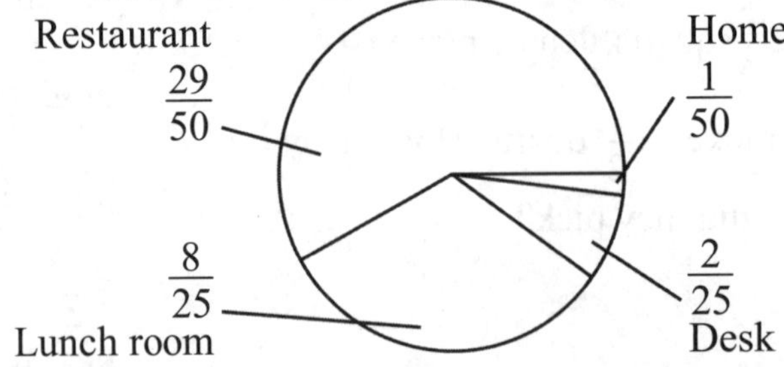

Review these examples for Objective 7:
11.

a. Which location has the largest share of workers? What was the share?

In the circle graph, the sector for Restaurant is the largest, so Restaurant had the largest share of workers, $\frac{29}{50}$.

b. How many actual workers ate in the Lunch Room?

Multiply the actual fraction from the graph of the Lunch Room by the number of workers surveyed.

$$\begin{aligned}\frac{8}{25}\cdot 1300 &= \frac{8}{25}\cdot\frac{1300}{1} \\ &= \frac{10{,}400}{25} \\ &= 416\end{aligned}$$

Thus, 416 workers ate in the Lunch Room.

Now Try:
11.

a. Which location had the smallest share of workers? What was the share?

____

b. How many actual workers ate at their Desk?

____

Name: Date:
Instructor: Section:

Objective 7 Practice Exercises

For extra help, see Example 11 on page 11 of your text.

In August 2014, 1300 workers were surveyed on where they eat during their lunch time. The circle graph shows the approximate fractions of locations.

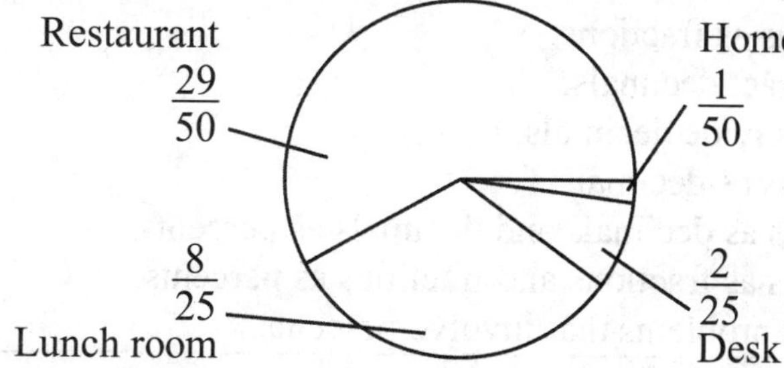

19. What location had the second-largest number of workers? **19.** ________

20. Estimate the number of workers who eat at a Restaurant. **20.** ________

21. How many actual workers ate at a Restaurant? **21.** ________

Name: Date:
Instructor: Section:

Chapter R PREALGEBRA REVIEW

R.2 Decimals and Percents

Learning Objectives

1	Write decimals as fractions.
2	Add and subtract decimals.
3	Multiply and divide decimals.
4	Write fractions as decimals.
5	Write percents as decimals and decimals as percents.
6	Write percents as fractions and fractions as percents.
7	Solve applied problems that involve percents.

Key Terms

Use the vocabulary terms listed below to complete each statement in exercises 1–3.

decimals **place value** **percent**

1. We use ____________________ to show parts of a whole.

2. A ____________________________ is assigned to each place to the left or right of the decimal point.

3. ____________________ means per one hundred.

Objective 1 Write decimals as fractions.

Video Examples

Review these examples for Objective 1:

1. Write each decimal as a fraction. Do not write in lowest terms.

a. 0.87

We read 0.87 as "eighty-seven hundredths," so the fraction form is $\frac{87}{100}$. Using the shortcut method, since there are two places to the right of the decimal point, there will be two zeros in the denominator.

$$0.87 = \frac{87}{100}$$

2 places 2 zeros

b. 0.043

We read 0.043 as "forty-three thousandths."

$$0.043 = \frac{43}{1000}$$

3 places 3 zeros

Now Try:

1. Write each decimal as a fraction. Do not write in lowest terms.

a. 0.72

____

b. 0.053

____

c. 5.3084

Here we have 4 places.

$$5.3084 = 5 + 0.3084$$

$$= \frac{50{,}000}{10{,}000} + \frac{3084}{10{,}000} \quad \text{The LCD is 10,000.}$$

$$= \frac{53{,}084}{10{,}000}$$

4 zeros

c. 3.7058

______

Objective 1 Practice Exercises

For extra help, see Example 1 on page 17 of your text.

Write each decimal as a fraction. Do not write in lowest terms.

1. 0.007   1. ______

2. 18.03   2. ______

3. 30.0005   3. ______

Objective 2 Add and subtract decimals.

Video Examples

Review these examples for Objective 2:

2. Add or subtract as indicated.

 a. 7.34 + 15.6 + 2.419

 Place the digits of the numbers in columns, so that tenths are in one column, hundredths in another column, and so on.

$$\begin{array}{r} 7.34 \\ 15.6 \\ +\ 2.419 \\ \hline 25.359 \end{array} \quad \text{Align decimal points.}$$

 To avoid errors, attach zeros to make all the numbers the same length.

$$\begin{array}{r} 7.34 \\ 15.6 \\ +\ 2.419 \\ \hline \end{array} \quad \text{becomes} \quad \begin{array}{r} 7.340 \\ 15.600 \\ +\ 2.419 \\ \hline 25.359 \end{array}$$

Now Try:

2. Add or subtract as indicated.

 a. 5.23 + 28.9 + 3.741

______

b. 56.8 – 42.307

$$\begin{array}{r} 56.8 \\ -\ 42.307 \\ \hline \end{array} \quad \text{becomes} \quad \begin{array}{r} 56.800 \\ -\ 42.307 \\ \hline 14.493 \end{array}$$

b. 64.5 – 37.218 __________

Objective 2 Practice Exercises

For extra help, see Example 2 on pages 17–18 of your text.

Add or subtract as indicated.

4. 45.83 + 20.923 + 5.7 **4.** __________

5. 768.5 – 13.402 **5.** __________

6. 689 – 79.832 **6.** __________

Objective 3 Multiply and divide decimals.

Video Examples

Review these examples for Objective 3:

3. Multiply.

a. 37.4×5.26

There is 1 decimal place in the first number and 2 decimal places in the second number. Therefore, there are 1 + 2 = 3 decimal places in the answer.

$$\begin{array}{r} 37.4 \\ \times\ 5.26 \\ \hline 2244 \\ 748 \\ 1870 \\ \hline 196.724 \end{array}$$

b. 0.08×0.6

Here $8 \times 6 = 48$. There are 2 decimal places in the first number and 1 decimal place in the second number. Therefore, there are 2 + 1 = 3 decimal places in the answer.

$0.08 \times 0.6 = 0.048$

Now Try:

3. Multiply.

a. 26.8×9.37 __________

b. 0.04×0.7 __________

Name: Date:
Instructor: Section:

4. Divide.

a. $1191.45 \div 23.5$

Write the problem as follows.

$23.5\overline{)1191.45}$

To change 23.5 into a whole number, move the decimal point one place to the right. Move the decimal point in 1191.45 the same number of places to the right, to get 11,914.5.

$235\overline{)11{,}914.5}$

Move the decimal point straight up and divide as with whole numbers.

$$\begin{array}{r} 50.7 \\ 235\overline{)11{,}914.5} \\ \underline{1175} \\ 1645 \\ \underline{1645} \\ 0 \end{array}$$

b. $9.581 \div 5.78$ (Round the answer to two decimal places.)

Move the decimal point two places to the right in 5.78, to get 578. Do the same thing with 9.581 to get 958.1.

$578\overline{)958.1}$

Move the decimal point straight up and divide as with whole numbers.

$$\begin{array}{r} 1.657 \\ 578\overline{)958.100} \\ \underline{578} \\ 3801 \\ \underline{3468} \\ 3330 \\ \underline{2890} \\ 4400 \\ \underline{4046} \\ 354 \end{array}$$

We carried out the division to three decimal places so that we could round to two decimal places, obtaining the quotient 1.66.

4. Divide.

a. $2472.12 \div 76.3$

____

b. $7.648 \div 5.36$ (Round the answer to two decimal places.)

____

Name: Date:
Instructor: Section:

Objective 3 Practice Exercises

For extra help, see Examples 3–5 on pages 18–20 of your text.

Multiply or divide as indicated.

7. 14.64×0.16 **7.** ________________

8. $498.624 \div 21.2$ **8.** ________________

9. $429.2 \div 1000$ **9.** ________________

Objective 4 Write fractions as decimals.

Video Examples

Review these examples for Objective 4:

6. Write each fraction as a decimal.

a. $\frac{27}{8}$

Divide 27 by 8. Add a decimal point and as many 0s as necessary.

$$\begin{array}{r} 3.375 \\ 8\overline{)27.000} \\ \underline{24} \\ 3\,0 \\ \underline{2\,4} \\ 60 \\ \underline{56} \\ 40 \\ \underline{40} \\ 0 \end{array}$$

$\frac{27}{8} = 3.375$

Now Try:

6. Write each fraction as a decimal.

a. $\frac{7}{20}$

____

b. $\frac{26}{9}$

$$\begin{array}{r} 2.888... \\ 9\overline{)26.000...} \\ \underline{18} \\ 80 \\ \underline{72} \\ 80 \\ \underline{72} \\ 80 \\ \underline{72} \\ 8 \end{array}$$

$\frac{26}{9} = 2.888...$

The remainder is never 0. Because 8 is always left after the subtraction, this quotient is a repeating decimal. A convenient notation for a repeating decimal is a bar over the digit (or digits) that repeats.

$\frac{26}{9} = 2.888...$ or $2.\overline{8}$

b. $\frac{32}{9}$

____

Objective 4 Practice Exercises

For extra help, see Example 6 on page 20 of your text.

Write each fraction as a decimal. For repeating decimals, write the answer two ways: using the bar notation and rounding to the nearest thousandth.

10. $\frac{3}{7}$

10. ________________

11. $\frac{4}{9}$

11. ________________

12. $\frac{151}{200}$

12. ________________

Name: Date:
Instructor: Section:

Objective 5 Write percents as decimals and decimals as percents.

Video Examples

Review these examples for Objective 5:

9. Convert each percent to a decimal and each decimal to a percent.

a. 54%

54% = 0.54

b. 3%

3% = 0.03

c. 0.29

0.29 = 29%

d. 4.6

4.6 = 460%

Now Try:

9. Convert each percent to a decimal and each decimal to a percent.

a. 91% __________

b. 6% __________

c. 0.43 __________

d. 5.2 __________

Objective 5 Practice Exercises

For extra help, see Examples 7–9 on pages 21–22 of your text.

Convert each percent to a decimal and each decimal to a percent.

13. 362% **13.** __________

14. 0.4% **14.** __________

15. 0.084 **15.** __________

Objective 6 Write percents as fractions and fractions as percents.

Video Examples

Review these examples for Objective 6:

10. Write each percent as a fraction. Give answers in lowest terms.

a. 12%

Recall writing 12% as a decimal.

$12\% = 12 \div 100 = 0.12$

Because 0.12 means 12 hundredths,

$$0.12 = \frac{12}{100} = \frac{12 \div 4}{100 \div 4} = \frac{3}{25}$$

Now Try:

10. Write each percent as a fraction. Give answers in lowest terms.

a. 30% __________

Name: Date:
Instructor: Section:

b. 350%

$$350\% = \frac{350}{100} = \frac{350 \div 50}{100 \div 50} = \frac{7}{2} = 3\frac{1}{2}$$

11. Write each fraction as a percent. Round to the nearest tenth as necessary.

a. $\frac{3}{5}$

$$\frac{3}{5} = \left(\frac{3}{5}\right)(100\%) = \left(\frac{3}{5}\right)\left(\frac{100}{1}\%\right) = \left(\frac{3}{5}\right)\left(\frac{20 \cdot 5}{1}\%\right)$$
$$= \frac{60}{1}\% = 60\%$$

b. $\frac{1}{15}$

$$\frac{1}{15} = \left(\frac{1}{15}\right)(100\%) = \left(\frac{1}{15}\right)\left(\frac{100}{1}\%\right)$$
$$= \left(\frac{1}{3 \cdot 5}\right)\left(\frac{5 \cdot 20}{1}\%\right)$$
$$= \frac{20}{3}\% = 6\frac{2}{3}\%$$

$6\frac{2}{3}\%$ rounds to 6.7%.

b. 125% ____________

11. Write each fraction as a percent. Round to the nearest tenth as necessary.

a. $\frac{3}{20}$ ____________

b. $\frac{1}{18}$ ____________

Objective 6 Practice Exercises

For extra help, see Examples 10–11 on pages 22–23 of your text.

Convert each percent to a fraction and each fraction to a percent.

16. 140% **16.** ____________

17. 55.6% **17.** ____________

18. $\frac{11}{40}$ **18.** ____________

Name: Date:
Instructor: Section:

Objective 7 Solve applied problems that involve percents.

Video Examples

Review this example for Objective 7:

12. A calendar with a regular price of \$12 is on sale at 46% off. Find the amount of the discount and the sale price of the calendar.

The discount is 46% of 12.

$$\begin{array}{ccc} 46\% & \text{of} & 12 \\ \downarrow & \downarrow & \downarrow \\ 0.46 & \cdot & 12 \end{array}$$

$$= 5.52$$

The discount is \$5.52. The sale price is found by subtracting.

Original price – discount = sale price

\$12 – \$5.52 = \$6.48

Now Try:

12. Olympic t-shirts are on sale at 56% off. The regular price is \$19. Find the amount of the discount and the sale price?

Objective 7 Practice Exercises

For extra help, see Example 12 on page 23 of your text.

Solve each problem.

19. Ranee bought a pair of shoes with a regular price of \$70, on sale at 20% off. Find the amount of the discount and the sale price?

19. ________________

20. Geishe's Shoes sells shoes at $33\frac{1}{3}\%$ off the regular price. Find the price of a pair of shoes normally priced at \$54, after the discount is given.

20. ________________

21. At the end of the season, a swim suit is on sale at 75% off. The regular price is \$56. Find the amount of the discount and the sale price?

21. ________________

Name: Date:
Instructor: Section:

Chapter 1 THE REAL NUMBER SYSTEM

1.1 Exponents, Order of Operations, and Inequality

Learning Objectives

1. Use exponents.
2. Use the rules for order of operations.
3. Use more than one grouping symbol.
4. Know the meanings of $\neq$, $<$, $>$, $\leq$, and $\geq$.
5. Translate word statements to symbols.
6. Write statements that change the direction of inequality symbols.

Key Terms

Use the vocabulary terms listed below to complete each statement in exercises 1–3.

exponent **base** **exponential expression**

1. A number written with an exponent is an ____________________.
2. The ____________ is the number that is a repeated factor when written with an exponent.
3. An ____________ is a number that indicates how many times a factor is repeated.

Objective 1 Use exponents.

Video Examples

Review this example for Objective 1:

1. Find the value of the exponential expression.

6^2

6^2 means $6 \cdot 6$, which equals 36.

Now Try:

1. Find the value of the exponential expression.

7^2

____

Objective 1 Practice Exercises

For extra help, see Example 1 on page 28 of your text.

Find the value of each exponential expression.

1. 3^3 1. __________

2. $\left(\frac{2}{3}\right)^4$ 2. __________

3. $(0.4)^2$ 3. __________

Name: Date:
Instructor: Section:

Objective 2 Use the rules for order of operations.

Video Examples

Review this example for Objective 2:

2. Find the value of the expression.

$6+7\cdot 3$

Multiply, then add.

$$6+7\cdot 3=6+21$$
$$=27$$

Now Try:

2. Find the value of the expression.

$5+2\cdot 9$

____

Objective 2 Practice Exercises

For extra help, see Example 2 on pages 29–30 of your text.

Find the value of each expression.

4. $20\div 5-3\cdot 1$

4. __________

5. $3\cdot 5^2-3\cdot 7-9$

5. __________

6. $6^2\div 3^2-4\cdot 3-2\cdot 5$

6. __________

Objective 3 Use more than one grouping symbol.

Video Examples

Review this example for Objective 3:

3. Find the value of the expression.

$3[9+4(7+8)]$

Start by adding inside the parentheses.

$$3[9+4(7+8)]=3[9+4(15)] \quad \text{Add.}$$
$$=3[9+60] \quad \text{Multiply.}$$
$$=3[69] \quad \text{Add.}$$
$$=207 \quad \text{Multiply.}$$

Now Try:

3. Find the value of the expression.

$5[3+4(8+2)]$

____

Name: Date:
Instructor: Section:

Objective 3 Practice Exercises

For extra help, see Example 3 on pages 30–31 of your text.

Find the value of each expression.

7. $\dfrac{10(5-3)-9(6-2)}{2(4-1)-2^2}$

7. ______________

8. $19-3[8(5-2)+6]$

8. ______________

9. $4[5+2(8-6)]+12$

9. ______________

Objective 4 Know the meanings of ≠, <, >, ≤, and ≥.

Video Examples

Review this example for Objective 4:

4. Determine whether the statement is true or false.

$$5\cdot 6-12\le 22$$
$$5\cdot 6-12\le 22$$
$$30-12\le 22$$
$$18\le 22$$

The statement is true.

Now Try:

4. Determine whether the statement is true or false.
$7\cdot 4-15\le 13$

____

Objective 4 Practice Exercises

For extra help, see Example 4 on page 32 of your text.

Tell whether each statement is true *or* false.

10. $3\cdot 4\div 2^2\ne 3$

10. ______________

11. $3.25>3.52$

11. ______________

12. $2[7(4)-3(5)]\le 45$

12. ______________

Name: Date:
Instructor: Section:

Objective 5 Translate word statements to symbols.

Video Examples

Review this example for Objective 5:

5. Write each word statement in symbols.

Thirteen is greater than or equal to nine plus four.

$13 \geq 9+4$

Now Try:

5. Write each word statement in symbols.
Nineteen is less than or equal to eleven plus 8.

____

Objective 5 Practice Exercises

For extra help, see Example 5 on page 32 of your text.

Write each word statement in symbols.

13. Seven equals thirteen minus six. **13.** __________

14. Five times the sum of two and nine is less than one hundred six. **14.** __________

15. Twenty is greater than or equal to the product of two and seven. **15.** __________

Objective 6 Write statements that change the direction of inequality symbols.

Video Examples

Review this example for Objective 6:

6. Write each statement as another true statement with the inequality symbol reversed.

a. $9>7$

$7<9$

Now Try:

6. Write each statement as another true statement with the inequality symbol reversed.
a. $15>11$

____

Objective 6 Practice Exercises

For extra help, see Example 6 on page 32 of your text.

Write each statement with the inequality symbol reversed.

16. $\frac{3}{4}>\frac{2}{3}$ **16.** __________

17. $12 \geq 8$ **17.** __________

18. $0.002>0.0002$ **18.** __________

Name: Date:
Instructor: Section:

Chapter 1 THE REAL NUMBER SYSTEM

1.2 Variables, Expressions, and Equations

Learning Objectives

1. Evaluate algebraic expressions, given values for the variables.
2. Translate word phrases to algebraic expressions.
3. Identify solutions of equations.
4. Identify solutions of equations from a set of numbers.
5. Distinguish between *equations* and *expressions*.

Key Terms

Use the vocabulary terms listed below to complete each statement in exercises 1−7.

variable	**constant**	**algebraic expression**	
equation	**solution**	**set**	**elements**

1. A(n) ____________________ is a statement that says two expressions are equal.

2. The objects that belong to a set are its ____________________ .

3. A ____________________ is a symbol, usually a letter, used to represent an unknown number.

4. A collection of numbers, variables, operation symbols, and grouping symbols is an____________________.

5. A ____________________ is collection of objects.

6. Any value of a variable that makes an equation true is a(n) ____________________ of the equation.

7. A ____________________ is a fixed, unchanging number.

Objective 1 Evaluate algebraic expressions, given values for the variables.

Video Examples

Review these examples for Objective 1:

1. Find the value of each algebraic expression for $x = 4$ and then $x = 7$.

$5x^2$

For $x = 4$,

$$\begin{aligned} 5x^2 &= 5 \cdot 4^2 && \text{Let } x = 4. \\ &= 5 \cdot 16 && \text{Square 4.} \\ &= 80 && \text{Multiply.} \end{aligned}$$

Now Try:

1. Find the value of each algebraic expression for $x = 6$ and then $x = 9$.

$7x^2$

____

Name: Date:
Instructor: Section:

For $x = 7$,

$$5x^2 = 5 \cdot 7^2 \quad \text{Let } x = 7.$$
$$= 5 \cdot 49 \quad \text{Square 7.}$$
$$= 245 \quad \text{Multiply.}$$

2. Find the value of each expression for $x = 7$ and $y = 6$.

$3x + 4y + 2$

Replace x with 7 and y with 6.

$$3x + 4y + 2 = 3 \cdot 7 + 4 \cdot 6 + 2$$
$$= 21 + 24 + 2 \quad \text{Multiply.}$$
$$= 47 \quad \text{Add.}$$

2. Find the value of each expression for $x = 8$ and $y = 4$.
$5x + 6y + 1$

____

Objective 1 Practice Exercises

For extra help, see Examples 1–2 on page 37 of your text.

Find the value of each expression if $x = 2$ *and* $y = 4$.

1. $9x - 3y + 2$

1. ______________

2. $\dfrac{2x + 3y}{3x - y + 2}$

2. ______________

3. $\dfrac{3y^2 + 2x^2}{5x + y^2}$

3. ______________

Name: Date:
Instructor: Section:

Objective 2 Translate word phrases to algebraic expressions.

Video Examples

Review this example for Objective 2:

3. Write the word phrase as an algebraic expression, using x as the variable.

The product of 15 and a number

$15 \cdot x$, or $15x$

Now Try:

3. Write the word phrase as an algebraic expression, using x as the variable.
The product of 20 and a number

____

Objective 2 Practice Exercises

For extra help, see Example 3 on page 38 of your text.

Write each word phrase as an algebraic expression. Use x as the variable.

4. Ten times a number, added to 21 — **4.** ____________

5. 11 fewer than eight times a number — **5.** ____________

6. Half a number subtracted from two-thirds of the number — **6.** ____________

Objective 3 Identify solutions of equations.

Video Examples

Review this example for Objective 3:

4. Decide whether the given number is a solution of the equation.

$8n - 7(n-4) = 41;\ \ 9$

$8n - 7(n-4) = 41$

$8 \cdot 9 - 7(9-4) \stackrel{?}{=} 41$

$8 \cdot 9 - 7 \cdot 5 \stackrel{?}{=} 41$

$72 - 35 \stackrel{?}{=} 41$

$37 = 41$ False – the left side does not equal the right side.

The number 9 is not a solution of the equation.

Now Try:

4. Decide whether the given number is a solution of the equation.
$9m - 4(m-3) = 41;\ \ 5$

____

Name: Date:
Instructor: Section:

Objective 3 Practice Exercises

For extra help, see Example 4 on pages 38–39 of your text.

Decide whether the given number is a solution of the equation.

7. $5+3x^2=19;\ 2$ **7.** ________________

8. $\dfrac{m+2}{3m-10}=1;\ 8$ **8.** ________________

9. $3y+5(y-5)=7;\ 4$ **9.** ________________

Objective 4 Identify solutions of equations from a set of numbers.

Video Examples

Review this example for Objective 4:

5. Write the word sentence as an equation. Use x as the variable. Then find the solution of the equation from the following set.
$\{0, 2, 4, 6, 8, 10\}$

The sum of a number and five is eleven.

the sum of a number and five	is	eleven.
↓	↓	↓
$x+5$	$=$	11

Because 6 + 5 = 11 is true, 6 is the only solution.

Now Try:

5. Write the word sentence as an equation. Use x as the variable. Then find the solution of the equation from the following set.
$\{0, 2, 4, 6, 8, 10\}$
The sum of a number and seven is eleven.

____

Objective 4 Practice Exercises

For extra help, see Example 5 on page 39 of your text.

Write each word sentence as an equation. Use x as the variable. Then find the solution of the equation from the following set. $\{1, 3, 5, 7, 9, 11\}$

10. Ten divided by a number is nine more than the number. **10.** ________________

11. Five more than a number is fourteen. **11.** ________________

12. Five times a number is 12 plus the number. **12.** ________________

Name: Date:
Instructor: Section:

Objective 5 Distinguish between *equations* and *expressions*.

Video Examples

Review this example for Objective 5:

6. Decide whether each is an equation or an expression.

$5x-4(x-6)$

Ask, "Is there an equality symbol?" The answer is no, so this is an expression.

Now Try:

6. Decide whether each is an equation or an expression.

$2(x-4)-4x$

____

Objective 5 Practice Exercises

For extra help, see Example 6 on page 40 of your text.

Identify each as an **expression** *or an* **equation**.

13. y^2-4y-3 **13.** _______________

14. $\frac{x+4}{5}$ **14.** _______________

15. $8x=2y$ **15.** _______________

Name: Date:
Instructor: Section:

Chapter 1 THE REAL NUMBER SYSTEM

1.3 Real Numbers and the Number Line

Learning Objectives

1 Classify numbers and graph them on number lines.
2 Tell which of two real numbers is less than the other.
3 Find the additive inverse of a real number.
4 Find the absolute value of a real number.
5 Interpret meanings of real numbers from a table of data.

Key Terms

Use the vocabulary terms listed below to complete each statement in exercises 1–14.

natural numbers	**whole numbers**	**number line**	**additive inverse**
integers	**negative number**	**positive number**	**signed numbers**
rational number	**set-builder notation**		**coordinate**
irrational number	**real numbers**	**absolute value**	

1. The set {0, 1, 2, 3, …} is called the set of ______________________________.

2. The __________________ of a number is the same distance from 0 on the number line as the original number, but located on the opposite side of 0.

3. The whole numbers together with their opposites and 0 are called __________________.

4. The set { 1, 2, 3, …} is called the set of ______________________________.

5. The ____________________________ of a number is the distance between 0 and the number on the number line.

6. A ______________________ shows the ordering of the real numbers on a line.

7. A real number that is not a rational number is called a(n) __________________.

8. The number that corresponds to a point on the number line is the __________________ of that point.

9. A number located to the left of 0 on a number line is a __________________.

10. A number located to the right of 0 on a number line is a __________________.

11. Numbers that can be represented by points on the number line are ____________________.

Name: Date:
Instructor: Section:

12. ______________________ uses a variable and a description to describe a set.

13. A number that can be written as the quotient of two integers is a ____________________.

14. Positive numbers and negative numbers are __________________.

Objective 1 Classify numbers and graph them on number lines.

Video Examples

Review these examples for Objective 1:

1. Use an integer to express the boldface italic number in the application.

In August, 2012, the National Debt was approximately $***16*** trillion.

Use –\$16 trillion because "debt" indicates a negative number.

2. Graph each number on a number line.

$-3\frac{1}{2}, -\frac{3}{2}, 0, \frac{7}{2}, 1$

To locate the improper fractions on the number line, write them as mixed numbers or decimals.

–3.5 –1.5 0 1 3.5

–5 –4 –3 –2 –1 0 1 2 3 4 5

3. List the numbers in the following set that belong to each set of numbers.

$\left\{-6, -\frac{5}{6}, 0, 0.\overline{3}, \sqrt{3}, 4\frac{1}{5}, 6, 6.7\right\}$

a. Whole numbers

Answer: 0 and 6

b. Integers

Answer: –6, 0, and 6

c. Rational numbers

Answer: $-6, -\frac{5}{6}, 0, 0.\overline{3}, 4\frac{1}{5}, 6, 6.7$

d. Irrational numbers

Answer: $\sqrt{3}$

Now Try:

1. Use an integer to express the boldface italic number in the application.
Death Valley is ***282*** feet below sea level.

____

2. Graph each number on a number line.

$\frac{1}{2}, 0, -3, -\frac{5}{2}$

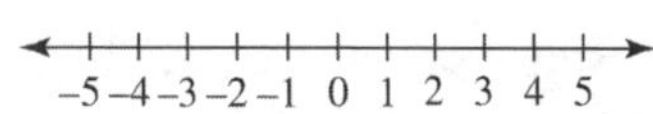

3. List the numbers in the following set that belong to each set of numbers.

$\left\{-10, -\frac{5}{8}, 0, 0.\overline{4}, \sqrt{5}, 5\frac{1}{2}, 7, 9.9\right\}$

a. Whole numbers

____

b. Integers

____

c. Rational numbers

____

d. Irrational numbers

____

Name: Date:
Instructor: Section:

Objective 1 Practice Exercises

For extra help, see Examples 1–3 on pages 43–45 of your text.

Use a real number to express each number in the following applications.

1. Last year Nina lost 75 pounds. **1.** ______________

2. Between 1970 and 1982, the population of Norway increased by 279,867. **2.** ______________

Graph the group of rational numbers on a number line.

3. –4.5, –2.3, 1.7, 4.2 **3.**

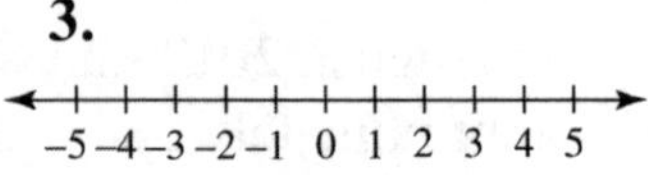

Objective 2 Tell which of two real numbers is less than the other.

Video Examples

Review this example for Objective 2:

4. Is the statement $-4 < -2$ true or false?

Because –4 is to the left of –2 on the number line, –4 is less than –2. The statement $-4 < -2$ is true.

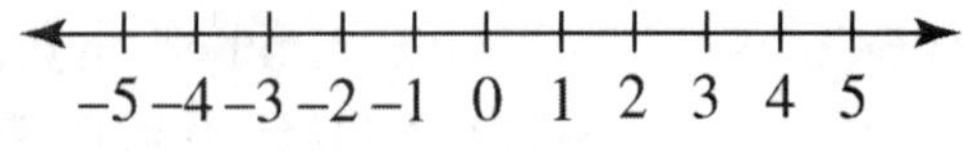

Now Try:

4. Is the statement $-10 < -8$ true or false.

____

Objective 2 Practice Exercises

For extra help, see Example 4 on page 46 of your text.

Decide whether each statement is **true** *or* **false**.

4. $-76 < 45$ **4.** ______________

5. $-5 > -5$ **5.** ______________

6. $-12 > -10$ **6.** ______________

Name: Date:
Instructor: Section:

Objective 3 Find the additive inverse of a real number.

Objective 3 Practice Exercises

For extra help, see pages 46–47 of your text.

Find the additive inverse of each number.

7. -25 7. ______

8. $\frac{3}{8}$ 8. ______

9. 4.5 9. ______

Objective 4 Find the absolute value of a real number.

Video Examples

Review these examples for Objective 4:

5. Simplify by finding the absolute value.

a. $|16|$

$|16| = 16$

b. $|-16|$

$|-16| = -(-16) = 16$

c. $-|-16|$

$-|-16| = -(16) = -16$

Now Try:

5. Simplify by finding the absolute value.

a. $|-10|$ ______

b. $-|10|$ ______

c. $|10-7|$ ______

Objective 4 Practice Exercises

For extra help, see Example 5 on page 48 of your text.

Simplify.

10. $-|49-39|$ 10. ______

11. $|-7.52+6.3|$ 11. ______

12. $|16-14|$ 12. ______

Name: Date:
Instructor: Section:

Objective 5 Interpret meanings of real numbers from a table of data.

Video Examples

Review this example for Objective 5:

6. In the table, which category represents a decrease for both years?

Category	Change from 2012 to 2013	Change from 2013 to 2014
Eggs	−0.3	3.9
Milk	1.6	0.7
Orange Juice	−8.8	−3.9
Electricity	0.8	3.9

Source: U.S. Bureau of Labor and Statistics

Since a decrease implies a negative number, the category Orange Juice has a negative number for both years. So the answer is Orange Juice.

Now Try:

6. In the table to the left, which category represents an increase for both years?

____

Objective 5 Practice Exercises

For extra help, see Example 6 on page 48 of your text.

The Consumer Price Index (CPI) measures the average change in prices of goods and services purchased by urban consumers in the United States. The table shows the percent change in CPI for selected categories of goods and services from 2012 to 2013 and from 2013 to 2014. Use the table to answer each question.

Category	Change from 2012 to 2013	Change from 2013 to 2014
Gasoline	−1.2	−0.9
Eggs	−0.3	3.9
Milk	1.6	0.7
Electricity	0.8	3.9

13. Which category represents a decrease for both years? **13.** _______________

14. Which category in which year represents the greatest percent decrease? **14.** _______________

15. Which category in which year represents the least change? **15.** _______________

Name: Date:
Instructor: Section:

Chapter 1 THE REAL NUMBER SYSTEM

1.4 Adding and Subtracting Real Numbers

Learning Objectives

1. Add two numbers with the same sign.
2. Add numbers with different signs.
3. Use the definition of subtraction.
4. Use the rules for order of operations when adding and subtracting signed numbers.
5. Translate words and phrases involving addition and subtraction.
6. Use signed numbers to interpret data.

Key Terms

Use the vocabulary terms listed below to complete each statement in exercises 1–2.

sum **addends** **minuend**

subtrahend **difference**

1. The number from which another number is being subtracted is called the ____________________.
2. The ______________________ is the number being subtracted.
3. The answer to a subtraction problem is called the ______________________.
4. The answer to an addition problem is called the ___________________________.
5. In an addition problem, the numbers being added are the ____________________.

Objective 1 Add two numbers with the same sign.

Video Examples

Review these examples for Objective 1:

1. Use a number line to find the sum.

$-3 + (-5)$.

Step 1 Start at 0 and draw an arrow 3 units to the left.

Step 2 From the left end of that arrow, draw another arrow 5 units to the left.

The number below the end of this second arrow is -5, so $-3 + (-5) = -8$.

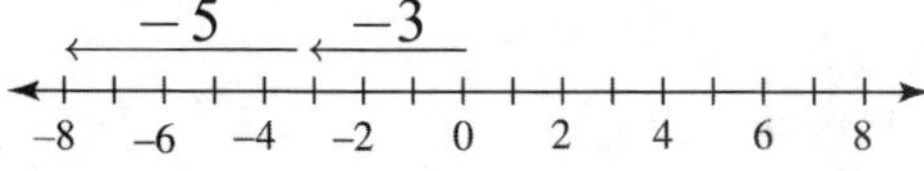

Now Try:

1. Use a number line to find the sum.
$-4 + (-1)$

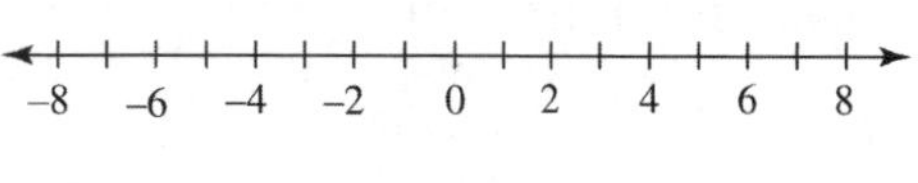

____

Name: Date:
Instructor: Section:

2. Find the sum.

$-3+(-7)$

$-3+(-7)=-10$

2. Find the sum.

$-8+(-4)$

______

Objective 1 Practice Exercises

For extra help, see Examples 1–2 on pages 52–53 of your text.

Find each sum.

1. $-7+(-11)$ **1.** ______

2. $-9+(-9)$ **2.** ______

3. $-2\frac{3}{8}+\left(-3\frac{1}{4}\right)$ **3.** ______

Objective 2 Add numbers with different signs.

Video Examples

Review these examples for Objective 2:

3. Use the number line to find the sum $-3+4$.

Step 1 Start at 0 and draw an arrow 3 units to the left.

Step 2 From the left end of that arrow, draw a second arrow 4 units to the right.

The number below the end of this second arrow is 1, so $-3+4=1$.

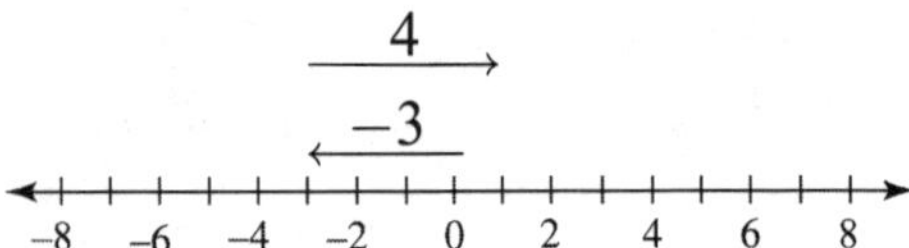

4. Find the sum.

$\frac{5}{8}+\left(-1\frac{1}{4}\right)$

$$\begin{aligned}\frac{5}{8}+\left(-1\frac{1}{4}\right)&=\frac{5}{8}+\left(-\frac{5}{4}\right)\\&=\frac{5}{8}+\left(-\frac{10}{8}\right)=+\left(\frac{5}{8}-\frac{10}{8}\right)\\&=-\frac{5}{8}\end{aligned}$$

Now Try:

3. Use the number line to find the sum $7+(-4)$.

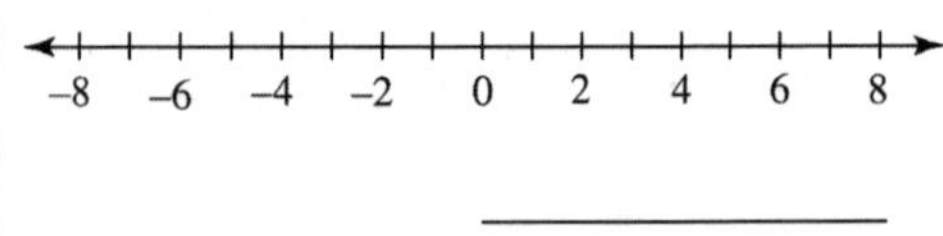

______

4. Find the sum.

$\frac{3}{5}+\left(-1\frac{3}{10}\right)$

______

Name: Date:
Instructor: Section:

Objective 2 Practice Exercises

For extra help, see Examples 3–4 on pages 53–54 of your text.

Use a number line to find the sum.

4. $-8+5$

4. 

−8 −6 −4 −2 0 2 4 6 8

Find each sum.

5. $\frac{7}{12}+\left(-\frac{3}{4}\right)$

5. ______________

6. $-\frac{4}{7}+\frac{3}{5}$

6. ______________

Objective 3 Use the definition of subtraction.

Video Examples

Review these examples for Objective 3:

6. Subtract.

a. $2-8$

$2-8=2+(-8)=-6$

b. $-6-(-9)$

$-6-(-9)=-6+(9)=3$

Now Try:

6. Subtract.

a. $3-6$

____

b. $-5-(-7)$

____

Objective 3 Practice Exercises

For extra help, see Examples 5–6 on pages 55–56 of your text.

Subtract.

7. $-14-11$

7. ______________

8. $15-(-2)$

8. ______________

9. $-\frac{3}{10}-\left(-\frac{3}{10}\right)$

9. ______________

Name: Date:
Instructor: Section:

Objective 4 Use the rules for order of operations with real numbers.

Video Examples

Review this example for Objective 4:

7. Perform each operation.

$(9+6)-7$

$(9+6)-7=15-7$
$=8$

Now Try:

7. Perform each operation.

$(7+4)-10$

____

Objective 4 Practice Exercises

For extra help, see Example 7 on pages 56–57 of your text.

Find each sum.

10. $-2+[4+(-18+13)]$

10. ____________

11. $[(-7)+14]+[(-16)+3]$

11. ____________

12. $-8.9+[6.8+(-4.7)]$

12. ____________

Objective 5 Translate words and phrases that indicate addition.

Video Examples

Review these examples for Objective 5:

8. Write a numerical expression for the phrase, and simplify the expression.

The sum of –9 and 5 and 3

–9 + 5 + 3 simplifies to –4 + 3, which equals –1.

9. Write a numerical expression for the phrase, and simplify the expression.

The difference between –10 and 7

–10 –7 simplifies to $-10+(-7)$, which equals –17

Now Try:

8. Write a numerical expression for the phrase, and simplify the expression.
The sum of –10 and 11 and 2

____

9. Write a numerical expression for the phrase, and simplify the expression.
The difference between –17 and 9

____

Name: Date:
Instructor: Section:

10. The early morning temperature on a mountain in California was –8°F. At noon the temperature was 38°F. What was the rise in temperature? We must subtract the lowest temperature from the highest temperature. $38-(-8)=38+8=46$ The rise was 46°F.	**10.** The floor of Death Valley is 282 ft below sea level. A nearby mountain has an elevation of 5182 ft above sea level. Find the difference between the highest and lowest elevations. __________

Objective 5 Practice Exercises

For extra help, see Examples 8–10 on pages 57–59 of your text.

Write a numerical expression for each phrase, and then simplify the expression.

13. 4 less than −4 **13.** __________

14. The sum of –4 and 12, decreased by 9 **14.** __________

Solve the problem.

15. Dr. Somers runs an experiment at –43.3°C. He then lowers the temperature by 7.9°C. What is the new temperature for the experiment? **15.** __________

Objective 6 Use signed numbers to interpret data.

Video Examples

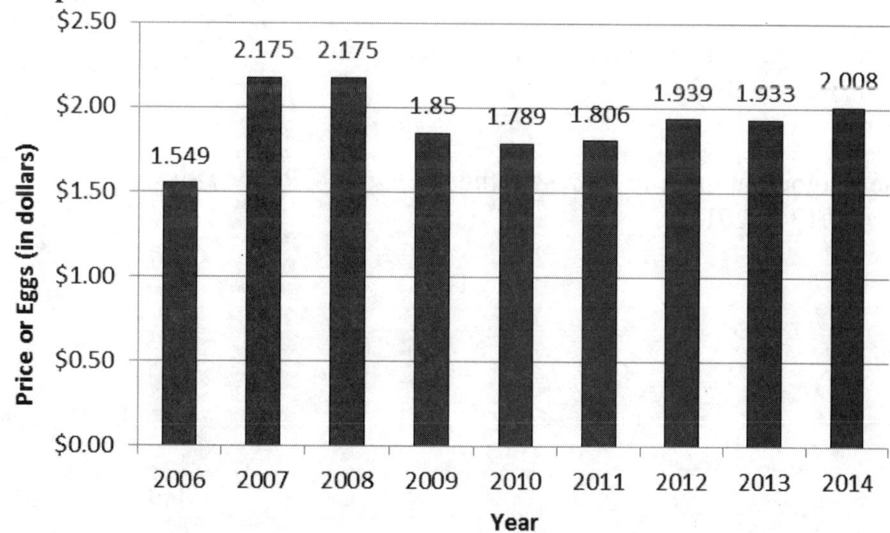

The bar graph above shows the Consumer Price Index (CPI) for a dozen of grade A large Eggs between 2006 and 2014. Source: U.S. Bureau of Labor and Statistics

Review this example for Objective 6: **11.** Use a signed number to represent the change in CPI from 2008 to 2009. $1.85 – $2.175 = –$0.325	**Now Try:** **11.** Use a signed number to represent the change in CPI from 2012 to 2013. __________

Name: Date:
Instructor: Section:

Objective 6 Practice Exercises

For extra help, see Example 11 on page 59 of your text.

The bar graph below shows the Consumer Price Index (CPI) for a dozen of grade A large Eggs between 2006 and 2014. Source: U.S. Bureau of Labor and Statistics

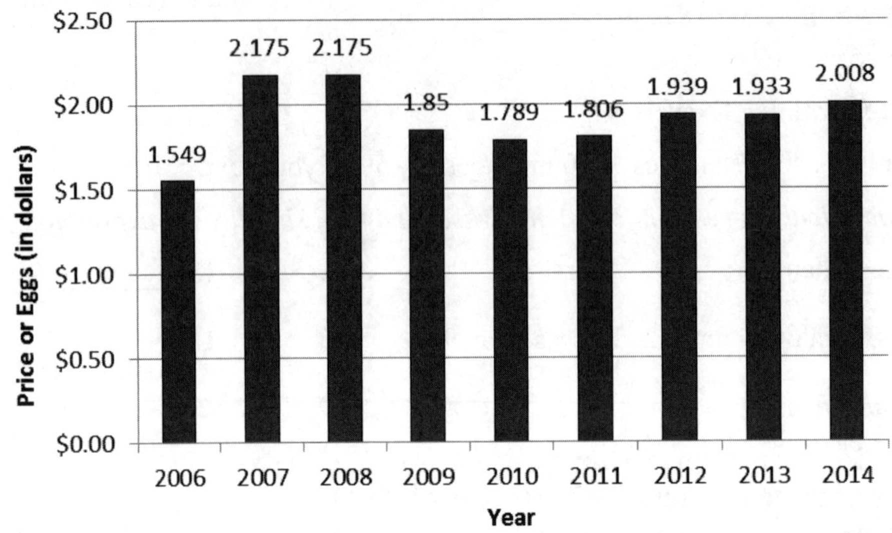

16. Use a signed number to represent the change in CPI from 2006 to 2007. **16.** ________________

17. Use a signed number to represent the change in CPI from 2009 to 2010. **17.** ________________

18. Use a signed number to represent the change in CPI from 2013 to 2014. **18.** ________________

Name: Date:
Instructor: Section:

Chapter 1 THE REAL NUMBER SYSTEM

1.5 Multiplying and Dividing Real Numbers

Learning Objectives

1. Find the product of a positive number and a negative number.
2. Find the product of two negative numbers.
3. Identify factors of integers.
4. Use the reciprocal of a number to apply the definition of division.
5. Use the rules for order of operations when multiplying and dividing signed numbers.
6. Evaluate expressions given values for the variables.
7. Translate words and phrases involving multiplication and division.
8. Translate simple sentences into equations.

Key Terms

Use the vocabulary terms listed below to complete each statement in exercises 1−3.

product **quotient** **reciprocals**

1. The answer to a division problem is called the ________________.
2. Pairs of numbers whose product is 1 are called ________________.
3. The answer to a multiplication problem is called the ________________.

Objective 1 Find the product of a positive number and a negative number.

Video Examples

Review this example for Objective 1:

1. Find the product using the multiplication rule.

$9(-6)$

$9(-6) = -(9 \cdot 6) = -54$

Now Try:

1. Find the product using the multiplication rule.
$8(-7)$

____

Objective 1 Practice Exercises

For extra help, see Example 1 on page 66 of your text.

Find each product.

1. $7(-4)$ 1. ________________

2. $\left(\frac{1}{5}\right)\left(-\frac{2}{3}\right)$ 2. ________________

3. $(-3.2)(4.1)$ 3. ________________

Name: Date:
Instructor: Section:

Objective 2 Find the product of two negative numbers.

Video Examples

Review this example for Objective 2:

2. Find the product using the multiplication rule.

a. $-7(-3)$

$-7(-3) = 21$

Now Try:

2. Find the product using the multiplication rule.

a. $-5(-6)$

______

Objective 2 Practice Exercises

For extra help, see Example 2 on page 67 of your text.

Find each product.

4. $(-4)(-10)$ **4.** ______

5. $\left(-\frac{2}{7}\right)\left(-\frac{14}{5}\right)$ **5.** ______

6. $(-0.4)(-3.4)$ **6.** ______

Objective 3 Identify factors of integers.

Video Examples

Review this example for Objective 3:

2b. Find all integer factors of the number 52.

$1 \times 52 = 52$
$2 \times 26 = 52$
$4 \times 13 = 52$
$(-1) \times (-52) = 52$
$(-2) \times (-26) = 52$
$(-4) \times (-13) = 52$

The integer factors of 52 are –52, –26, –13, –4, –2, –1, 1, 2, 4, 13, 26, and 52.

Now Try:

2b. Find all integer factors of the number 12.

______

Objective 3 Practice Exercises

For extra help, see page 68 of your text.

Find all integer factors of each number.

7. 8 **7.** ______

Name: Date:
Instructor: Section:

8. 38

8. ____________

9. 42

9. ____________

Objective 4 Use the reciprocal of a number to apply the definition of division.

Video Examples

Review these examples for Objective 4:

3. Find each quotient.

a. $\frac{30}{6}$

$\frac{30}{6} = 5$

b. $\frac{15}{-3}$

$\frac{15}{-3} = -5$

Now Try:

3. Find each quotient.

a. $\frac{16}{8}$

____

b. $\frac{-18}{-6}$

____

Objective 4 Practice Exercises

For extra help, see Example 3 on page 69 of your text.

Find each quotient.

10. $\frac{-120}{-20}$

10. ____________

11. $\frac{0}{-2}$

11. ____________

12. $\frac{10}{0}$

12. ____________

Name: Date:
Instructor: Section:

Objective 5 **Use the rules for order of operations when multiplying and dividing signed numbers.**

Video Examples

Review these examples for Objective 5:

4. Simplify.

a. $-6(-1-4)$

$$-6(-1-4)=-6(-5)$$
$$=30$$

b. $\dfrac{6(-4)-5(3)}{3(2-7)}$

$$\frac{6(-4)-5(3)}{3(2-7)}=\frac{-24-15}{3(-5)}$$
$$=\frac{-39}{-15}$$
$$=\frac{13}{5}$$

Now Try:

4. Simplify.

a. $-4(-5-2)$

____

b. $\dfrac{-9(-3)+4(-8)}{-4(5-6)}$

____

Objective 5 Practice Exercises

For extra help, see Example 4 on pages 70–71 of your text.

Perform the indicated operations.

13. $-4[(-2)(7)-2]$

13. ____________

14. $\dfrac{-7(2)-(-3)}{5+(-3)}$

14. ____________

15. $\dfrac{-4[8-(-3+7)]}{-6[3-(-2)]-3(-3)}$

15. ____________

Name: Date:
Instructor: Section:

Objective 6 Evaluate algebraic expressions given values for the variables.

Video Examples

Review this example for Objective 6:

5. Evaluate the expression for $x=-2,\ y=-4,$ and $m=-5.$

$(5x+6y)(-3m)$

Substitute the given values for the variables. Then simplify.

$(5x+6y)(-3m)$

$=[5(-2)+6(-4)][-3(-5)]$

$=[-10+(-24)][15]$

$=[-34]15$

$=-510$

Now Try:

5. Evaluate the expression for $x=-5,\ y=-3,$ and $p=-4.$

$(6x+2y)(-3p)$

____

Objective 6 Practice Exercises

For extra help, see Example 5 on page 71 of your text.

Evaluate the following expressions if $x=-3, y=2,$ *and* $a=4.$

16. $-x+[(-a+y)-2x]$

16. ____________

17. $(-4+x)(-a)-|x|$

17. ____________

18. $\dfrac{4a-x}{y^2}$

18. ____________

Objective 7 Translate words and phrases involving multiplication and division.

Video Examples

Review this example for Objective 7:

6. Write a numerical expression for the phrase, and simplify the expression.

Three fifths of the sum of –6 and –7

$\frac{3}{5}[-6+(-7)]$ simplifies to $\frac{3}{5}[-13],$

which equals $-\dfrac{39}{5}.$

Now Try:

6. Write a numerical expression for the phrase, and simplify the expression.
Five-sixths of the sum of –8 and –4

____

Name: Date:
Instructor: Section:

Objective 7 Practice Exercises

For extra help, see Examples 6–7 on pages 72–73 of your text.

Write a numerical expression for each phrase and simplify.

19. The product of –7 and 3, added to –7 **19.** ________________

20. Three-tenths of the difference between 50 and –10, subtracted from 85 **20.** ________________

21. The sum of –12 and the quotient of 49 and –7 **21.** ________________

Objective 8 Translate simple sentences into equations.

Video Examples

Review this example for Objective 8:

8. Write the sentence in symbols, using x to represent the number.

The quotient of 27 and a number is –3.

$$\frac{27}{x} = -3$$

Now Try:

8. Write the sentence in symbols, using x to represent the number.

The quotient of 36 and a number is –4

____

Objective 8 Practice Exercises

For extra help, see Example 8 on page 73 of your text.

Write each statement in symbols, using x as the variable.

22. Two-thirds of a number is –7. **22.** ________________

23. –8 times a number is 72. **23.** ________________

24. When a number is divided by –4, the result is 1. **24.** ________________

Name: Date:
Instructor: Section:

Chapter 1 THE REAL NUMBER SYSTEM

1.6 Properties of Real Numbers

Learning Objectives
1. Use the commutative properties.
2. Use the associative properties.
3. Use the identity properties.
4. Use the inverse properties.
5. Use the distributive property.

Key Terms

Use the vocabulary terms listed below to complete each statement in exercises 1−2.

identity element for addition

identity element for multiplication

1. When the ______________________________, which is 0, is added to a number, the number is unchanged.

2. When a number is multiplied by the ______________________________, which is 1, the number is unchanged.

Objective 1 Use the commutative properties.

Video Examples

Review these examples for Objective 1:

1. Use a commutative property to complete each statement.

   a. $-7+6=6+$ ______

   Using the commutative property of addition,
   $-7+6=6+(-7)$

   b. $(-3)5=$ ____(-3)

   Using the commutative property of multiplication,
   $(-3)5=5(-3)$

Now Try:

1. Use a commutative property to complete each statement.

   a. $-12+8=8+$ ______

   ____

   b. $(-4)2=$ ____(-4)

   ____

Objective 1 Practice Exercises

For extra help, see Example 1 on page 79 of your text.

Complete each statement. Use a commutative property.

1. $y+4=$ _____$+y$

1. ______________

2. $5(2) = \_\_\_\_(5)$ **2.** ________

3. $-4(4+z) = \_\_\_\_(-4)$ **3.** ________

Objective 2 Use the associative properties.

Video Examples

Review these examples for Objective 2:

2. Use an associative property to complete each statement.

a. $-5+(3+7)=(-5+\_\_\_\_)+7$

Using the associative property of addition,
$-5+(3+7)=(-5+3)+7$

b. $[4\cdot(-9)]\cdot 2 = 4\cdot\_\_\_\_$

Using the associative property of multiplication,
$[4\cdot(-9)]\cdot 2 = 4\cdot[(-9)\cdot 2]$

3. Decide whether each statement is an example of a commutative property, an associative property, or both.

a. $(5+9)+11=5+(9+11)$

The order of the three numbers is the same, but the change is in grouping. This is an example of the associative property.

b. $7\cdot(9\cdot 11)=7\cdot(11\cdot 9)$

The only change involves the order of the number, so this is an example of the commutative property.

c. $(12+3)+6=12+(6+3)$

Both the order and the grouping are changed. This is an example of both the associative and commutative properties.

Now Try:

2. Use an associative property to complete each statement.

a. $-8+(4+6)=(-8+\_\_\_)+6$

____

b. $[8\cdot(-3)]\cdot 4 = 8\cdot\_\_\_\_$

____

3. Decide whether each statement is an example of a commutative property, an associative property, or both.

a. $(13+8)+25=13+(8+25)$

____

b. $4\cdot(15\cdot 30)=4\cdot(30\cdot 15)$

____

c. $(21+19)+4=21+(4+19)$

____

Name: Date:
Instructor: Section:

4. Find the sum.

$21x + 3 + 17x + 29$

$$\begin{aligned} &21x+3+17x+29 \\ &= 21x+(3+17x)+29 \\ &= (21x+17x)+(3+29) \\ &= 38x+32 \end{aligned}$$

4. Find the sum

$15x + 12 + 24x + 8$

____

Objective 2 Practice Exercises

For extra help, see Examples 2–4 on pages 79–80 of your text.

Complete each statement. Use an associative property.

4. $4(ab) = \_\_\_\_ \cdot b$

4. __________

5. $[x + (-4)] + 3y = x + \_\_\_\_$

5. __________

6. $4r + (3s + 14t) = \_\_\_\_ + 14t$

6. __________

Objective 3 Use the identity properties.

Video Examples

Review these examples for Objective 3:

5. Use an identity property to complete each statement.

a. $-6 + \_\_\_\_ = -6$

Use the identity property for addition.

$-6 + 0 = -6$

b. $\_\_\_\_ \cdot \frac{1}{6} = \frac{1}{6}$

Use the identity property for multiplication.

$1 \cdot \frac{1}{6} = \frac{1}{6}$

Now Try:

5. Use an identity property to complete each statement.

a. $8 + \_\_\_\_ = 8$

____

b. $-9 \cdot \_\_\_\_ = -9$

____

Name: Date:
Instructor: Section:

6.

a. Write $\frac{56}{35}$ in lowest terms.

$$\frac{56}{35}=\frac{8\cdot 7}{5\cdot 7}$$
$$=\frac{8}{5}\cdot\frac{7}{7}$$
$$=\frac{8}{5}\cdot 1$$
$$=\frac{8}{5}$$

b. Perform the operation: $\frac{5}{6}-\frac{7}{18}$

$$\frac{5}{6}-\frac{7}{18}=\frac{5}{6}\cdot 1-\frac{7}{18}$$
$$=\frac{5}{6}\cdot\frac{3}{3}-\frac{7}{18}$$
$$=\frac{15}{18}-\frac{7}{18}$$
$$=\frac{8}{18}$$
$$=\frac{4}{9}$$

6.

a. Write $\frac{49}{63}$ in lowest terms.

______

b. Perform the operation:
$\frac{3}{7}+\frac{5}{21}$

______

Objective 3 Practice Exercises

For extra help, see Examples 5–6 on pages 80–81 of your text.

Use an identity property to complete each statement.

7. $4+0=$ ____ **7.** ______

8. ____ $\cdot 1=12$ **8.** ______

Use an identity property to simplify the expression.

9. $\frac{30}{35}$ **9.** ______

Name: Date:
Instructor: Section:

Objective 4 Use the inverse properties.

Video Examples

Review these examples for Objective 4:

7. Use an inverse property to complete each statement.

a. $\_\_\_\_\cdot\frac{6}{7}=1$

Use the inverse property of multiplication.

$\frac{7}{6}\cdot\frac{6}{7}=1$

b. $5+\_\_\_\_=0$

Use the inverse property of addition.

$5+(-5)=0$

c. $-9(\_\_\_\_)=1$

Use the inverse property of multiplication.

$-9\left(-\frac{1}{9}\right)=1$

d. $\_\_\_\_+\frac{1}{4}=0$

Use the inverse property of addition.

$-\frac{1}{4}+\frac{1}{4}=0$

8. Simplify $-4x+1+4x$.

$-4x+1+4x$

$=(-4x+1)+4x$	Order of operations
$=[1+(-4x)]+4x$	Commutative property
$=1+[(-4x)+4x]$	Associative property
$=1+0$	Inverse property
$=1$	Identity property

Now Try:

7. Use an inverse property to complete each statement.

a. $\frac{8}{5}\cdot\_\_\_\_=1$

____

b. $8+\_\_\_\_=0$

____

c. $-\frac{1}{10}(\_\_\_\_)=1$

____

d. $-11+\_\_\_\_=0$

____

8. Simplify $-\frac{1}{4}x+6+\frac{1}{4}x$.

____

Objective 4 Practice Exercises

For extra help, see Examples 7–8 on pages 81–82 of your text.

Complete the statements so that they are examples of either an identity property or an inverse property. Identify which property is used.

10. $-4+\_\_\_\_=0$ **10.** ________

11. $-9+\_\_\_\_=-9$ **11.** ________

Name: Date:
Instructor: Section:

12. $-\frac{3}{5}\cdot$ ______ $=1$

12. ______________

Objective 5 Use the distributive property.

Video Examples

Review these examples for Objective 5:

9. Use the distributive property to rewrite each expression.

a. $7(p-6)$

$$\begin{aligned}7(p-6)&=7[p+(-6)]\\&=7p+7(-6)\\&=7p-42\end{aligned}$$

b. $-3(5x-2)$

$$\begin{aligned}-3(5x-2)&=-3[5x+(-2)]\\&=-3(5x)+(-3)(-2)\\&=(-3\cdot 5)x+(-3)(-2)\\&=-15x+6\end{aligned}$$

c. $4\cdot 8+4\cdot 7$

$$4\cdot 8+4\cdot 7=4(8+7)$$

d. $5\cdot 3+5x+5m$

$$5\cdot 3+5x+5m=5(3+x+m)$$

10. Rewrite each expression.

a. $-(5x+7)$

$$\begin{aligned}-(5x+7)&=-1\cdot(5x+7)\\&=-1\cdot 5x+(-1)\cdot 7\\&=-5x-7\end{aligned}$$

b. $-(-p-5r+9x)$

$$\begin{aligned}&-(-p-5r+9x)\\&=-1\cdot(-1p-5r+9x)\\&=-1\cdot(-1p)-1\cdot(-5r)-1\cdot(9x)\\&=p+5r-9x\end{aligned}$$

Now Try:

9. Use the distributive property to rewrite each expression.

a. $17(x-6)$

____

b. $-4(2x-5)$

c. $3\cdot 11+3\cdot 7$

____

d. $12y+12\cdot 6+12x$

____

10. Rewrite each expression.

a. $-(3x+4)$

b. $-(-4x-5y+z)$

____

c. $6a+6b+6$

$$6a+6b+6=6a+6b+6\cdot 1$$
$$=6(a+b+1)$$

c. $3x+3y+3$

____

Objective 5 Practice Exercises

For extra help, see Examples 9–10 on pages 83–84 of your text.

Use the distributive property to rewrite each expression. Simplify if possible.

13. $n(2a-4b+6c)$ **13.** ______________

14. $-2(5y-9z)$ **14.** ______________

15. $-(-2k+7)$ **15.** ______________

Name: Date:
Instructor: Section:

Chapter 1 THE REAL NUMBER SYSTEM

1.7 Simplifying Expressions

Learning Objectives

1. Simplify expressions.
2. Identify terms and numerical coefficients.
3. Identify like terms.
4. Combine like terms.
5. Simplify expressions from word phrases.

Key Terms

Use the vocabulary terms listed below to complete each statement in exercises 1–3.

term **numerical coefficient** **like terms**

1. In the term $4x^2$, "4" is the________________________________.

2. A number, a variable, or a product or quotient of a number and one or more variables raised to powers is called a ________________________________.

3. Terms with exactly the same variables, including the same exponents, are called ________________________________.

Objective 1 Simplify expressions.

Video Examples

Review these examples for Objective 1:

1. Simplify each expression.

a. $8(4m-6n)$

Use the distributive property.

$$8(4m-6n)=8(4m)+8(-6n)$$
$$=32m-48n$$

b. $9-(4y-6)$

$$9-(4y-6)=9-1(4y-6)$$
$$=9-4y+6$$
$$=15-4y$$

Now Try:

1. Simplify each expression.

a. $7(5x-3y)$

____

b. $8-(7x-3)$

____

Name: Date:
Instructor: Section:

Objective 1 Practice Exercises

For extra help, see Example 1 on page 88 of your text.

Simplify each expression.

1. $4(2x+5)+7$ **1.** ______________

2. $-4+s-(12-21)$ **2.** ______________

3. $-2(-5x+2)+7$ **3.** ______________

Objective 2 Identify terms and numerical coefficients.

Objective 2 Practice Exercises

For extra help, see pages 88–89 of your text.

Give the numerical coefficient of each term.

4. $-2y^2$ **4.** ______________

5. $\frac{7x}{9}$ **5.** ______________

6. $5.6r^5$ **6.** ______________

Objective 3 Identify like terms.

Objective 3 Practice Exercises

For extra help, see page 89 of your text.

Identify each group of terms as **like** *or* **unlike**.

7. $4x^2, -7x^2$ **7.** ______________

8. $-8m, -8m^2$ **8.** ______________

9. $7xy, -6xy^2$ **9.** ______________

Name: Date:
Instructor: Section:

Objective 4 Combine like terms.

Video Examples

Review these examples for Objective 4:

2. Combine like terms in each expression.

a. $8r+5r+4r$

$$8r+5r+4r=(8+5+4)r$$
$$=17r$$

b. $9x+x$

$$9x+x=9x+1x$$
$$=(9+1)x$$
$$=10x$$

c. $15x^2-8x^2$

$$15x^2-8x^2=(15-8)x^2$$
$$=7x^2$$

3. Simplify each expression.

a. $8k-5-4(7-3k)$

$$8k-5-4(7-3k)=8k-5-4(7)-4(-3k)$$
$$=8k-5-28+12k$$
$$=20k-33$$

b. $-\frac{3}{5}(x-10)-\frac{1}{10}x$

$$-\frac{3}{5}(x-10)-\frac{1}{10}x=-\frac{3}{5}x-\frac{3}{5}(-10)-\frac{1}{10}x$$
$$=-\frac{3}{5}x+6-\frac{1}{10}x$$
$$=-\frac{6}{10}x+6-\frac{1}{10}x$$
$$=-\frac{7}{10}x+6$$

Now Try:

2. Combine like terms in each expression.

a. $14r+7r+2r$ __________

b. $18x+x$ __________

c. $17x^2-9x^2$ __________

3. Simplify each expression.

a. $7k-9-5(3-6k)$ __________

b. $-\frac{3}{4}(x-8)-\frac{1}{2}x$ __________

Objective 4 Practice Exercises

For extra help, see Examples 2–3 on pages 89–91 of your text.

Simplify.

10. $12y-7y^2+4y-3y^2$ **10.** __________

Name: Date:
Instructor: Section:

11. $-4(x+4)+2(3x+1)$ **11.** ________________

12. $2.5(3y+1)-4.5(2y-3)$ **12.** ________________

Objective 5 Simplify expressions from word phrases.

Video Examples

Review this example for Objective 5:

4. Translate the phrase into a mathematical expression and simplify.

The sum of 8, three times a number, nine times a number, and seven times a number.

Use x for the number.

$8+3x+9x+7x$ simplifies to $8+19x$.

Now Try:

4. Translate the phrase into a mathematical expression and simplify.

The sum of 11, ten times a number, eight times a number, and four times a number

____

Objective 5 Practice Exercises

For extra help, see Example 4 on page 91 of your text.

Write each phrase as a mathematical expression and simplify by combining like terms. Use x as the variable.

13. The sum of six times a number and 12, added to four times the number. **13.** ________________

14. The sum of seven times a number and 2, subtracted from three times the number. **14.** ________________

15. Four times the difference between twice a number and six times the number, added to six times the sum of the number and 9. **15.** ________________

Name: Date:
Instructor: Section:

Chapter 2 LINEAR EQUATIONS AND INEQUALITIES IN ONE VARIABLE

2.1 The Addition Property of Equality

Learning Objectives
1 Identify linear equations.
2 Use the addition property of equality.
3 Simplify, and then use the addition property of equality.

Key Terms

Use the vocabulary terms listed below to complete each statement in exercises 1–3.

linear equation **solution set** **equivalent equations**

1. Equations that have exactly the same solutions sets are called ____________________.

2. An equation that can be written in the form $Ax + B = C$, where A, B, and C are real numbers and $A \neq 0$, is called a ____________________.

3. The set of all numbers that satisfy an equation is called its ____________________.

Objective 1 Identify linear equations.

Objective 1 Practice Exercises

For extra help, see page 104 of your text.

Tell whether each of the following is a linear equation.

1. $3x^2 + 4x + 3 = 0$ 1. ____________

2. $\frac{5}{x} - \frac{3}{2} = 0$ 2. ____________

3. $4x - 2 = 12x + 9$ 3. ____________

Name: Date:
Instructor: Section:

Objective 2 Use the addition property of equality.

Video Examples

Review these examples for Objective 2:

1. Solve $x-15=8$.

$$x-15=8$$
$$x-15+15=8+15$$
$$x=23$$

Check $x-15=8$

$$23-15\stackrel{?}{=}8$$
$$8=8 \quad \text{True}$$

The solution is 23, and the solution set is $\{23\}$.

3. Solve $-5=x+17$.

$$-5=x+17$$
$$-5-17=x+17-17$$
$$-22=x$$

Check $-5=x+17$

$$-5\stackrel{?}{=}-22+17$$
$$-5=-5 \quad \text{True}$$

The solution set is $\{-22\}$.

5. Solve $\frac{5}{6}k+9=\frac{11}{6}k$.

$$\frac{5}{6}k+9=\frac{11}{6}k$$
$$\frac{5}{6}k+9-\frac{5}{6}k=\frac{11}{6}k-\frac{5}{6}k$$
$$9=1k$$
$$9=k$$

Check by substituting 9 in the original equation. The solution set is $\{9\}$.

6. Solve $9-5p=-6p+3$.

$$9-5p=-6p+3$$
$$9-5p+6p=-6p+3+6p$$
$$9+p-9=3-9$$
$$p=-6$$

Check by substituting –6 in the original equation. The solution set is $\{-6\}$.

Now Try:

1. Solve $x-12=9$.

____

3. Solve $-10=x+9$.

____

5. Solve $\frac{4}{7}k+13=\frac{11}{7}k$.

____

6. Solve $10-8p=-9p+7$.

____

Name: Date:
Instructor: Section:

Objective 2 Practice Exercises

For extra help, see Examples 1–6 on pages 105–107 of your text.

Solve each equation by using the addition property of equality. Check each solution.

4. $y-4=16$ **4.** ________________

5. $\frac{9}{8}p-\frac{1}{2}=\frac{1}{8}p$ **5.** ________________

6. $9.5y-2.4=10.5y$ **6.** ________________

Objective 3 Simplify, and then use the addition property of equality.

Video Examples

Review these examples for Objective 3:

7. Solve $5t-16+t+4=9+5t+6$.

$$5t-16+t+4=9+5t+6$$
$$6t-12=15+5t$$
$$6t-12-5t=15+5t-5t$$
$$t-12=15$$
$$t-12+12=15+12$$
$$t=27$$

Check by substituting 27 in the original equation. The solution set is $\{27\}$.

8. Solve $4(3+6x)-(5+23x)=19$.

$$4(3+6x)-(5+23x)=19$$
$$4(3)+4(6x)-1(5)-1(23x)=19$$
$$12+24x-5-23x=19$$
$$x+7=19$$
$$x+7-7=19-7$$
$$x=12$$

Check by substituting 12 in the original equation. The solution set is $\{12\}$.

Now Try:

7. Solve $8t-9+t+7=12+8t+15$.

____

8. Solve $5(7+8x)-(29+39x)=14$.

____

Name: Date:
Instructor: Section:

Objective 3 Practice Exercises

For extra help, see Examples 7–8 on page 108 of your text.

Solve each equation. First simplify each side of the equation as much as possible. Check each solution.

7. $3(t+3)-(2t+7)=9$ **7.** ________________

8. $-4(5g-7)+3(8g-3)=15-4+3g$ **8.** ________________

9. $3.6p+4.8+4.0p=8.6p-3.1+0.7$ **9.** ________________

Name: Date:
Instructor: Section:

Chapter 2 LINEAR EQUATIONS AND INEQUALITIES IN ONE VARIABLE

2.2 The Multiplication Property of Equality

Learning Objectives
1 Use the multiplication property of equality.
2 Simplify, and then use the multiplication property of equality.

Key Terms

Use the vocabulary terms listed below to complete each statement in exercises 1−2.

multiplication property of equality **addition property of equality**

1. The ____________________ states that multiplying both sides of an equation by the same nonzero number will not change the solution.

2. When the same quantity is added to both sides of an equation, the ____________________ is being applied.

Objective 1 Use the multiplication property of equality.

Video Examples

Review these examples for Objective 1:

1. Solve $6x = 78$.

$$6x = 78$$
$$\frac{6x}{6} = \frac{78}{6}$$
$$x = 13$$

Check $6x = 78$
$$6(13) \stackrel{?}{=} 78$$
$$78 = 78 \quad \text{True}$$
The solution set is $\{13\}$.

3. Solve $4.3x = 10.32$.

$$4.3x = 10.32$$
$$\frac{4.3x}{4.3} = \frac{10.32}{4.3}$$
$$x = 2.4$$
Check by substituting 2.4 in the original equation. The solution set is $\{2.4\}$.

Now Try:

1. Solve $4x = 56$.

____

3. Solve $3.6x = 20.52$.

____

Name: Date:
Instructor: Section:

4. Solve $\frac{x}{7}=5$.

$$\frac{x}{7}=5$$
$$\frac{1}{7}x=5$$
$$7\cdot\frac{1}{7}x=7\cdot 5$$
$$x=35$$

Check by substituting 35 in the original equation. The solution set is $\{35\}$.

4. Solve $\frac{x}{8}=3$.

____

5. Solve $\frac{5}{6}x=15$.

$$\frac{5}{6}x=15$$
$$\frac{6}{5}\cdot\frac{5}{6}x=\frac{6}{5}\cdot 15$$
$$1\cdot x=\frac{6}{5}\cdot\frac{15}{1}$$
$$x=18$$

Check by substituting 18 in the original equation. The solution set is $\{18\}$.

5. Solve $\frac{7}{9}h=28$.

____

6. Solve $-x=5$.

$$-x=5$$
$$-1\cdot x=5$$
$$-1(-1\cdot x)=-1(5)$$
$$[-1(-1)]\cdot x=-5$$
$$1\cdot x=-5$$
$$x=-5$$

Check by substituting –5 in the original equation. The solution set is {–5}.

6. Solve $-x=3$.

____

Objective 1 Practice Exercises

For extra help, see Examples 1–6 on pages 113–115 of your text.

Solve each equation and check your solution.

1. $-3w=51$

1. ______________

Name: Date:
Instructor: Section:

2. $\frac{3p}{7}=-6$

2. ______________

3. $-2.7v=-17.28$

3. ______________

Objective 2 Simplify, and then use the multiplication property of equality.

Video Examples

Review this example for Objective 2:

7. Solve $9m+4m=39$.

$$9m+4m=39$$
$$13m=39$$
$$\frac{13m}{13}=\frac{39}{13}$$
$$m=3$$

Check by substituting 3 in the original equation.
The solution set is $\{3\}$.

Now Try:

7. Solve $12m+8m=80$.

____

Objective 2 Practice Exercises

For extra help, see Example 7 on page 115 of your text.

Solve each equation and check your solution.

4. $-7b+12b=125$

4. ______________

5. $3w-7w=20$

5. ______________

6. $-11h-6h+14h=-21$

6. ______________

Name: Date:
Instructor: Section:

Chapter 2 LINEAR EQUATIONS AND INEQUALITIES IN ONE VARIABLE

2.3 More on Solving Linear Equations

Learning Objectives
1. Learn and use the four steps for solving a linear equation.
2. Solve equations that have no solution or infinitely many solutions.
3. Solve equations with fractions or decimals as coefficients.
4. Write expressions for two related unknown quantities.

Key Terms

Use the vocabulary terms listed below to complete each statement in exercises 1–3.

conditional equation **identity** **contradiction**

1. An equation with no solution is called a(n) ____________________.

2. A(n) ____________________ is an equation that is true for some values of the variable and false for other values.

3. An equation that is true for all values of the variable is called a(n) ____________.

Objective 1 Learn and use the four steps for solving a linear equation.

Video Examples

Review these examples for Objective 1:

1. Solve $-5x+8=23$.

Step 1 There are no parentheses, fractions, or decimals in this equation, so this step is not necessary.

$$-5x+8=23$$

Step 2 $$-5x+8-8=23-8$$

$$-5x=15$$

Step 3 $$\frac{-5x}{-5}=\frac{15}{-5}$$

$$x=-3$$

Step 4 Check by substituting –3 for x in the original equation.

$$-5x+8=23$$
$$-5(-3)+8\overset{?}{=}23$$
$$15+8\overset{?}{=}23$$
$$23=23 \quad \text{True}$$

The solution, –3, checks, so the solution set is $\{-3\}$.

Now Try:

1. Solve $-8x+11=59$.

____

Name: Date:
Instructor: Section:

2. Solve $4x+3=6x-11$.

Step 1 There are no parentheses, fractions, or decimals in this equation, so begin with Step 2.

$$4x+3=6x-11$$

Step 2 $4x+3-4x=6x-11-4x$

$$3=2x-11$$

$$3+11=2x-11+11$$

$$14=2x$$

Step 3 $\frac{14}{2}=\frac{2x}{2}$

$$7=x$$

Step 4 Check by substituting 7 for x in the original equation.

$$4x+3=6x-11$$

$$4(7)+3\overset{?}{=}6(7)-11$$

$$28+3\overset{?}{=}42-11$$

$$31=31 \quad \text{True}$$

The solution, 7, checks, so the solution set is $\{7\}$.

2. Solve $5x+4=8x-20$.

____

4. Solve $9a-(4+3a)=2a+5$.

$$9a-(4+3a)=2a+5$$

Step 1 $9a-4-3a=2a+5$

$$6a-4=2a+5$$

Step 2 $6a-4-2a=2a+5-2a$

$$4a-4=5$$

$$4a-4+4=5+4$$

$$4a=9$$

Step 3 $\frac{4a}{4}=\frac{9}{4}$

$$a=\frac{9}{4}$$

Step 4 Check that the solution set is $\left\{\frac{9}{4}\right\}$.

4. Solve $10a-(11+3a)=5a+4$.

____

Objective 1 Practice Exercises

For extra help, see Examples 1–5 on pages 117–120 of your text.

Solve each equation and check your solution.

1. $7t+6=11t-4$

1. ____________

2. $3a-6a+4(a-4)=-2(a+2)$ **2.** ________

3. $3(t+5)=6-2(t-4)$ **3.** ________

Objective 2 Solve equations that have no solution or infinitely many solutions.

Video Examples

Review these examples for Objective 2:

6. Solve $6x-18=6(x-3)$.

$$\begin{aligned} 6x-18&=6(x-3)\\ 6x-18&=6x-18\\ 6x-18-6x&=6x-18-6x\\ -18&=-18\\ -18+18&=-18+18\\ 0&=0 \end{aligned}$$

The solution set is {all real numbers}.

7. Solve $3x+4(x-5)=7x+5$.

$$\begin{aligned} 3x+4(x-5)&=7x+5\\ 3x+4x-20&=7x+5\\ 7x-20&=7x+5\\ 7x-20-7x&=7x+5-7x\\ -20&=5 \quad \text{False} \end{aligned}$$

There is no solution. The solution set is $\varnothing$.

Now Try:

6. Solve $3x+4(x-5)=7x-20$.

____

7. Solve $-5x+17=x-6(x+3)$.

____

Objective 2 Practice Exercises

For extra help, see Examples 6–7 on page 121 of your text.

Solve each equation and check your solution.

4. $3(6x-7)=2(9x-6)$ **4.** ________

5. $6y-3(y+2)=3(y-2)$ **5.** ________

Name: Date:
Instructor: Section:

6. $3(r-2)-r+4=2r+6$ **6.** ________

Objective 3 Solve equations with fractions or decimals as coefficients.

Video Examples

Review these examples for Objective 3:

9. Solve $\frac{1}{4}(x+3)-\frac{2}{5}(x+1)=2$.

To clear fractions, multiply by 20, the LCD.

$$\frac{1}{4}(x+3)-\frac{2}{5}(x+1)=2$$

Step 1 $$20\left[\frac{1}{4}(x+3)-\frac{2}{5}(x+1)\right]=20(2)$$

$$20\left[\frac{1}{4}(x+3)\right]+20\left[-\frac{2}{5}(x+1)\right]=20(2)$$

$$5(x+3)-8(x+1)=40$$

$$5x+15-8x-8=40$$

$$-3x+7=40$$

Step 2 $$-3x+7-7=40-7$$

$$-3x=33$$

Step 3 $$\frac{-3x}{-3}=\frac{33}{-3}$$

$$x= \quad 11$$

Step 4 Check to confirm that {–11} is the solution set.

10. Solve $0.2x+0.04(10-x)=0.06(4)$.

To clear decimals, multiply by 100.

$$0.2x+0.04(10-x)=0.06(4)$$

Step 1 $$100[0.2x+0.04(10-x)]=100[0.06(4)]$$

$$100(0.2x)+100[0.04(10-x)]=100[0.06(4)]$$

$$20x+4(10)+4(-x)=24$$

$$20x+40-4x=24$$

$$16x+40=24$$

Step 2 $$16x+40-40=24-40$$

$$16x=-16$$

Step 3 $$\frac{16x}{16}=\frac{-16}{16}$$

$$x=-1$$

Step 4 Check to confirm that {–1} is the solution set.

Now Try:

9. Solve
$\frac{1}{7}(x+5)-\frac{1}{2}(x+4)=-2$.

____

10. Solve
$0.5x+0.04(5-8x)=0.07(8)$.

____

Name: Date:
Instructor: Section:

Objective 3 Practice Exercises

For extra help, see Examples 8–10 on pages 122–124 of your text.

Solve each equation and check your solution.

7. $\frac{3}{8}x - \frac{1}{3}x = \frac{1}{12}$ **7.** ______

8. $\frac{1}{3}(2m-1) - \frac{3}{4}m = \frac{5}{6}$ **8.** ______

9. $0.45a - 0.35(20-a) = 0.02(50)$ **9.** ______

Objective 4 Write expressions for two related unknown quantities.

Video Examples

Review this example for Objective 4:

11. Two numbers have a sum of 51. If one of the numbers is represented by x, find an expression for the other number.

If one number is x, then the other number is obtained by subtracting x from 51.

$51 - x$.

To check, we find the sum of the two numbers.

$x + (51 - x) = 51$

Now Try:

11. Two numbers have a sum of 67. If one of the numbers is represented by t, find an expression for the other number.

______

Objective 4 Practice Exercises

For extra help, see Example 11 on page 124 of your text.

Write an expression for the two related unknown quantities.

10. Two numbers have a sum of 36. One is m. Find the other number. **10.** ______

11. The product of two numbers is 17. One number is p. What is the other number? **11.** ______

12. Admission to the circus costs x dollars for an adult and y dollars for a child. Find the total cost of 6 adults and 4 children. **12.** ______

Name: Date:
Instructor: Section:

Chapter 2 LINEAR EQUATIONS AND INEQUALITIES IN ONE VARIABLE

2.4 Applications of Linear Equations

Learning Objectives

1. Learn the six steps for solving applied problems.
2. Solve problems involving unknown numbers.
3. Solve problems involving sums of quantities.
4. Solve problems involving consecutive integers.
5. Solve problems involving complementary and supplementary angles.

Key Terms

Use the vocabulary terms listed below to complete each statement in exercises 1−5.

complementary angles **right angle** **supplementary angles**

straight angle **consecutive integers**

1. Two angles whose measures sum to 180° are ______________________.
2. Two angles whose measures sum to 90° are ______________________.
3. An angle whose measure is exactly 90° is a ______________________.
4. An angle whose measure is exactly 180° is a ______________________.
5. Two integers that differ by 1 are ______________________.

Objective 1 Learn the six steps for solving applied problems.

Objective 1 Practice Exercises

For extra help, see page 128 of your text.

1. Write the six problem-solving steps. **1.** ______________

____

____

____

____

____

Name: Date:
Instructor: Section:

Objective 2 Solve problems involving unknown numbers.

Video Examples

Review this example for Objective 2:

2. The product of 5, and a number decreased by 8, is 150. What is the number?

Step 1 Read the problem carefully. We are asked to find a number.

Step 2 Assign a variable to represent the unknown quantity.
Let x = the number.

Step 3 Write an equation.

The product of 5,	and	a number	decreased by	8,	is	150.
↓		↓	↓	↓	↓	↓
$5\cdot$		$(x$	$-$	$8)$	$=$	150

Step 4 Solve the equation.

$$5(x-8)=150$$
$$5x-40=150$$
$$5x-40+40=150+40$$
$$5x=190$$
$$\frac{5x}{5}=\frac{190}{5}$$
$$x=38$$

Step 5 State the answer. The number is 38.

Step 6 Check. The number 38 decreased by 8 is 30. The product of 5 and 30 is 150. The answer, 38, is correct.

Now Try:

2. The product of 8, and a number decreased by 11, is 40. What is the number?

____

Objective 2 Practice Exercises

For extra help, see Examples 1–2 on pages 128–129 of your text.

Write an equation for each of the following and then solve the problem. Use x as the variable.

2. If 4 is added to 3 times a number, the result is 7. Find the number.

2. ____________

____

Name: Date:
Instructor: Section:

3. If −2 is multiplied by the difference between 4 and a number, the result is 24. Find the number.

3. ______________

____

4. If four times a number is added to 7, the result is five less than six times the number. Find the number.

4. ______________

____

Objective 3 Solve problems involving sums of quantities.

Video Examples

Review these examples for Objective 3:

3. George and Al were opposing candidates in the school board election. George received 21 more votes than Al, with 439 votes cast. How many votes did Al receive?

Step 1 Read the problem carefully. We are given total votes and asked to find the number of votes Al received.

Step 2 Assign a variable.
Let x = the number of votes Al received.
Then $x + 21$ = the number of votes George received.

Step 3 Write an equation.

The total	is	votes for Al	plus	votes for George
↓	↓	↓	↓	↓
439	=	x	+	$(x+21)$

Step 4 Solve the equation.

$$439 = x + (x + 21)$$
$$439 = 2x + 21$$
$$439 - 21 = 2x + 21 - 21$$
$$418 = 2x$$
$$\frac{418}{2} = \frac{2x}{2}$$
$$209 = x \quad \text{or} \quad x = 209$$

Step 5 State the answer. Al received 209 votes.

Now Try:

3. On a psychology test, the highest grade was 38 points more than the lowest grade. The sum of the two grades was 142. Find the lowest grade.

____

Step 6 Check. George won 209 + 21 = 230 votes. The total number of votes is 209 + 230 = 439. The answer checks.

6. Penny is making punch for a party. The recipe requires twice as much orange juice as cranberry juice and 8 times as much ginger ale as cranberry juice. If she plans to make 176 ounces of punch, how much of each ingredient should she use?

Step 1 Read the problem. The three amounts of ingredients must be found.

Step 2 Assign a variable.
Let x = the number of ounces of cranberry juice.
Then $2x$ = the number of ounces of orange juice, and $8x$ = the number of ounces of ginger ale.

Step 3 Write an equation.

Cranberry	plus	orange	plus	ginger ale	is	total
↓	↓	↓	↓	↓	↓	↓
x	$+$	$2x$	$+$	$8x$	$=$	176

Step 4 Solve the equation.

$$x+2x+8x=176$$
$$11x=176$$
$$\frac{11x}{11}=\frac{176}{176}$$
$$x=16$$

Step 5 State the answer. There are 16 ounces of cranberry juice, 2(16) = 32 ounces of orange juice, and 8(16) = 128 ounces of ginger ale.

Step 6 Check. The sum is 176. All conditions of the problem are satisfied.

6. Linda wishes to build a rectangular dog pen using 52 feet of fence and the back of her house, which is 36 feet long to enclose the pen. How wide will the dog pen be if the pen is 36 feet long?

____

Objective 3 Practice Exercises

For extra help, see Examples 3–6 on pages 130–133 of your text.

Write an equation for each of the following and then solve the problem. Use x as the variable.

5. Mount McKinley in Alaska is 5910 feet higher than Mount Rainier in Washington. Together, their heights total 34,730 feet. How high is each mountain?

5. ________________

Mt. Rainier ____________

Mt. McKinley __________

Name: Date:
Instructor: Section:

6. Charles bought five general admission tickets and four student tickets for a movie. He paid \$35.25. If each student ticket cost \$3.50, how much did each general admission ticket cost?

6. ______________

____

7. Pablo, Faustino, and Mark swim at a public pool each day for exercise. One day Pablo swam five more than three times as many laps as Mark, and Faustino swam four times as many laps as Mark. If the men swam 29 laps altogether, how many laps did each one swim?

7. ______________

Mark ______________

Pablo ______________

Faustino ______________

Objective 4 Solve problems involving consecutive integers.

Video Examples

Review this example for Objective 4:

8. Find two consecutive odd integers such that if three times the smaller is added to twice the larger, the sum is 69.

Step 1 Read the problem. We must find two consecutive odd integers.

Step 2 Assign a variable.
Let x = the lesser consecutive odd integer.
Then $x + 2$ = the greater consecutive odd integer.

Step 3 Write an equation.

Three times the smaller	is added to	twice the larger	is	69.
↓	↓	↓	↓	↓
$3x$	$+$	$2(x+2)$	$=$	69

Step 4 Solve the equation.

$$3x + 2x + 4 = 69$$
$$5x + 4 = 69$$
$$5x = 65$$
$$x = 13$$

Step 5 State the answer. The lesser integer is 13. The greater is $13 + 2 = 15$.

Now Try:

8. The sum of four consecutive even integers is 4. Find the integers.

____

Step 6 Check. Three times the smaller is 39, added to twice the larger, 30, is a sum of 69. The answers check.

Objective 4 Practice Exercises

For extra help, see Examples 7–8 on pages 133–135 of your text.

Solve each problem.

8. Find two consecutive even integers such that the smaller, added to twice the larger, is 292.

8. ________________

9. Find two consecutive integers such that the larger, added to three times the smaller, is 109.

9. ________________

10. Find three consecutive odd integers whose sum is 363.

10. ________________

Objective 5 Solve problems involving complementary and supplementary angles.

Video Examples

Review this example for Objective 5:

10. Find the measure of an angle if its supplement measures 4° less than three times its complement.

Step 1 Read the problem. We must find the measure of an angle.

Step 2 Assign a variable.
Let $x =$ the degree measure of the angle
Then $90 - x =$ the degree measure of its complement,
and $180 - x =$ the degree measure of its supplement.

Step 3 Write an equation.

The supplement	is	Three times the complement	minus	4
↓	↓	↓	↓	↓
$180-x$	$=$	$3(90-x)$	$-$	4

Now Try:

10. Find the measure of an angle such that the sum of the measures of its complement and its supplement is 138°.

____

Name: Date:
Instructor: Section:

Step 4 Solve the equation.

$$180 - x = 270 - 3x - 4$$
$$180 - x = 266 - 3x$$
$$180 - x + 3x = 266 - 3x + 3x$$
$$180 + 2x = 266$$
$$180 + 2x - 180 = 266 - 180$$
$$2x = 86$$
$$x = 43$$

Step 5 State the answer. The angle is 43°.

Step 6 Check. If the angle measures 43°, then its complement measures 90° – 43° = 47°, and the supplement measures 180° – 43° = 137°. Also, 137 is equal to 4 less than 3 times 47° (that is, 137° = 3(47) – 4). The answer is correct.

Objective 5 Practice Exercises

For extra help, see Examples 9–10 on pages 135–136 of your text.

Solve each problem.

11. Find the measure of an angle if the measure of the angle is 8° less than three times the measure of its supplement. **11.** ______________

12. Find the measure of an angle whose supplement measures 20° more than twice its complement. **12.** ______________

13. Find the measure of an angle whose complement is 9° more than twice its measure. **13.** ______________

Name: Date:
Instructor: Section:

Chapter 2 LINEAR EQUATIONS AND INEQUALITIES IN ONE VARIABLE

2.5 Formulas and Additional Applications from Geometry

Learning Objectives

1. Solve a formula for one variable, given the values of the other variables.
2. Use a formula to solve an applied problem.
3. Solve problems involving vertical angles and straight angles.
4. Solve a formula for a specified variable.

Key Terms

Use the vocabulary terms listed below to complete each statement in exercises 1−4.

formula **area** **perimeter** **vertical angles**

1. The nonadjacent angles formed by two intersecting lines are called ______________________________.
2. An equation in which variables are used to describe a relationship is called a(n) ______________________.
3. The distance around a figure is called its ______________________.
4. A measure of the surface covered by a figure is called its ____________________.

Objective 1 Solve a formula for one variable, given the values of the other variables.

Video Examples

Review this example for Objective 1:

1. Find the value of the remaining variable in each formula.

$A = LW$; $A = 54, L = 8$

Substitute the given values for A and L into the formula.

$$A = LW$$
$$54 = 8W$$
$$\frac{54}{8} = \frac{8W}{8}$$
$$6.75 = W$$

The width is 6.75. Since 8(6.75) = 54, the answer checks.

Now Try:

1. Find the value of the remaining variable in each formula.

$A = LW$; $A = 88, L = 16$

____

Name: Date:
Instructor: Section:

Objective 1 Practice Exercises

For extra help, see Example 1 on page 142 of your text.

In the following exercises, a formula is given, along with the values of all but one of the variables in the formula. Find the value of the variable that is not given.

1. $S = \frac{a}{1-r}; S = 60,\ r = 0.4$

1. ________________

2. $I = prt;\ I = 288,\ r = 0.04,\ t = 3$

2. ________________

3. $A = \frac{1}{2}(b+B)h;\ b = 6,\ B = 16,\ A = 132$

3. ________________

Objective 2 Use a formula to solve an applied problem.

Video Examples

Review these examples for Objective 2:

2. Find the dimensions of a rectangle. The length is 4 m less than three times the width. The perimeter is 96 m.

Step 1 Read the problem. We must find the dimensions of the rectangle.

Step 2 Assign a variable.
Let W = the width of the rectangle, in meters.
Then $L = 3W - 4$ is the length, in meters.

Step 3 Write an equation. Use the formula for the perimeter of a rectangle. Substitute $3W - 4$ for the length.

$$P = 2L + 2W$$

$$96 = 2(3W - 4) + 2W$$

Step 4 Solve.

$$96 = 6W - 8 + 2W$$

$$96 = 8W - 8$$

$$96 + 8 = 8W - 8 + 8$$

$$104 = 8W$$

$$\frac{104}{8} = \frac{8W}{8}$$

$$13 = W$$

Now Try:

2. Ruth has 42 feet of binding for a rectangular rug that she is weaving. If the rug is 9 feet wide, how long can she make the rug if she wishes to use all the binding on the perimeter of the rug?

____

Step 5 State the answer. The width is 13 m. The length is 3(13) – 4 = 35 m.

Step 6 Check. The perimeter is 2(13) + 2(35) = 96 m. The answer checks.

3. The longest side of a triangle is 4 feet longer than the shortest side. The medium side is 2 feet longer than the shortest side. If the perimeter is 36 feet, what are the lengths of the three sides?

Step 1 Read the problem. We must find the lengths of the sides.

Step 2 Assign a variable.
Let s = the length of the shortest side, in feet.
Then $s + 2$ = the length of the medium side, in feet,
and $s + 4$ = the length of the longest side, in feet.

Step 3 Write an equation. Use the formula for the perimeter of a triangle.

$$P = a + b + c$$

$$36 = s + (s + 2) + (s + 4)$$

Step 4 Solve.

$$36 = 3s + 6$$

$$30 = 3s$$

$$10 = s$$

Step 5 State the answer. Since s represents the length of the shortest side, its measure is 10 ft.
$s + 2 = 10 + 2 = 12$ ft is the length of the medium side.
$s + 4 = 10 + 4 = 14$ ft is the length of the longest side.

Step 6 Check. The perimeter is $10 + 12 + 14 = 36$ ft, as required.

3. The longest side of a triangle is twice as long as the shortest side. The medium side is 5 feet longer than the shortest side. If the perimeter is 65 feet, what are the lengths of the three sides?

____

Objective 2 Practice Exercises

For extra help, see Examples 2–4 on pages 143–144 of your text.

Use a formula to write an equation for each of the following applications; then solve the application. (Use 3.14 as an approximation for π.)

4. Find the height of a triangular banner whose area is 48 square inches and base is 12 inches.

4. ____________

Name: Date:
Instructor: Section:

5. Linda invests \$5000 at 6% simple interest and earns \$450. How long did Linda invest her money? 5. ______________

6. The circumference of a circular garden is 628 feet. Find the area of the garden. (Hint: First find the radius of the garden.) 6. ______________

Objective 3 Solve problems involving vertical angles and straight angles.

Video Examples

Review this example for Objective 3:

5.

Find the measure of the marked angles in the figure below.

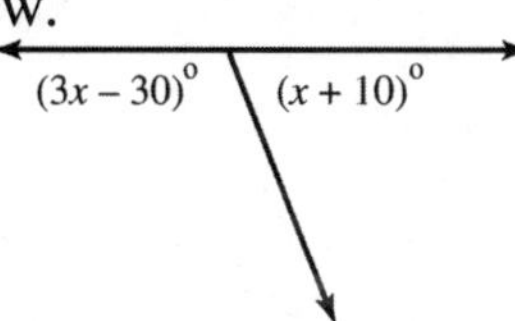

The measures of the marked angles must add to 180° because together they form a straight angle. The angles are supplements of each other.

$$(3x-30)+(x+10)=180$$
$$4x-20=180$$
$$4x=200$$
$$x=50$$

Replace x with 50 in the measure of each marked angle.

$$3x-30=3(50)-30=150-30=120$$
$$x+10=50+10=60$$

The two angles measure 120° and 60°.

Now Try:

5.

Find the measure of the marked angles in the figure below.

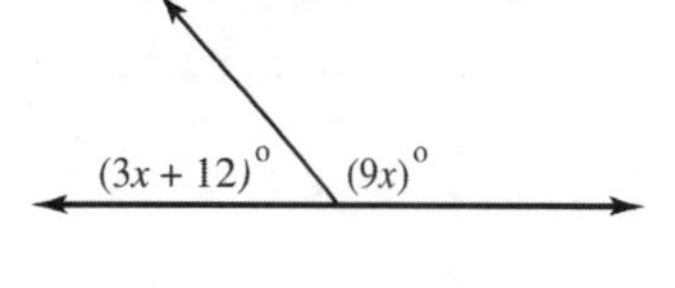

____

Name: Date:
Instructor: Section:

Objective 3 Practice Exercises

For extra help, see Example 5 on page 145 of your text.

Find the measure of each marked angle.

7.

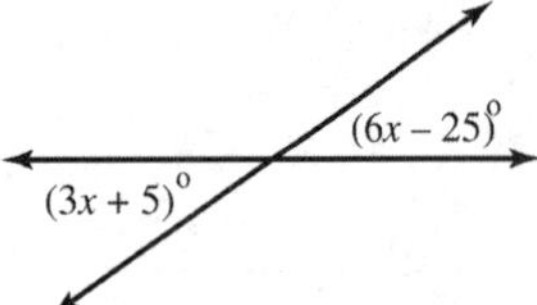

7. ______________

8.

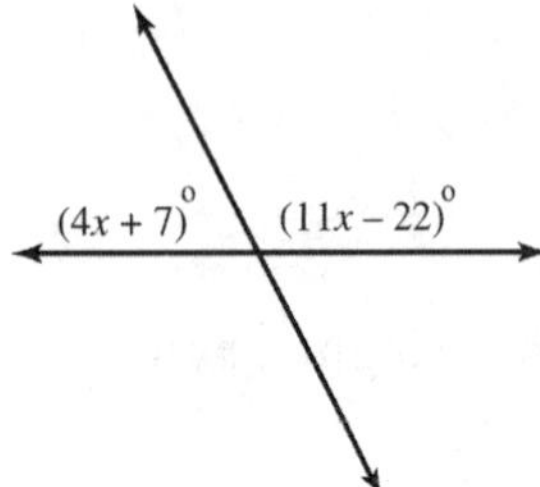

8. ______________

9.

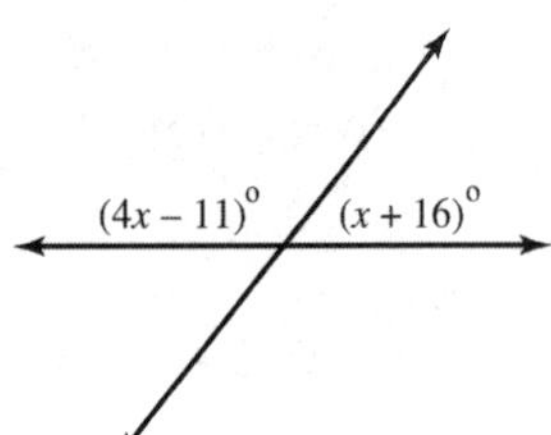

9. ______________

Objective 4 Solve a formula for a specified variable.

Video Examples

Review these examples for Objective 4:

6. Solve $A = \frac{1}{2}bh$ for h.

$$A = \frac{1}{2}bh$$

$$2A = bh$$

$$\frac{2A}{b} = \frac{bh}{b}$$

$$\frac{2A}{b} = h \quad \text{or} \quad h = \frac{2A}{b}$$

Now Try:

6. Solve $d = rt$ for r.

____

Name: Date:
Instructor: Section:

7. Solve $A = p + prt$ for r.

$$A = p + prt$$
$$A - p = p + prt - p$$
$$A - p = prt$$
$$\frac{A-p}{pt} = \frac{prt}{pt}$$
$$\frac{A-p}{pt} = r \quad \text{or} \quad r = \frac{A-p}{pt}$$

8. Solve $V = k + gt$ for t.

$$V = k + gt$$
$$V - k = k + gt - k$$
$$V - k = gt$$
$$\frac{V-k}{g} = \frac{gt}{g}$$
$$\frac{V-k}{g} = t$$
$$t = \frac{V-k}{g}$$

7. Solve $P = a + b + c$ for a.

____

8. Solve $A = \frac{1}{2}h(b + B)$ for h.

____

Objective 4 Practice Exercises

For extra help, see Examples 6–9 on pages 146–147 of your text.

Solve each formula for the specified variable.

10. $V = LWH$ for H

10. ____________

11. $S = (n - 2)180$ for n

11. ____________

12. $V = \frac{1}{3}\pi r^2 h$ for h

12. ____________

Name: Date:
Instructor: Section:

Chapter 2 LINEAR EQUATIONS AND INEQUALITIES IN ONE VARIABLE

2.6 Ratio, Proportion, and Percent

Learning Objectives

1. Write ratios.
2. Solve proportions.
3. Solve applied problems using proportions.
4. Find percents and percentages.

Key Terms

Use the vocabulary terms listed below to complete each statement in exercises 1–4.

ratio **proportion** **cross products** **terms**

1. A ______________________________ is a statement that two ratios are equal.

2. A ______________________ is a comparison of two quantities using a quotient.

3. In the proportion, $\frac{a}{b} = \frac{c}{d}$, a, b, c, and d are called the ________________.

4. To see whether a proportion is true, determine if the ___________________ are equal.

Objective 1 Write ratios.

Video Examples

Review these examples for Objective 1:

1. Write a ratio for each word phrase.

a. 7 hr to 9 hr

$$\frac{7 \text{ hr}}{9 \text{ hr}} = \frac{7}{9}$$

b. 15 hr to 4 days

First convert 4 days to hours.

$$4 \text{ days} = 4 \cdot 24 = 96 \text{ hr}$$

Now write the ratio using the common unit of measure, hours.

$$\frac{15 \text{ hr}}{4 \text{ days}} = \frac{15 \text{ hr}}{96 \text{ hr}} = \frac{15}{96}, \quad \text{or} \quad \frac{5}{32}$$

Now Try:

1. Write a ratio for each word phrase.

a. 11 hr to 17 hr

____

b. 32 hr to 5 days

____

Name: Date:
Instructor: Section:

2. The local grocery store charges the following prices for a bottle of olive oil.
16-ounce bottle: \$6.99
25.5-ounce bottle: \$9.99
32-ounce bottle: \$12.99
44-ounce bottle: \$14.99
Which size is the best buy? That is, which size has the lowest unit price?

To find the best buy, write ratios comparing the price for each size bottle to the number of units (ounces) per bottle.

Size	Unit price (dollars per ounce)
16 oz	$\frac{\$6.99}{16} = \$0.437$
25.5 oz	$\frac{\$9.99}{25.5} = \$0.392$
32 oz	$\frac{\$12.99}{32} = \$0.406$
44 oz	$\frac{\$14.99}{44} = \$0.341$

Because the 44-oz size has the lowest unit price, \$0.341, it is the best buy.

2. The local grocery store charges the following prices for a jar of applesauce.
16-ounce jar: \$1.19
24-ounce jar: \$1.29
48-ounce jar: \$2.69
64-ounce jar: \$3.49
Which size is the best buy? That is, which size has the lowest unit price?

____

Objective 1 Practice Exercises

For extra help, see Examples 1–2 on pages 152–153 of your text.

Write a ratio for each word phrase. Write fractions in lowest terms.

1. 8 men to 3 men **1.** __________

2. 9 dollars to 48 quarters **2.** __________

A supermarket was surveyed and the following prices were charged for items in various sizes. Find the best buy (based on price per unit) for each of the following items.

3. Trash bags **3.** __________
10-count box: \$2.89
20-count box: \$5.29
45-count box: \$6.69
85-count box: \$13.99

Name: Date:
Instructor: Section:

Objective 2 Solve proportions.

Video Examples

Review these examples for Objective 2:

3. Decide whether the proportion is *true* or *false*.

$$\frac{2}{5}=\frac{12}{30}$$

Check to see whether the cross product are equal.

$$5\cdot 12=60$$

$$\frac{2}{5}=\frac{12}{30}$$

$$2\cdot 30=60$$

The cross products are equal, so the proportion is true.

4. Solve the proportion $\frac{6}{11}=\frac{x}{88}$.

Solve for x.

$$\frac{6}{11}=\frac{x}{88}$$
$$6\cdot 88=11\cdot x$$
$$528=11x$$
$$48=x$$

Check by substituting 48 for x in the proportion. The solution set is $\{48\}$.

5. Solve the equation $\frac{n-1}{3}=\frac{2n+1}{4}$.

$$\frac{n-1}{3}=\frac{2n+1}{4}$$
$$3(2n+1)=4(n-1)$$
$$6n+3=4n-4$$
$$2n+3=-4$$
$$2n=-7$$
$$n=-\frac{7}{2}$$

A check confirms that the solution is $-\frac{7}{2}$, so the solution set is $\left\{-\frac{7}{2}\right\}$.

Now Try:

3. Decide whether the proportion is *true* or *false*.
$\frac{5}{6}=\frac{20}{24}$

____

4. Solve the proportion $\frac{x}{15}=\frac{42}{90}$.

____

5. Solve the equation
$\frac{2x+1}{2}=\frac{7x+3}{9}$.

____

Name: Date:
Instructor: Section:

Objective 2 Practice Exercises

For extra help, see Examples 3–5 on pages 154–155 of your text.

Solve each equation.

4. $\frac{z}{20}=\frac{25}{125}$

4. ______________

5. $\frac{m}{5}=\frac{m-2}{2}$

5. ______________

6. $\frac{z+1}{4}=\frac{z+7}{2}$

6. ______________

Objective 3 Solve applied problems using proportions.

Video Examples

Review this example for Objective 3:

6. If four pounds of fertilizer will cover 50 square feet of garden, how many pounds would be needed for 125 square feet?

To solve this problem, set up a proportion, with pounds in the numerator and square feet in the denominator.

$$\frac{4}{50}=\frac{x}{125}$$
$$4(125)=50x$$
$$500=50x$$
$$10=x$$

10 lb of fertilizer are needed.

Now Try:

6. Margie earns \$168.48 in 26 hours. How much does she earn in 40 hours?

____

Objective 3 Practice Exercises

For extra help, see Example 6 on page 156 of your text.

Solve each problem.

7. On a road map, 6 inches represents 50 miles. How many inches would represent 125 miles?

7. ______________

Name: Date:
Instructor: Section:

8. If 12 rolls of tape cost \$4.60, how much will 15 rolls cost? **8.** ________________

9. A garden service charges \$30 to install 50 square feet of sod. Find the charge to install 225 square feet. **9.** ________________

Objective 4 Find percents and percentages.

Video Examples

Review these examples for Objective 4:

7. Solve each problem.

a. What is 18% of 700?

Let n = the number. The word of indicates multiplication.

$$\begin{array}{ccccc} \text{What} & \text{is} & 18\% & \text{of} & 700 \\ \downarrow & \downarrow & \downarrow & \downarrow & \downarrow \\ n & = & 0.18 & \cdot & 700 \end{array}$$

$$n = 126$$

Thus, 126 is 18% of 700.

b. 54% of what number is 162?

$$\begin{array}{ccccc} 54\% & \text{of} & \text{what number} & \text{is} & 162 \\ \downarrow & \downarrow & \downarrow & \downarrow & \downarrow \\ 0.54 & \cdot & n & = & 162 \end{array}$$

$$n = \frac{162}{0.54}$$

$$n = 300$$

54% of 300 is 162.

c. 75 is what percent of 500?

$$\begin{array}{ccccc} 75 & \text{is} & \text{what percent} & \text{of} & 500 \\ \downarrow & \downarrow & \downarrow & \downarrow & \downarrow \\ 75 & = & p & \cdot & 500 \end{array}$$

$$\frac{75}{500} = p$$

$$0.15 = p$$

Thus, 75 is 15% of 500.

Now Try:

7. Solve each problem.

a. What is 35% of 400?

____

b. 42% of what number is 399?

____

c. 102 is what percent of 120?

____

Name: Date:
Instructor: Section:

8. An advertisement for a BluRay player gives a sale price of \$175.50. The regular price is \$225. Find the percent discount on this BluRay player.

The savings amounted to \$225 – \$175.50 = \$49.50. We can now restate the problem: What percent of 225 is 49.50?

$$\begin{array}{ccccc} \text{What percent} & \text{of} & 225 & \text{is} & 49.50 \\ \downarrow & \downarrow & \downarrow & \downarrow & \downarrow \\ p & \cdot & 225 & = & 49.50 \end{array}$$

$$p = \frac{49.50}{225}$$

$$p = 0.22 \quad \text{or} \quad 22\%$$

The discount is 22%.

8. The number of students enrolled in a calculus course is 145. If 40% of these students are female, how many are male?

____

Objective 4 Practice Exercises

For extra help, see Examples 7–8 on pages 156–157 of your text.

Answer each question about percent.

10. What is 2.5% of 3500?

10. __________

11. What percent of 5200 is 104?

11. __________

Solve the problem.

12. Paul recently bought a duplex for \$144,000. He expects to earn \$6120 per year on this investment. What percent of the purchase price will he earn?

12. __________

Name: Date:
Instructor: Section:

Chapter 2 LINEAR EQUATIONS AND INEQUALITIES IN ONE VARIABLE

2.7 Further Applications of Linear Equations

Learning Objectives

1. Use percent in solving problems involving rates.
2. Solve problems involving mixtures
3. Solve problems involving simple interest.
4. Solve problems involving denominations of money.
5. Solve problems involving distance, rate, and time.

Objective 1 Use percent solving problems involving rates.

Video Examples

Review this example for Objective 1:

1. The purchase price of a new car is $12,500. In order to finance the car a purchaser is required to make a minimum down payment of 20% of the purchase price. What is the minimum down payment required?

$$12,500 \cdot 0.20 = 2500$$

The down payment is $2500

Now Try:

1. A certificate of deposit pays 2.6% simple interest in one year on a principal of $4500. What interest is being paid on this deposit?

____

Objective 1 Practice Exercises

For extra help, see Example 1 on page 162 of your text.

Solve each problem.

1. Twelve percent of a college student body has a grade point average of 3.0 or better. If there are 1250 students enrolled in the college, how many have a grade point average of less than 3.0?

1. ____________

2. In a class of freshmen and sophomores only, there are 85 students. If 60% of the class is sophomores, how many students are sophomores?

2. ____________

Name: Date:
Instructor: Section:

3. At a large university, 40% of the student body is from out of state. If the total enrollment at the university is 25,000 students, how many are from in-state?

3. ______________

Objective 2 Solve problems involving mixtures.

Video Examples

Review this example for Objective 2:

2. How many gallons of a 20% alcohol solution must be mixed with 15 gallons of a 12% alcohol solution to obtain a 14% alcohol solution?

Step 1 Read the problem. Find the amount of the 20% alcohol solution.

Step 2 Assign a variable. Let x = the amount of the 20% alcohol solution.

Liters of Solution	Rate (as a decimal)	Liters of Pure Acid
x	0.20	$0.2x$
15	0.12	0.12(15)
$x+15$	0.14	$0.14(x+15)$

Step 3 Write an equation.

$$0.20x+0.12(15)=0.14(x+15)$$

Step 4 Solve.

$$0.20x+1.8=0.14x+2.1$$
$$0.06x=0.3$$
$$x=5$$

Step 5 State the answer. 5 gallons of the 20% alcohol solution must be added.

Step 6 Check.

$$0.20(5)+0.12(15)\stackrel{?}{=}0.14(5+15)$$
$$2.8=2.8$$

The answer checks.

Now Try:

2. How many ounces of a 35% solution of acid must be mixed with a 60% solution to get 20 ounces of a 50% solution?

____

Name: Date:
Instructor: Section:

Objective 2 Practice Exercises

For extra help, see Examples 2–3 on pages 163–164 of your text.

Solve each problem.

4. How many pounds of peanuts worth \$3 per pound must be mixed with mixed nuts worth \$5.50 per pound to make 40 pounds of a mixture worth \$5 per pound?

4. ________________

5. How many pounds of candy worth \$7 per pound must be mixed with candy worth \$4.50 per pound to make 100 pounds of candy worth \$6 per pound?

5. ________________

Objective 3 Solve problems involving simple interest.

Video Examples

Review this example for Objective 3:

4. Larry invested some money at 8% simple interest and \$700 less than this amount at 7%. His total annual income from the interest was \$584. How much was invested at each rate?

Step 1 Read the problem. Find the two amounts.

Step 2 Assign a variable.
Let x = the amount at 8%.
Let $x - 700$ = the amount at 7%.

Amount invested (in dollars)	Rate (as a decimal)	Interest for One Year (in dollars)
x	0.08	$0.08x$
$x-700$	0.07	$0.07(x-700)$

Step 3 Write an equation.
$0.08x + 0.07(x-700) = 584$

Now Try:

4. Tracy invested some money at 5% and \$300 more than twice this amount at 7%. Her total annual income from the two investments is \$325. How much is invested at each rate?

____

Step 4 Solve.

$$0.08x + 0.07x - 49 = 584$$
$$0.15x - 49 = 584$$
$$0.15x = 633$$
$$x = 4220$$

Step 5 State the answer. Larry must invest \$4220 at 8%, and 4220 – 700 = \$3520 at 7%.

Step 6 Check. The sum of the two amounts should be \$584.

$$0.08(4220) + 0.07(3520) = 584$$

Objective 3 Practice Exercises

For extra help, see Example 4 on page 165 of your text.

Solve each problem.

6. Louisa invested \$16,000 in bonds paying 7% simple interest. How much additional money should she invest at 4% simple interest so that the average return on the two investments is 6%?

6. ________________

7. Desiree invested some money at 9% and \$100 less than three times that amount at 7%. Her total annual interest was \$83. How much did she invest at each rate?

7. 7%: ____________

9%: ____________

8. Jacob has \$48,000 invested in stocks paying 6%. How much additional money should he invest in certificates of deposit paying 2.5% so that the average return on the two investments is 4%?

8. ________________

Name: Date:
Instructor: Section:

Objective 4 Solve problems involving denominations of money.

Video Examples

Review this example for Objective 4:

5. A collection of dimes and nickels has a total value of \$2.70. The number of nickels is 2 more than twice the number of dimes. How many of each type of coin are in the collection?

Step 1 Read the problem. Find the number of dimes and the number of nickels.

Step 2 Assign a variable.
Let d = the number of dimes.
Then $2d + 2$ = the number of nickels.

Number of coins	Denomination (in dollars)	Total Value (in dollars)
d	0.10	$0.10d$
$2d+2$	0.05	$0.05(2d+2)$

Step 3 Write an equation.
$$0.10d + 0.05(2d+2) = 2.70$$

Step 4 Solve.
$$10d + 5(2d+2) = 270$$
$$10d + 10d + 10 = 270$$
$$20d + 10 = 270$$
$$20d = 260$$
$$d = 13$$

Step 5 State the answer. There are 13 dimes and $2(13)+2 = 28$ nickels.

Step 6 Check.
$0.10(13) + 0.05(28) = 1.30 + 1.40 = 2.70$
The answer is correct.

Now Try:

5. Erica's piggy bank has quarters and nickels in it. The total number of coins in the piggy bank is 50. Their total value is \$8.90. How many of each type are in the piggy bank?

____

____

Name: Date:
Instructor: Section:

Objective 4 Practice Exercises

For extra help, see Example 5 on page 166 of your text.

Solve each problem.

9. Anthony sells two different size jars of peanut butter. The large size sells for \$2.60 and the small size sells for \$1.80. He has 80 jars worth \$164. How many of each size jar does he have?

9.
large ______________
small ______________

10. Stan has 14 bills in his wallet worth \$95 altogether. If the wallet contains only \$5 and \$10 bills, how many bills of each denomination does he have?

10.
\$5 bills ______________
\$10 bills ______________

11. Twice as many general admission tickets to a basketball game were sold as reserved seat tickets. General admission tickets cost \$10 and reserved seat tickets cost \$15. If the total value of both kinds of tickets was \$26,250, how many tickets of each kind were sold?

11.
general admission ______
reserved seats ________

Name: Date:
Instructor: Section:

Objective 5 Use percent involving distance, rate, and time.

Video Examples

Review these examples for Objective 5:

6. A driver averaged 58 mph and took 10 hours to drive from Little Rock to Indianapolis. What is the distance between Little Rock and Indianapolis?

We must find the distance, given the rate and time using $rt = d$.

$58 \cdot 10 = 580$ miles

The distance is 580 miles.

7. A car and a truck leave Oklahoma City at the same time and travel west on the same route. The car travels at a constant rate of 62 mph. The truck travels at a constant rate of 68 mph. In how many hours will the distance between them be 30 miles?

Step 1 Read the problem.

Step 2 Assign a variable. We are looking for time.
Let t = the number of hours until the distance between them is 30 miles.

	Rate	Time	Distance
Truck	68	t	$68t$
Car	62	t	$62t$

Step 3 Write an equation.

$68t - 62t = 30$

Step 4 Solve.

$6t = 30$

$t = 5$

Step 5 State the answer. It will take 5 hours for the truck and car to be 30 miles apart.

Step 6 Check. After 5 hours, the truck travels $68 \cdot 5 = 340$ miles and the car travels $62 \cdot 5 = 310$ miles. The difference is $340 - 310 = 30,$ as required.

Now Try:

6. A driver averaged 54 mph and took 5 hours to drive from Los Angeles to Las Vegas. What is the distance between Los Angeles and Las Vegas?

____

7. A car and a truck leave Dallas at the same time and travel north on the same route. The car travels at a constant rate of 67 mph. The truck travels at a constant rate of 72 mph. In how many hours will the distance between them be 25 miles?

____

Name: Date:
Instructor: Section:

Objective 5 Practice Exercises

For extra help, see Examples 6–8 on pages 167–169 of your text.

Solve each problem.

12. A driver averages 52 mph and took 10 hours to drive from Charlotte, North Carolina to Orlando, Florida. What is the distance between Charlotte and Orlando?

12. ________________

13. A driver averages 50 mph and took 9 hours to drive from Denver, Colorado to Albuquerque, New Mexico. What is the distance between Denver and Albuquerque?

13. ________________

14. A car and a truck leave Billings, Montana at the same time and travel east on the same route. The truck travels at a constant rate of 78 mph. The car travels at a constant rate of 67 mph. In how many hours will the distance between them be 33 miles?

14. ________________

Name: Date:
Instructor: Section:

Chapter 2 LINEAR EQUATIONS AND INEQUALITIES IN ONE VARIABLE

2.8 Solving Linear Inequalities

Learning Objectives

1. Graph intervals on a number line.
2. Use the addition property of inequality.
3. Use the multiplication property of inequality.
4. Solve linear inequalities using both properties of inequality.
5. Solve applied problems using inequalities.
6. Solve linear inequalities with three parts.

Key Terms

Use the vocabulary terms listed below to complete each statement in exercises 1−5.

inequalities **interval** **interval notation**

linear inequality **three-part inequality**

1. An inequality that says that one number is between two other numbers is a(n)________________________________.
2. A portion of a number line is called a(n) ____________________________.
3. A(n) ________________________ can be written in the form $Ax + B < C$, $Ax + B \leq C$, $Ax + B > C$, or $Ax + B \geq C$, where A, B, and C are real numbers with $A \neq 0$.
4. Algebraic expressions related by $<$, $\leq$, $>$, or $\geq$ are called ____________________.
5. The _________________________ for $a \leq x < b$ is $[a, b)$.

Objective 1 Graph intervals on a number line.

Video Examples

Review this example for Objective 1:

1. Write the inequality in interval notation, and graph the interval.

   $x > -3$

   The statement $x > -3$ says that x can represent any value greater than –3, but cannot equal –3, written $(-3, \infty)$. We graph this interval by placing a parenthesis at –3 and drawing an arrow to the right. The parenthesis indicates that –3 is not part of the graph.

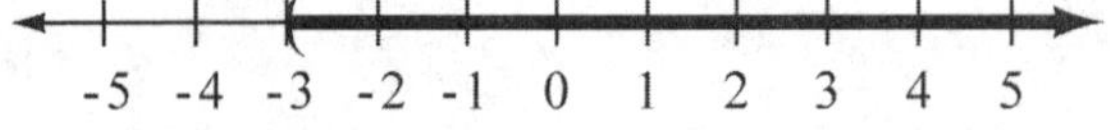

Now Try:

1. Write the inequality in interval notation, and graph the interval.

   $x > -1$

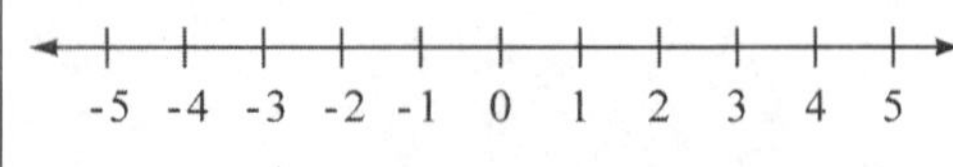

Name: Date:
Instructor: Section:

Objective 1 Practice Exercises

For extra help, see Example 1 on page 175 of your text.

Write each inequality in interval notation and graph the interval.

1. $3 < a$

1. ________________

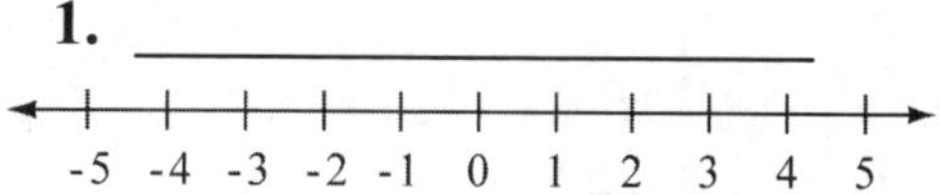

2. $y \geq -2$

2. ________________

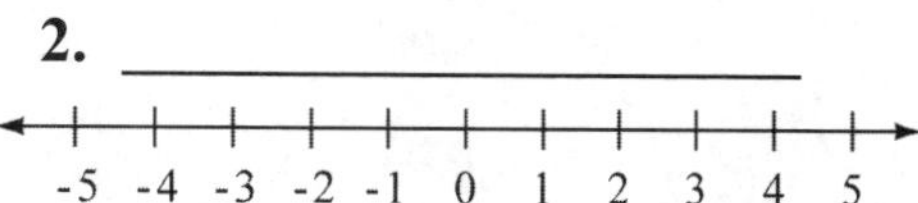

3. $x < -4$

3. ________________

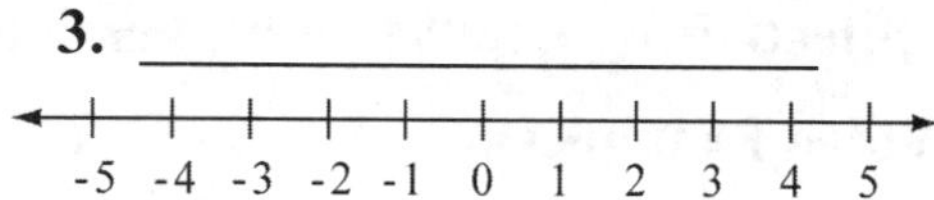

Objective 2 Use the addition property of inequality.

Video Examples

Review this example for Objective 2:

2. Solve $8+4x \geq 3x+3$ and graph the solution set.

$$8+4x \geq 3x+3$$
$$8+4x-3x \geq 3x+3-3x$$
$$8+x \geq 3$$
$$8+x-8 \geq 3-8$$
$$x \geq -5$$

The solution set, $[-5, \infty)$ is graphed below.

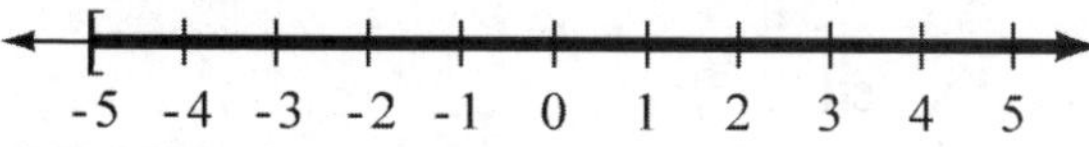

Now Try:

2. Solve $5+9x \geq 8x+2$ and graph the solution set.

____

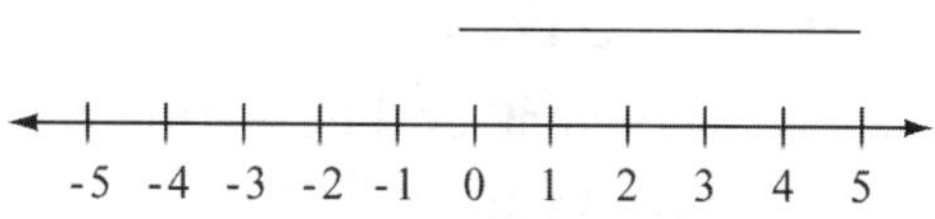

Objective 2 Practice Exercises

For extra help, see Example 2 on page 176 of your text.

Solve each inequality. Write the solution set in interval notation and then graph it.

4. $5a+3 \leq 6a$

4. ________________

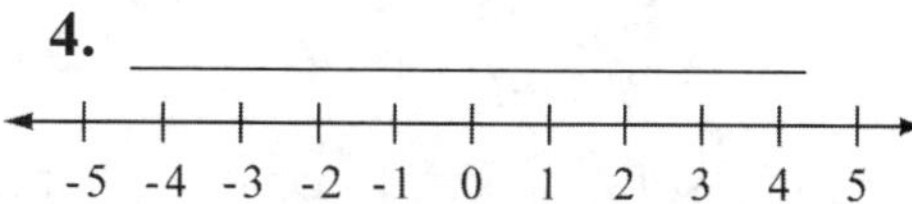

5. $6+3x<4x+4$

5. ______________

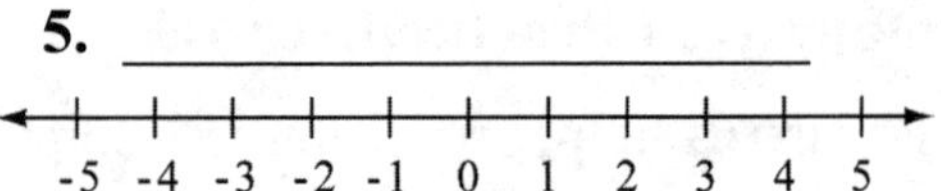

6. $3+5p\leq 4p+3$

6. ______________

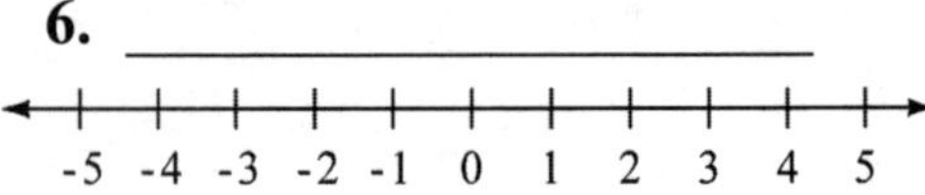

Objective 3 Use the multiplication property of inequality.

Video Examples

Review these examples for Objective 3:

3. Solve each inequality, and graph the solution set.

a. $6x<-24$

We divide each side by 6.

$$6x<-24$$

$$\frac{6x}{6}<\frac{-24}{6}$$

$$x<-4$$

The graph of the solution set $(-\infty,-4)$, is shown below.

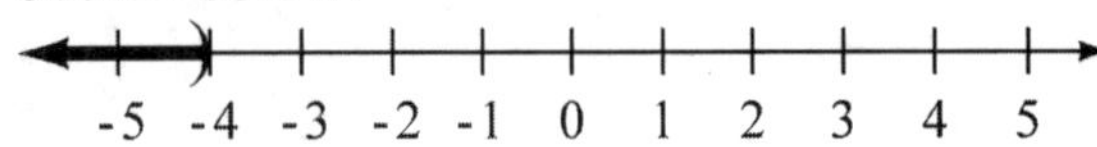

b. $-6x\geq 30$

Here each side of the inequality must be divided by –6, a negative number, which does require changing the direction of the inequality symbol.

$$-6x\geq 30$$

$$\frac{-6x}{-6}\leq\frac{30}{-6}$$

$$x\leq -5$$

The solution set, $(-\infty,-5]$, is graphed below.

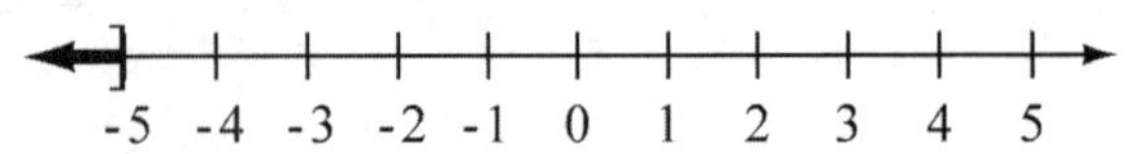

Now Try:

3. Solve each inequality, and graph the solution set.

a. $8x\leq -40$

____

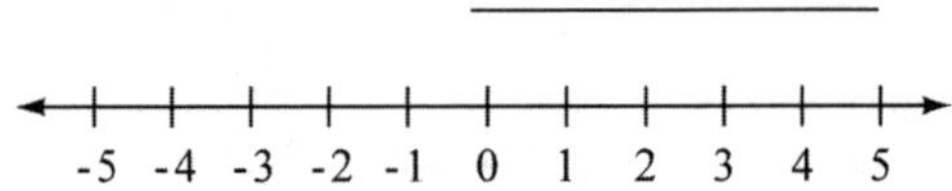

b. $-9t>36$

____

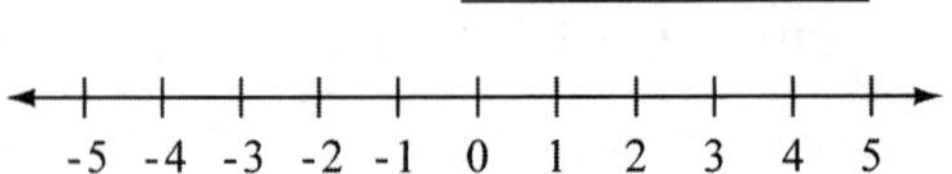

Name: Date:
Instructor: Section:

Objective 3 Practice Exercises

For extra help, see Example 3 on pages 178–179 of your text.

Solve each inequality. Write the solution set in interval notation and then graph it.

7. $-2s < 4$

7. ______________

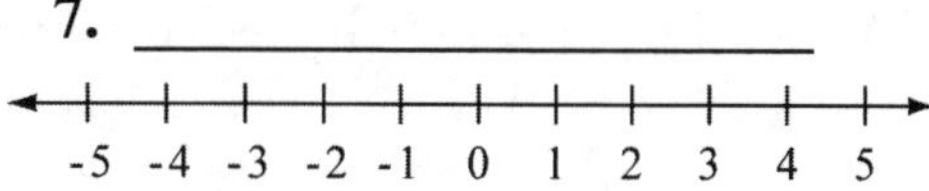

8. $4k \geq -16$

8. ______________

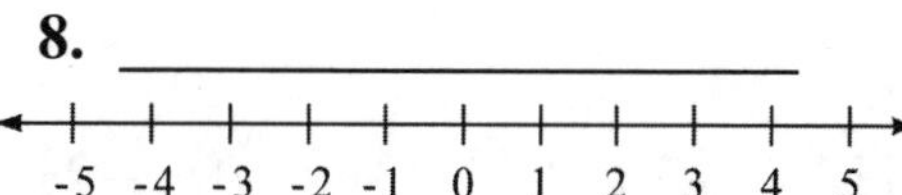

9. $-9m \geq -36$

9. ______________

-5 -4 -3 -2 -1 0 1 2 3 4 5

Objective 4 Solve linear inequalities using both properties of inequality.

Video Examples

Review this example for Objective 4:

4. Solve $4x+3-7>-2x+8+3x$. Graph the solution set.

Step 1 Combine like terms and simplify.

$$4x+3-7>-2x+8+3x$$

$$4x-4>x+8$$

Step 2 Use the addition property of inequality.

$$4x-4-x>x+8-x$$

$$3x-4>8$$

$$3x-4+4>8+4$$

$$3x>12$$

Step 3 Use the multiplication property of inequality.

$$\frac{3x}{3}>\frac{12}{3}$$

$$x>4$$

The solution set is $(4, \infty)$. The graph is shown below.

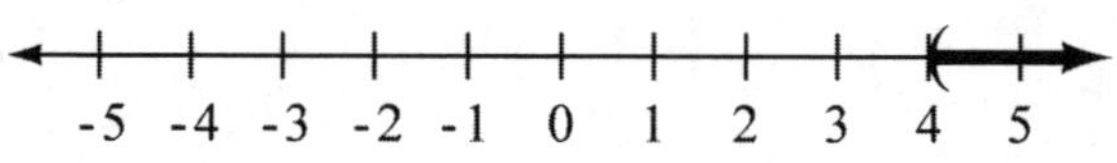

Now Try:

4. Solve $8x-5+4 \geq 6x-3x+9$. Graph the solution set.

____

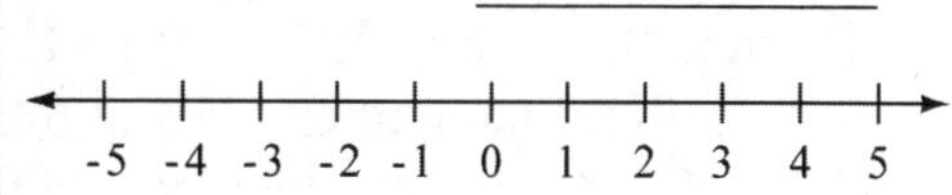

Name: Date:
Instructor: Section:

Objective 4 Practice Exercises

For extra help, see Example 4–6 on pages 179–180 of your text.

Solve each inequality. Write the solution set in interval notation and then graph it.

10. $4(y-3)+2>3(y-2)$

10. ______________________

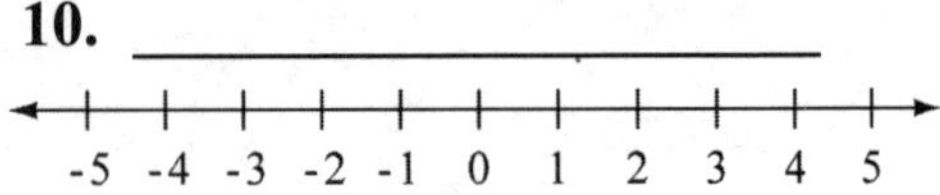

11. $-3(m+2)+3\le -4(m-2)-6$

11. ______________________

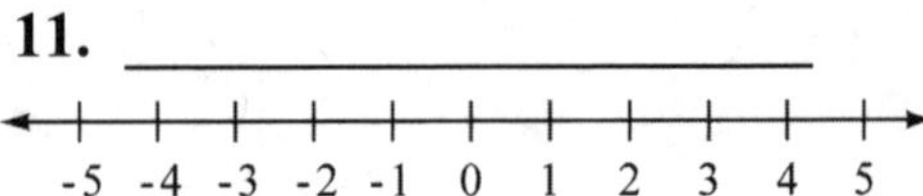

12. $7(2-x)\le -2(x-3)-x$

12. ______________________

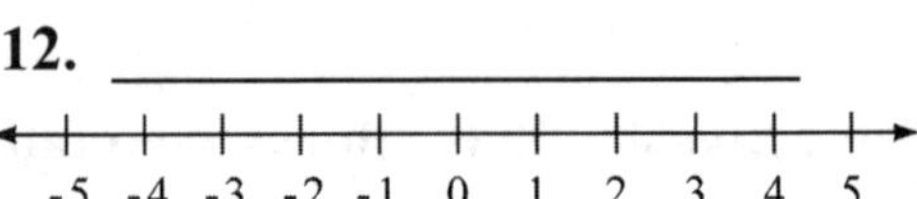

Objective 5 Solve applied problems using inequalities.

Video Examples

Review this example for Objective 5:

7. Ruth tutors mathematics in the evenings in an office for which she pays \$600 per month rent. If rent is her only expense and she charges each student \$40 per month, how many students must she teach to make a profit of at least \$1600 per month?

Step 1 Read the problem again.

Step 2 Assign a variable.
Let x = the number of students.

Step 3 Write an inequality.
$40x-600\ge 1600$

Step 4 Solve.

$$40x-600+600\ge 1600+600$$
$$40x\ge 2200$$
$$\frac{40x}{40}\ge\frac{2200}{40}$$
$$x\ge 55$$

Now Try:

7. Two sides of a triangle are equal in length, with the third side 8 feet longer than one of the equal sides. The perimeter of the triangle cannot be more than 38 feet. Find the largest possible value for the length of the equal sides.

____

Name: Date:
Instructor: Section:

Step 5 State the answer. Ruth must have 55 or more students to have at least \$1600 profit.

Step 6 Check. $40(55)-600=1600$ Also, any number greater than 55 makes the profit greater than \$1600.

Objective 5 Practice Exercises

For extra help, see Example 7 on page 181 of your text.

Solve each problem.

13. Lauren has grades of 98 and 86 on her first two chemistry quizzes. What must she score on her third quiz to have an average of at least 91 on the three quizzes?

13. ________________

14. Nina has a budget of \$230 for gifts for this year. So far she has bought gifts costing \$47.52, \$38.98, and \$26.98. If she has three more gifts to buy, find the average amount she can spend on each gift and still stay within her budget.

14. ________________

15. If twice the sum of a number and 7 is subtracted from three times the number, the result is more than –9. Find all such numbers.

15. ________________

Name: Date:
Instructor: Section:

Objective 6 Solve linear inequalities with three parts.

Video Examples

Review this example for Objective 6:

9. Solve the inequality, and graph the solution set.

$3 \le 4x - 5 < 7$

$$3 \le 4x - 5 < 7$$
$$3 + 5 \le 4x - 5 + 5 < 7 + 5$$
$$8 \le 4x < 12$$
$$\frac{8}{4} \le \frac{4x}{4} < \frac{12}{4}$$
$$2 \le x < 3$$

The solution set is $[2, 3)$. The graph is shown below.

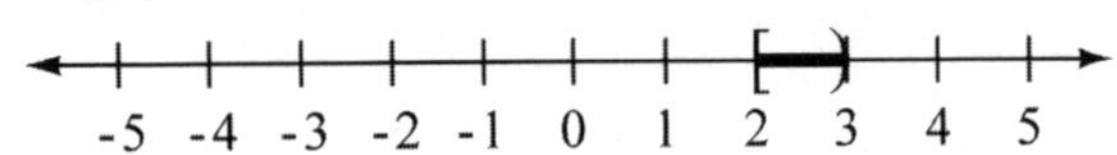

Now Try:

9. Solve the inequality, and graph the solution set.
$8 \le 6x - 4 < 20$

____

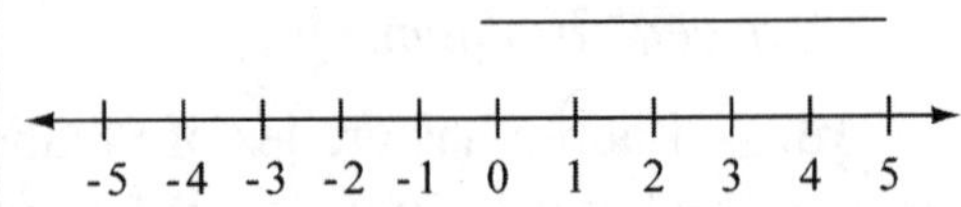

Objective 6 Practice Exercises

For extra help, see Examples 8–9 on pages 182–183 of your text.

Solve each inequality. Write the solution set in interval notation and then graph it.

16. $7 < 2x + 3 \le 13$

16. ______________

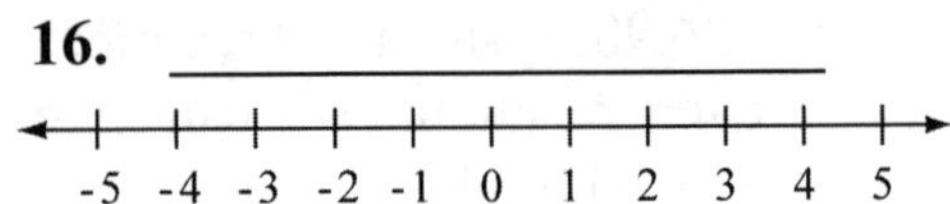

17. $-17 \le 3x - 2 < -11$

17. ______________

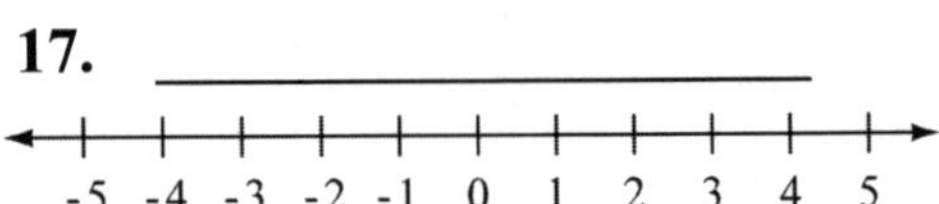

18. $1 < 3z + 4 < 19$

18. ______________

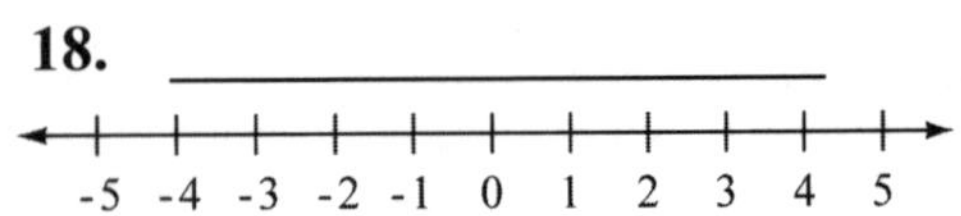

Name: Date:
Instructor: Section:

Chapter 3 LINEAR EQUATIONS AND INEQUALITIES IN TWO VARIABLES; FUNCTIONS

3.1 Linear Equations and Rectangular Coordinates

Learning Objectives

1 Interpret graphs.
2 Write a solution as an ordered pair.
3 Decide whether a given ordered pair is a solution of a given equation.
4 Complete ordered pairs for a given equation.
5 Complete a table of values.
6 Plot ordered pairs.

Key Terms

Use the vocabulary terms listed below to complete each statement in exercises 1−13.

line graph **linear equation in two variables**

ordered pair **table of values** ***x*-axis**

***y*-axis** **rectangular (Cartesian) coordinate system**

origin **quadrants** **plane** **coordinates**

plot **scatter diagram**

1. A ______________________________ uses dots connected by lines to show trends.

2. An equation that can be written in the form $Ax + By = C$, where A, B, and C are real numbers and $A, B \neq 0$, is called a ______________________________.

3. ______________________________ are the numbers in the ordered pair that specify the location of a point on a rectangular coordinate system.

4. In a coordinate system, the horizontal axis is called the ______________________.

5. In a coordinate system, the vertical axis is called the ________________________.

6. A pair of numbers written between parentheses in which order is important is called a(n) ______________________________.

7. Together, the x-axis and the y-axis form a ______________________.

8. A coordinate system divides the plane into four regions called ________________.

9. The axis lines in a coordinate system intersect at the ______________________.

10. To ____________________ an ordered pair is to find the corresponding point on a coordinate system.

Name: Date:
Instructor: Section:

11. A graph of ordered pairs is called a ____________________________.

12. A table showing selected ordered pairs of numbers that satisfy an equation is called a ________________________.

13. A flat surface determined by two intersecting lines is a __________________.

Objective 1 Interpret graphs.

Video Examples

The line graph shows the number of degrees awarded by a university for the years 2000–2005.

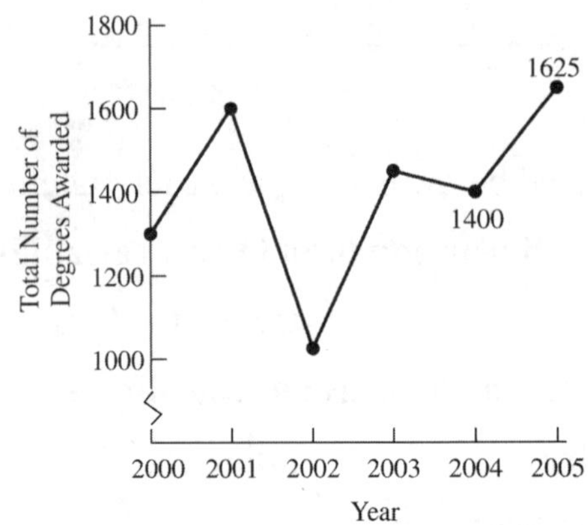

Review these examples for Objective 1:

1.

a. Between which years did the number of degrees awarded decrease?

The line between 2001 and 2002 and between 2003 and 2004 falls, so the number of degrees awarded decreased between 2001-2002 and 2003-2004.

b. Estimate the total number of degrees awarded in 2002 and in 2005. About how many more degrees were awarded in 2005?

Move up from 2002 on the horizontal scale to the point plotted for 2002. This point is about 1000. So about 1000 degrees were awarded in 2002.
Similarly, locate the point plotted for 2005. Moving across to the vertical scale, the graph indicates that the number of degrees awarded in 2005 was 1625.
Between 2002 and 2005, the increase was

$$1625 - 1000 = 625.$$

Now Try:

1.

a. Between which years did the number of degrees awarded increase?

____

b. Estimate the total number of degrees awarded in 2000 and in 2001. About how many more degrees were awarded in 2001?

____

Name: Date:
Instructor: Section:

Objective 1 Practice Exercises

For extra help, see Example 1 on page 200 of your text.

The line graph shows the number of degrees awarded by a university for the years 2000–2005. Use this graph to answer exercises 1–2.

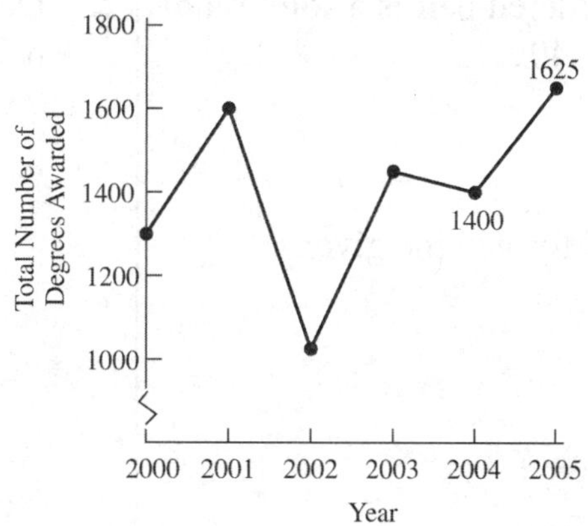

1. Between what pairs of consecutive years did the number of degrees decrease? **1.** ________________

2. If 20% of the degrees awarded in 2005 were MBA degrees, how many MBAs were awarded in 2005? **2.** ________________

Objective 2 Write a solution as an ordered pair.

Objective 2 Practice Exercises

For extra help, see page 201 of your text.

Write each solution as an ordered pair.

3. $x = 4$ and $y = 7$ **3.** ________________

4. $y = \frac{1}{3}$ and $x = 0$ **4.** ________________

5. $x = 0.2$ and $y = 0.3$ **5.** ________________

Name: Date:
Instructor: Section:

Objective 3 Decide whether a given ordered pair is a solution of a given equation.

Video Examples

Review these examples for Objective 3:

2. Decide whether each ordered pair is a solution of the equation $4x+5y=40$.

a. $(5,\ 4)$

Substitute 5 for x and 4 for y in the given equation.

$$\begin{aligned} 4x+5y &= 40 \\ 4(5)+5(4) &\stackrel{?}{=} 40 \\ 20+20 &\stackrel{?}{=} 40 \\ 40 &= 40 \quad \text{True} \end{aligned}$$

This result is true, so $(5,\ 4)$ is a solution of $4x+5y=40$.

b. $(-3,\ 6)$

Substitute –3 for x and 6 for y in the given equation.

$$\begin{aligned} 4x+5y &= 40 \\ 4(-3)+5(6) &\stackrel{?}{=} 40 \\ -12+30 &\stackrel{?}{=} 40 \\ 18 &= 40 \quad \text{False} \end{aligned}$$

This result is false, so $(-3,\ 6)$ is not a solution of $4x+5y=40$.

Now Try:

2. Decide whether each ordered pair is a solution of the equation $3x-4y=12$.

a. $(8,\ 3)$

____

b. $(5,-4)$

____

Objective 3 Practice Exercises

For extra help, see Example 2 on page 202 of your text.

Decide whether the given ordered pair is a solution of the given equation.

6. $4x-3y=10;\ (1,2)$

6. ____________

7. $2x-3y=1;\ \left(0,\frac{1}{3}\right)$

7. ____________

8. $x=-7;\ (-7,9)$

8. ____________

Name: Date:
Instructor: Section:

Objective 4 Complete ordered pairs for a given equation.

Video Examples

Review this example for Objective 4:

3. Complete the ordered pair for the equation $y = 5x + 8$.

$(3, \_\_\_\_)$

Replace x with 3.

$y = 5x + 8$

$y = 5(3) + 8$

$y = 15 + 8$

$y = 23$

The ordered pair is (3, 23).

Now Try:

3. Complete the ordered pair for the equation $y = 4x - 7$.

$(5, \_\_\_\_)$

____

Objective 4 Practice Exercises

For extra help, see Example 3 on pages 202–203 of your text.

For each of the given equations, complete the ordered pairs beneath it.

9. $y = 2x - 5$

(a) $(2, \ )$

(b) $(0, \ )$

(c) $(\ , 3)$

(d) $(\ , -7)$

(e) $(\ , 9)$

9.
(a) __________
(b) __________
(c) __________
(d) __________
(e) __________

10. $y = 3 + 2x$

(a) $(-4, \ )$

(b) $(2, \ )$

(c) $(\ , 0)$

(d) $(-2, \ )$

(e) $(\ , -7)$

10.
(a) __________
(b) __________
(c) __________
(d) __________
(e) __________

Name: Date:
Instructor: Section:

Objective 5 Complete a table of values.

Video Examples

Review this example for Objective 5:

4. Complete the table of values for the equation. Then write the results as ordered pairs.

$2x-3y=6$

x	y
9	
6	
	−2
	8

From the table, we can write the ordered pairs: (9, ____), (6, ____), (____, –2), (____, 8).

From the first row of the table, let $x = 9$ in the equation. From the second row of the table, let $x = 6$.

If $x=9$,	If $x=6$,
$2x-3y=6$	$2x-3y=6$
$2(9)-3y=6$	$2(6)-3y=6$
$18-3y=6$	$12-3y=6$
$-3y=-12$	$-3y=-6$
$y=4$	$y=2$

The first two ordered pairs are (9, 4) and (6, 2).

From the third and fourth rows of the table, let $y = -2$ and $y = 8$, respectively.

If $y=-2$,	If $y=8$,
$2x-3y=6$	$2x-3y=6$
$2x-3(-2)=6$	$2x-3(8)=6$
$2x+6=6$	$2x-24=6$
$2x=0$	$2x=30$
$x=0$	$x=15$

The last two ordered pairs are (0, –2) and (15, 8). The completed table and corresponding ordered pairs follow.

x	y	Ordered pairs
9	4 →	$(9,\ 4)$
6	2 →	$(6,\ 2)$
0	−2 →	$(0,-2)$
15	8 →	$(15,\ 8)$

Now Try:

4. Complete the table of values for the equation. Then write the results as ordered pairs.

$4x-y=8$

x	y
1	
5	
	0
	4

____

Name: Date:
Instructor: Section:

Objective 5 Practice Exercises

For extra help, see Example 4 on pages 203–204 of your text.

Complete each table of values. Write the results as ordered pairs.

11. $2x+5=7$ **11.** ________________

x	y
	-3
	0
	5

12. $y-4=0$ **12.** ________________

x	y
-4	
0	
6	

13. $4x+3y=12$ **13.** ________________

x	y
0	
	0
	-1

Objective 6 Plot ordered pairs.

Video Examples

Review these examples for Objective 6:

5. Plot the given points in a coordinate system.
(5, 4) (–2,–1) (–3, 5) (4,–2)
(1,–2.5) (3, 0) (0, 4)

Step 1 Move right or left the number of units that correspond to the x-coordinate in the ordered pair—right if the x-coordinate is positive and left if it is negative.

Step 2 Then turn and move up or down the number of units that corresponds to the y-coordinate—up if the y-coordinate is

Now Try:

5. Plot the given points in a coordinate system.
(2, 4) (–5, 1) (–4,–2)
(3,–5) (2,–1.5) (6, 0)
(0,–6)

positive or down if it is negative.

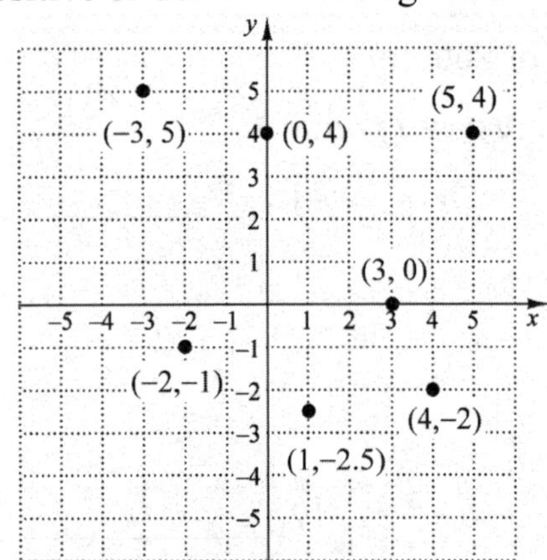

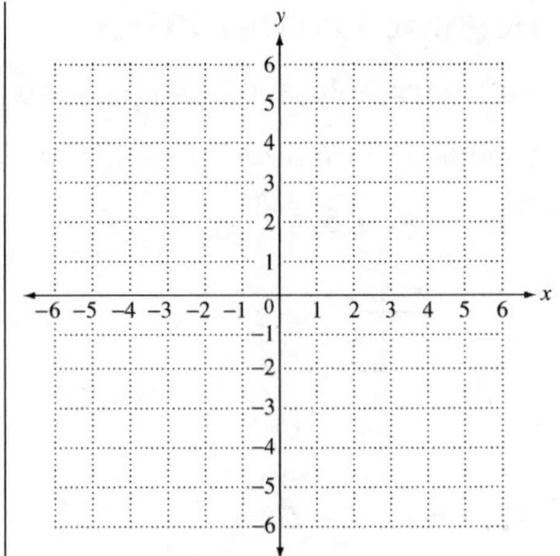

Objective 6 Practice Exercises

For extra help, see Examples 5–6 on pages 205–206 of your text.

Plot the each ordered pair on a coordinate system.

14. $(0,-2)$

14.

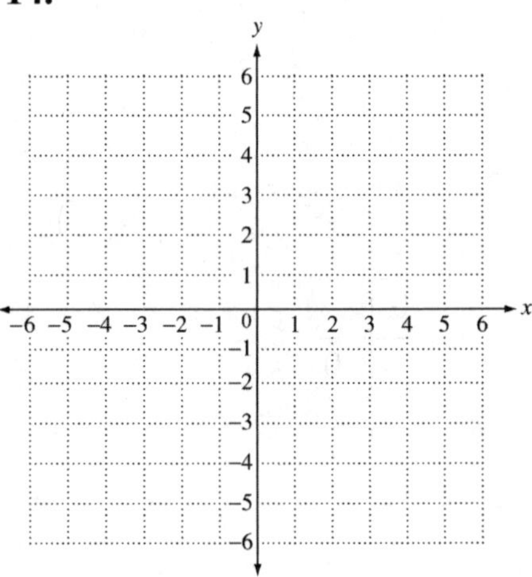

15. $(-3,\ 4)$

15.

Name: Date:
Instructor: Section:

16. $(2,-5)$

16.

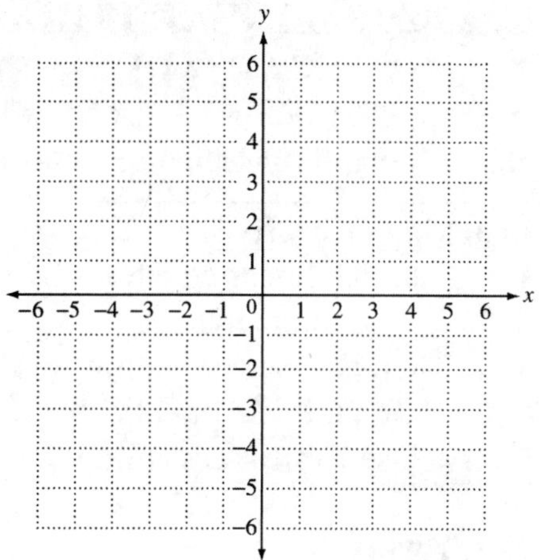

Name: Date:
Instructor: Section:

Chapter 3 LINEAR EQUATIONS AND INEQUALITIES IN TWO VARIABLES; FUNCTIONS

3.2 Graphing Linear Equations in Two Variables

Learning Objectives	
1	Graph linear equations by plotting ordered pairs.
2	Find intercepts.
3	Graph linear equations of the form $Ax + By = 0$.
4	Graph linear equations of the form $y = b$ or $x = a$.
5	Use a linear equation to model data.

Key Terms

Use the vocabulary terms listed below to complete each statement in exercises 1–4.

graph **graphing** ***y*-intercept** ***x*-intercept**

1. If a graph intersects the y-axis at k, then the ____________________ is $(0, k)$.

2. If a graph intersects the x-axis at k, then the ____________________ is $(k, 0)$.

3. The process of plotting the ordered pairs that satisfy a linear equation and drawing a line through them is called ____________________.

4. The set of all points that correspond to the ordered pairs that satisfy the equation is called the ____________________ of the equation.

Objective 1 Graph linear equations by plotting ordered pairs.

Video Examples

Review this example for Objective 1:

2. Graph $2x+3y=6$.

First let $x = 0$ and then let $y= 0$ to determine two ordered pairs.

$$\begin{array}{r|r} 2(0)+3y=6 & 2x+3(0)=6 \\ 0+3y=6 & 2x+0=6 \\ 3y=6 & 2x=6 \\ y=2 & x=3 \end{array}$$

The ordered pairs are (0, 2) and (3, 0). Find a third ordered pair by choosing a number other than 0 for x or y. We choose $y = 4$.

$$\begin{aligned} 2x+3(4)&=6 \\ 2x+12&=6 \\ 2x&=-6 \\ x&=-3 \end{aligned}$$

Now Try:

2. Graph $x+y=3$.

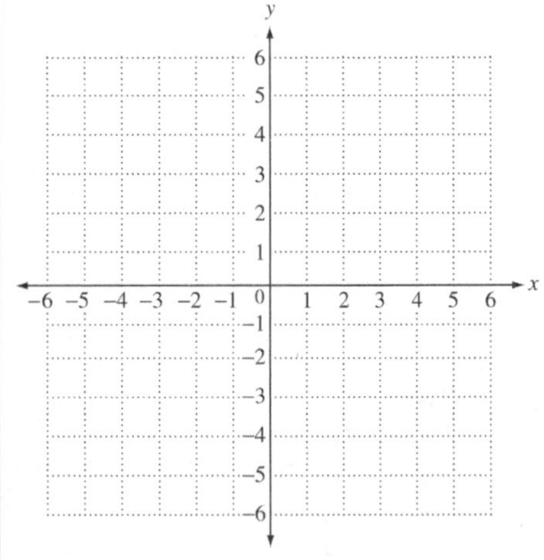

Name: Date:
Instructor: Section:

This gives the ordered pair (–3, 4). We plot the three ordered pairs (0, 2), (3, 0), and (–3, 4) and draw a line through them.

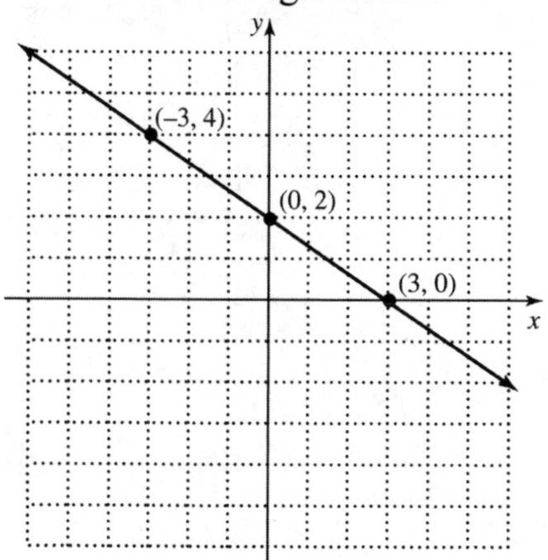

Objective 1 Practice Exercises

For extra help, see Examples 1–2 on pages 213–214 of your text.

Complete the ordered pairs for each equation. Then graph the equation by plotting the points and drawing a line through them.

1. $y = 3x - 2$

$(0,\ \ )$

$(\ \ ,0)$

$(2,\ \ )$

1.

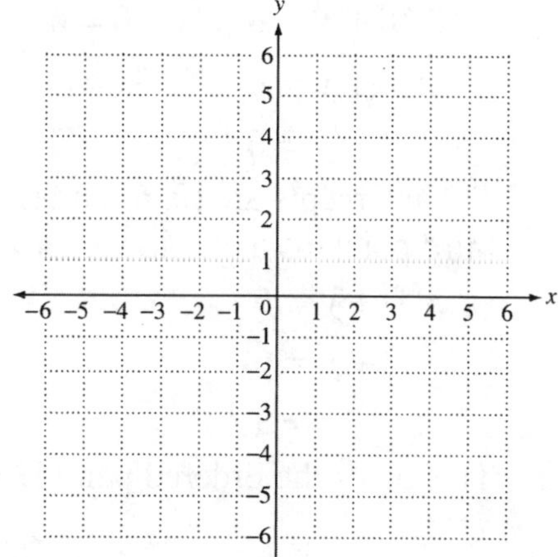

2. $x - y = 4$

$(0,\ \ )$

$(\ \ ,0)$

$(-2,\ \ )$

2.

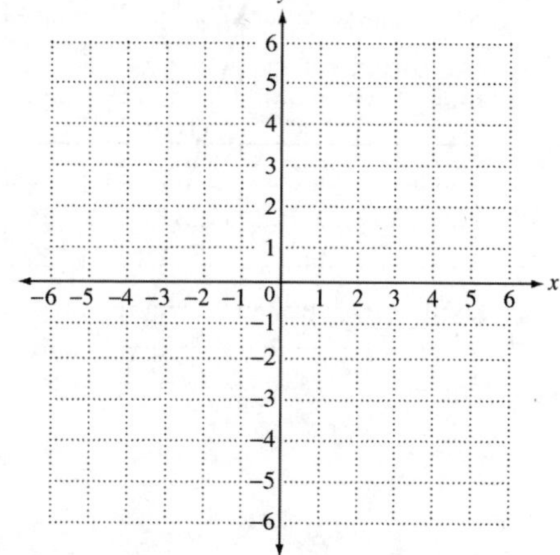

Name: Date:
Instructor: Section:

3. $x = 2y + 1$

$(0, \ )$

$(\ , 0)$

$(\ , -2)$

3.

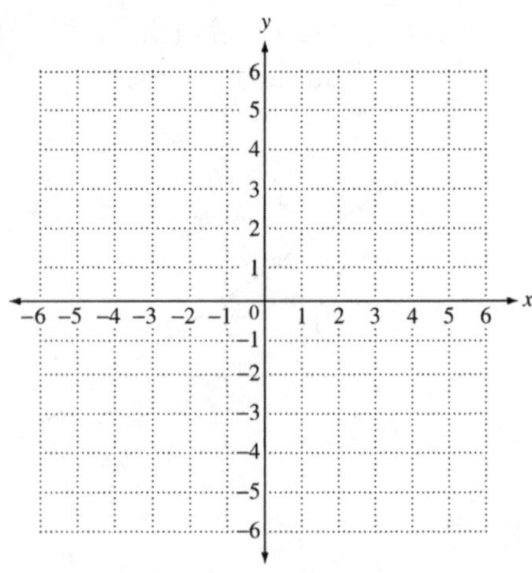

Objective 2 Find intercepts.

Video Examples

Review this example for Objective 2:

3. Graph $3x + y = 6$ using intercepts.

To find the y-intercept, let $x = 0$.
To find the x-intercept, let $y = 0$.

$3(0) + y = 6$	$3x + 0 = 6$
$0 + y = 6$	$3x = 6$
$y = 6$	$x = 2$

The intercepts are (0, 6) and (2, 0). To find a third point, as a check, we let $x = 1$.

$$3(1) + y = 6$$
$$3 + y = 6$$
$$y = 3$$

This gives the ordered pair (1, 3).

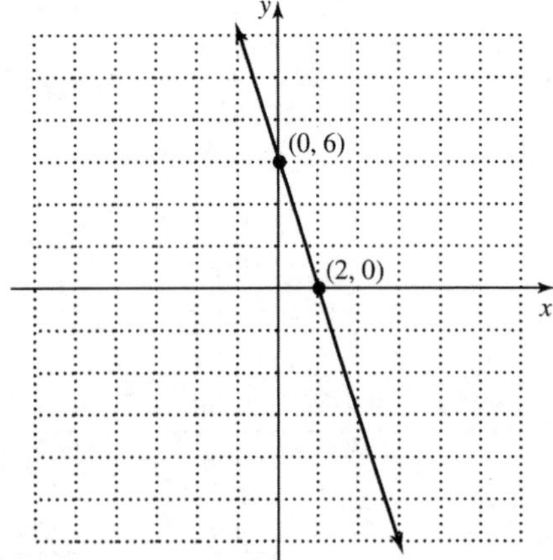

Now Try:

3. Graph $5x - 2y = -10$ using intercepts.

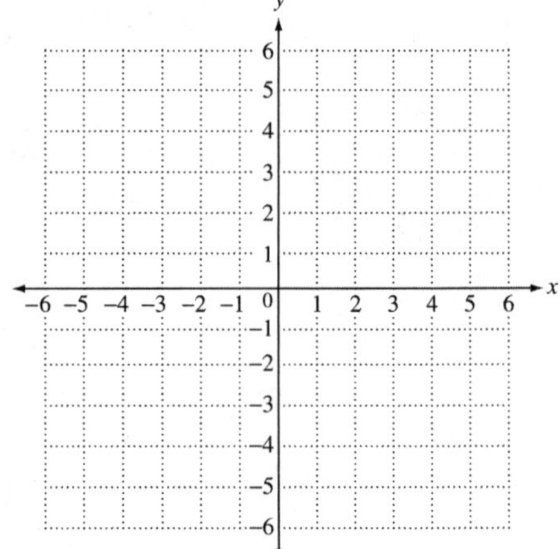

Name: Date:
Instructor: Section:

4. Graph $y=\frac{3}{2}x-4$.

To find the y-intercept, let $x = 0$.
To find the x-intercept, let $y = 0$.

$y=\frac{3}{2}(0)-4$	$0=\frac{3}{2}x-4$
$y=0-4$	$4=\frac{3}{2}x$
$y=-4$	$\frac{8}{3}=x$

The intercepts are (0, –4) and $\left(\frac{8}{3},\ 0\right)$. To find a third point, as a check, we let $x = 2$.

$y=\frac{3}{2}(2)-4$

$y=3-4$

$y=-1$

This gives the ordered pair (2, –1).

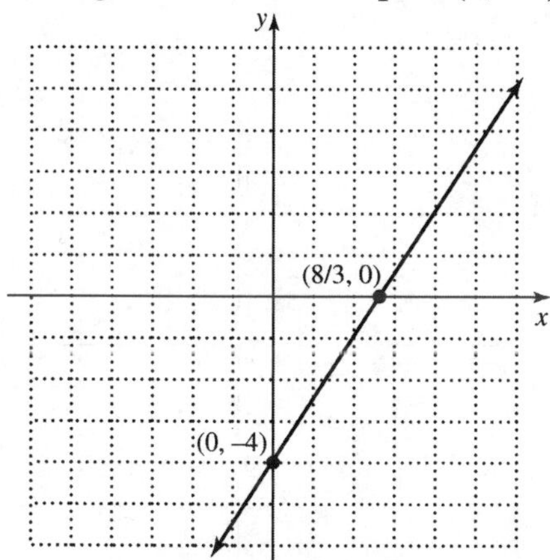

4. Graph $y=4x-4$.

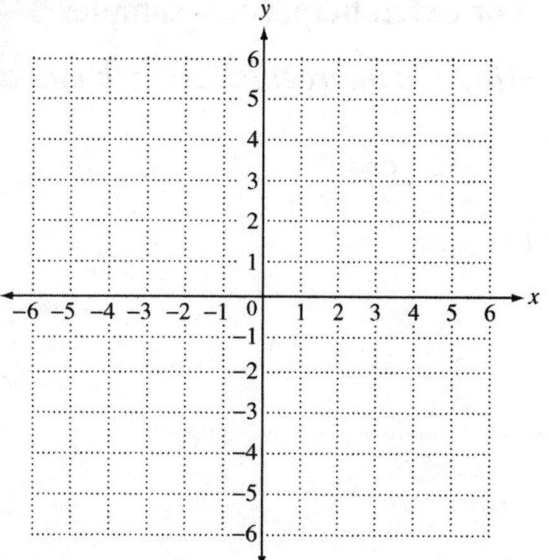

Name: Date:
Instructor: Section:

Objective 2 Practice Exercises

For extra help, see Examples 3–4 on pages 214–216 of your text.

Find the intercepts for each equation. Then graph the equation.

4. $y = \frac{2}{3}x - 2$

4.

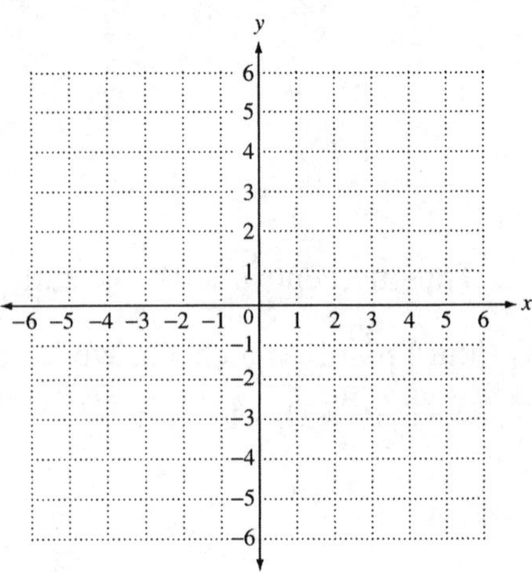

5. $4x - 7y = -8$

5.

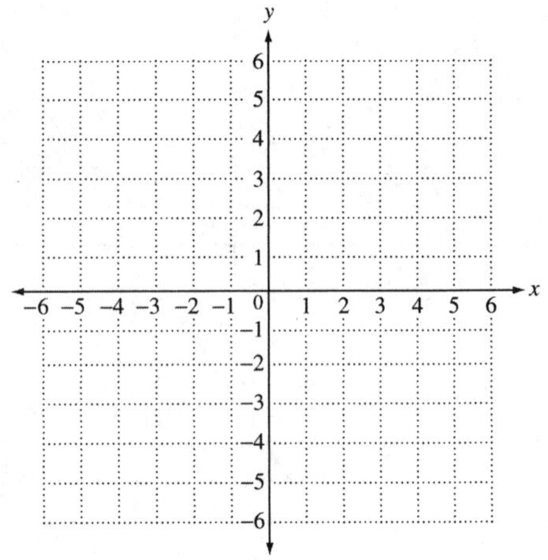

Name: Date:
Instructor: Section:

Objective 3 Graph linear equations of the form $Ax + By = 0$.

Video Examples

Review this example for Objective 3:

5. Graph $x+5y=0$.

To find the y-intercept, let $x = 0$.
To find the x-intercept, let $y = 0$.

$0+5y=0$	$x+5(0)=0$
$5y=0$	$x+0=0$
$y=0$	$x=0$

The x- and y-intercepts are the same point (0, 0). We must select two other values for x or y to find two other points. We choose $y = 1$ and $y = -1$.

$x+5(1)=0$	$x+5(-1)=0$
$x+5=0$	$x-5=0$
$x=-5$	$x=5$

We use (–5, 1), (0, 0), and (5, –1) to draw the graph.

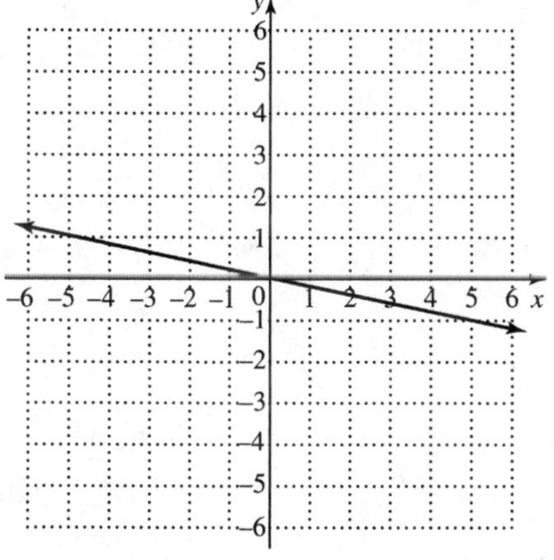

Now Try:

5. Graph $3x-y=0$.

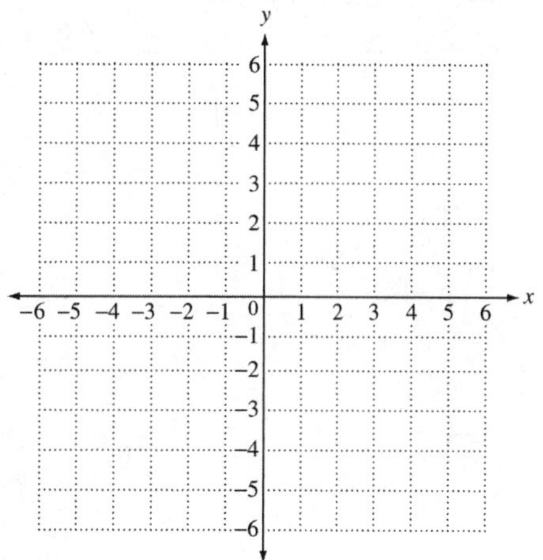

Name: Date:
Instructor: Section:

Objective 3 Practice Exercises

For extra help, see Example 5 on page 216 of your text.

Graph each equation.

6. $-3x-2y=0$

6.

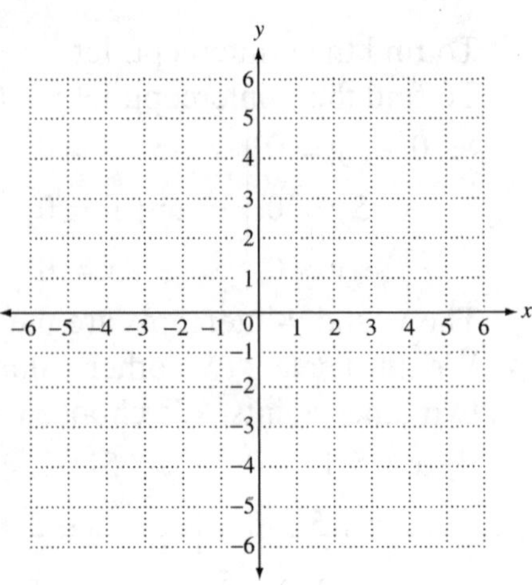

7. $x+y=0$

7.

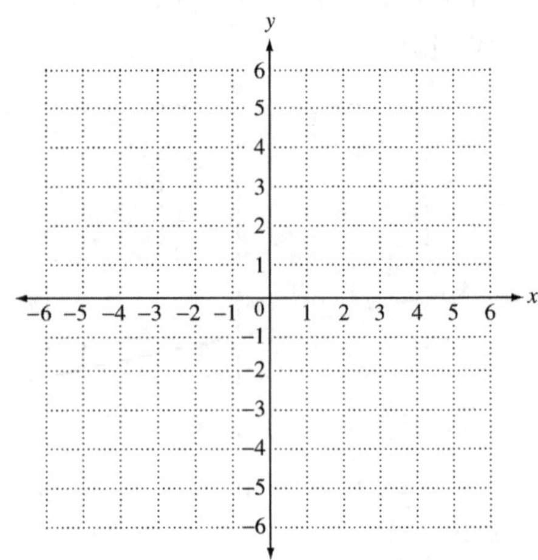

8. $y=2x$

8.

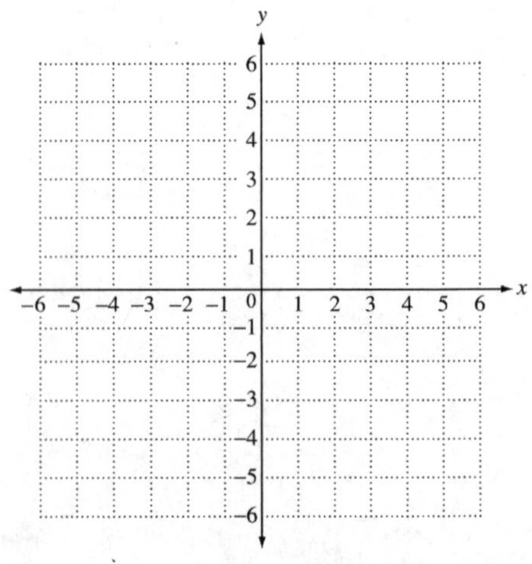

Name: Date:
Instructor: Section:

Objective 4 Graph linear equations of the form $y = b$ or $x = a$.

Video Examples

Review these examples for Objective 4:

6. Graph $y = -2$.

For any value of x, y is always –2. Three ordered pairs that satisfy the equation are (–4, –2), (0, –2) and (2, –2). Drawing a line through these points gives the horizontal line. The y-intercept is (0, –2). There is no x-intercept.

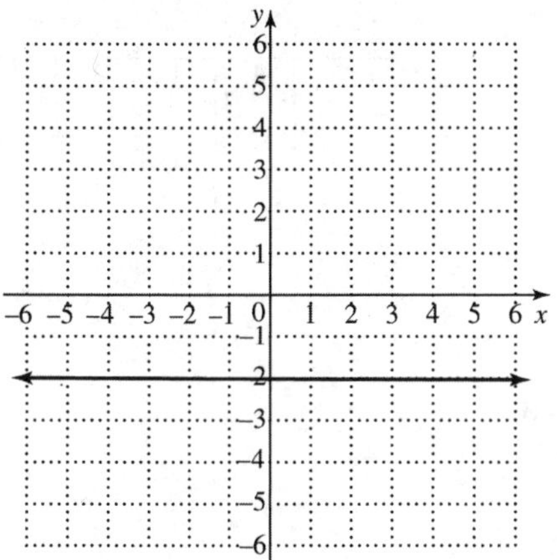

7. Graph $x + 4 = 0$.

First we subtract 4 from each side of the equation to get the equivalent equation $x = -4$. All ordered-pair solutions of this equation have x-coordinate –4.
Three ordered pairs that satisfy the equation are (–4, –1), (–4, 0), and (–4, 3). The graph is a vertical line. The x-intercept is (–4, 0). There is no y-intercept.

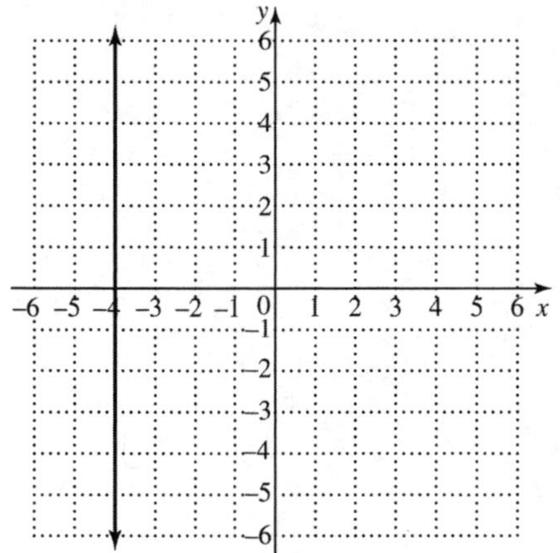

Now Try:

6. Graph $y = 4$.

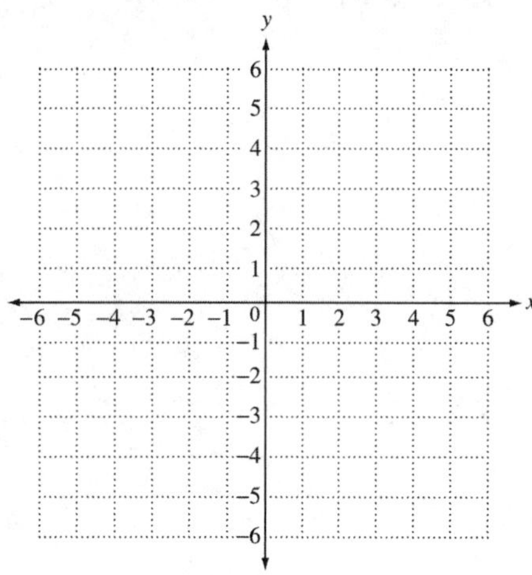

7. Graph $x = 0$.

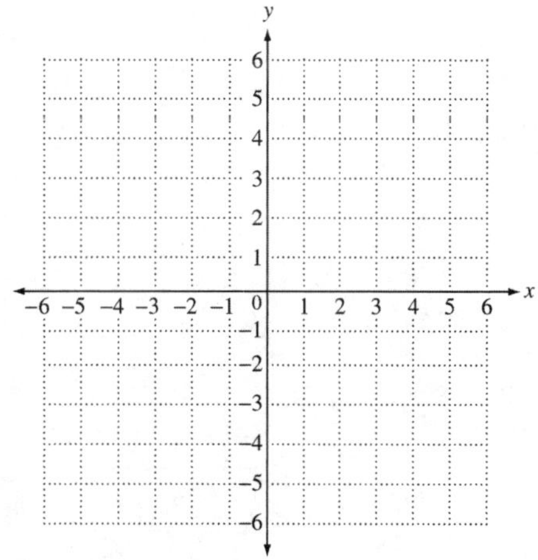

Name: Date:
Instructor: Section:

Objective 4 Practice Exercises

For extra help, see Examples 6–7 on page 217 of your text.

Graph each equation.

9. $x-1=0$

9.

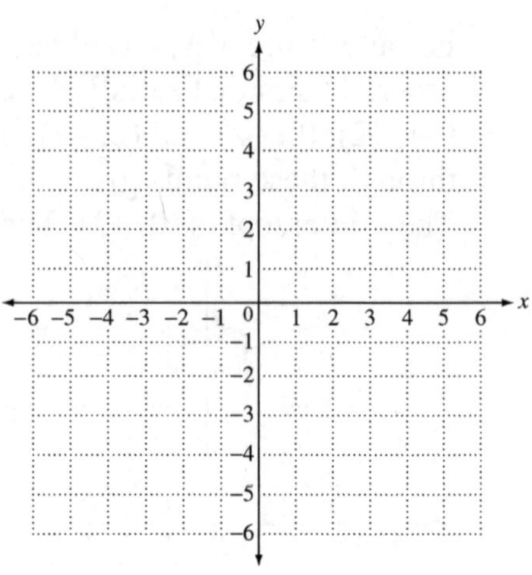

10. $y+3=0$

10.

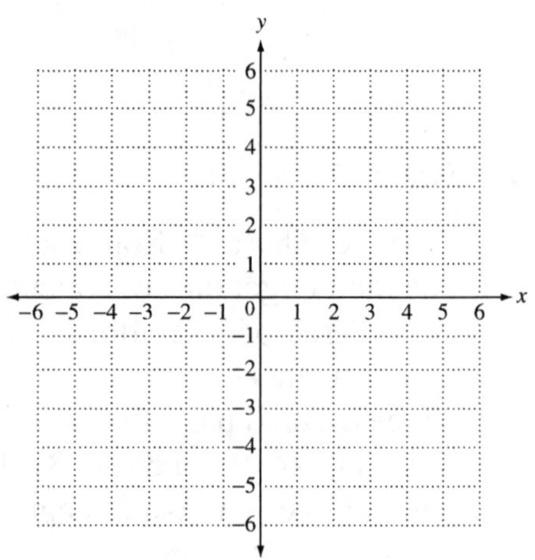

Name: Date:
Instructor: Section:

Objective 5 Use a linear equation to model data.

Video Examples

Review these examples for Objective 5:

8. Every year sea turtles return to a certain group of islands to lay eggs. The number of turtle eggs that hatch can be approximated by the equation $y = -70x + 3260$, where y is the number of eggs that hatch and $x = 0$ representing 1990.

a. Use this equation to find the number of eggs that hatched in 1995, 2000, and 2005, and 2015.

Substitute the appropriate value for each year x to find the number of eggs hatched in that year.

For 1995:

$y = -70(5) + 3260$ $\quad 1995 - 1990 = 5$

$y = 2910$ eggs $\quad$ Replace x with 5.

For 2000:

$y = -70(10) + 3260$ $\quad 2000 - 1990 = 10$

$y = 2560$ eggs $\quad$ Replace x with 10.

For 2005:

$y = -70(15) + 3260$ $\quad 2005 - 1990 = 15$

$y = 2210$ eggs $\quad$ Replace x with 15.

For 2015:

$y = -70(25) + 3260$ $\quad 2015 - 1990 = 25$

$y = 1510$ eggs $\quad$ Replace x with 25.

b. Write the information from part (a) as four ordered pairs, and use them to graph the given linear equation.

Since x represents the year and y represents the number of eggs, the ordered pairs are (5, 2910), (10, 2560), (15, 2210), and (25, 1510).

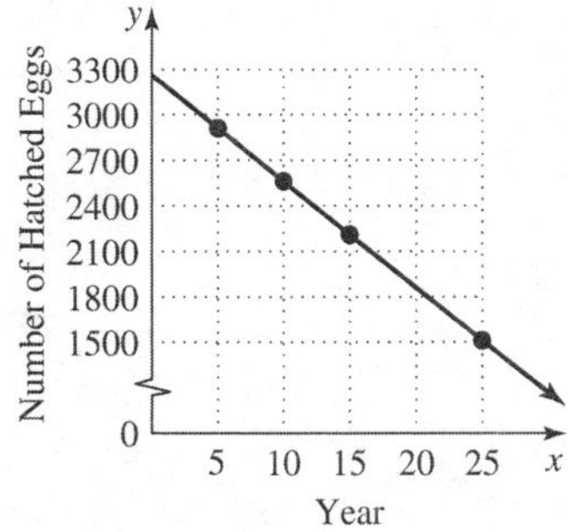

Now Try:

8. Suppose that the demand and price for a certain model of calculator are related by the equation $y = 45 - \frac{3}{5}x$, where y is the price (in dollars) and x is the demand (in thousands of calculators).

a. Assuming that this model is valid for a demand up to 50,000 calculators, use this equation to find the price of calculators at each level of demand.

0 calculators ____________

5000 calculators ____________

20,000 calculators ____________

45,000 calculators ____________

b. Write the information from part (a) as four ordered pairs, and use them to graph the given linear equation.

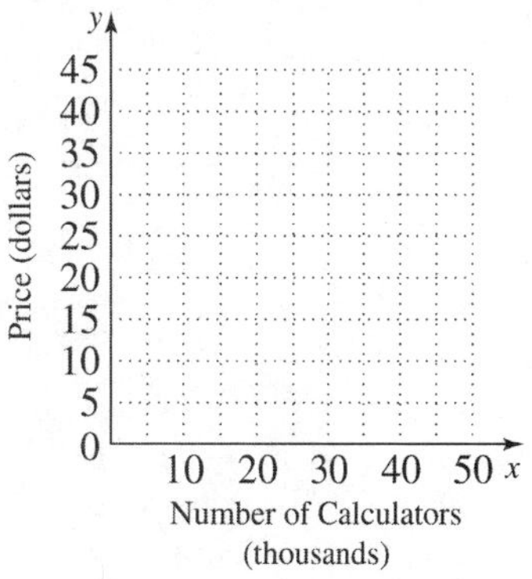

c. Use this graph and the equation to estimate the number of eggs that will hatch in 2010.

For 2010, $x = 20$. On the graph, find 20 on the horizontal axis, move up to the graphed line and then across to the vertical axis. It appears that in 2010, there were about 1900 eggs.

To use the equation, substitute 20 for x.

$$y = -70(20) + 3260$$

$$y = 1860 \text{ eggs}$$

This result for 2020 is close to our estimate of 1900 eggs from the graph.

c. Use this graph and the equation to estimate the price of 30,000 calculators.

____

Objective 5 Practice Exercises

For extra help, see Example 8 on pages 218–219 of your text.

Solve each problem. Then graph the equation.

11. The profit y in millions of dollars earned by a small computer company can be approximated by the linear equation $y = 0.63x + 4.9$, where $x = 0$ corresponds to 2004, $x = 1$ corresponds to 2005, and so on. Use this equation to approximate the profit in each year from 2004 through 2007.

11. 2004 ____________

2005 ____________

2006 ____________

2007 ____________

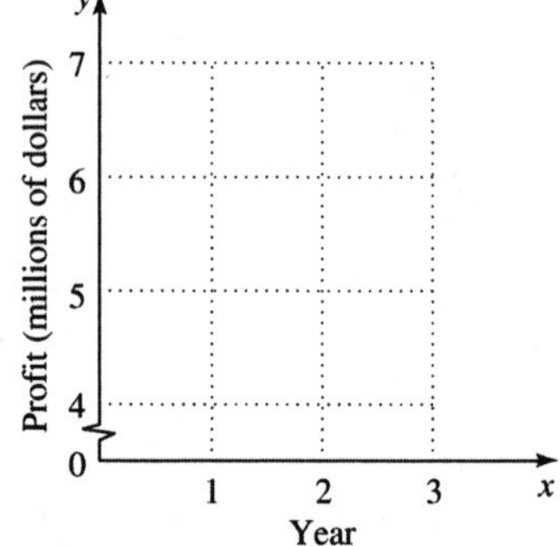

Name: Date:
Instructor: Section:

12. The number of band instruments sold by Elmer's Music Shop can be approximated by the equation $y = 325 + 42x$, where y is the number of instruments sold and x is the time in years, with $x = 0$ representing 2003. Use this equation to approximate the number of instruments sold in each year from 2003 through 2006.

12. 2003 ___________

2004 ___________

2005 ___________

2006 ___________

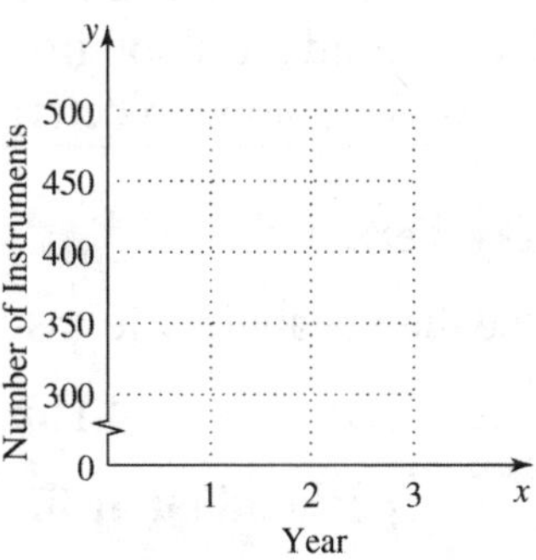

13. According to *The Old Farmer's Almanac*, the temperature in degrees Celsius can be determined by the equation $y = \frac{1}{3}x + 4$, where x is the number of cricket chirps in 25 seconds and y is the temperature in degrees Celsius. Use this equation to find the temperature when there are 48 chirps, 54 chirps, 60 chirps, and 66 chirps.

13. 48 ___________

54 ___________

60 ___________

66 ___________

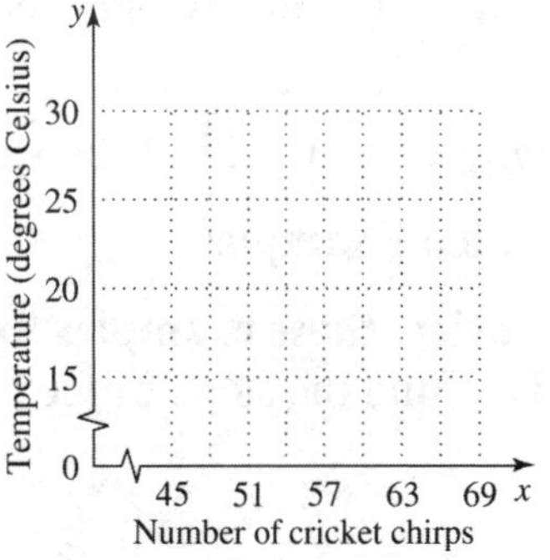

Name: Date:
Instructor: Section:

Chapter 3 LINEAR EQUATIONS AND INEQUALITIES IN TWO VARIABLES; FUNCTIONS

3.3 The Slope of a Line

Learning Objectives

1. Find the slope of a line given two points.
2. Find the slope from the equation of a line.
3. Use slope to determine whether two lines are parallel, perpendicular, or neither.

Key Terms

Use the vocabulary terms listed below to complete each statement in exercises 1–5.

rise **run** **slope** **parallel lines**

perpendicular lines

1. Two lines that intersect in a 90° angle are called ________________________.
2. The ______________ of a line is the ratio of the change in y compared to the change in x when moving along the line from one point to another.
3. The vertical change between two different points on a line is called the ____________________.
4. Two lines in a plane that never intersect are called ____________________.
5. The horizontal change between two different points on a line is called the ____________________.

Objective 1 Find the slope of a line given two points.

Video Examples

Review these examples for Objective 1:

1. Find the slope of the line.

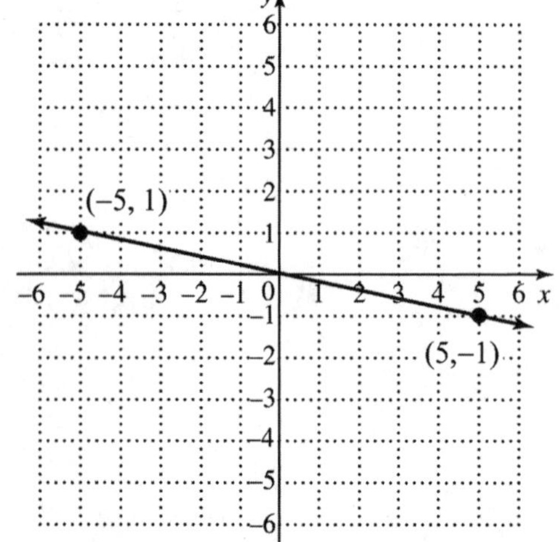

Now Try:

1. Find the slope of the line.

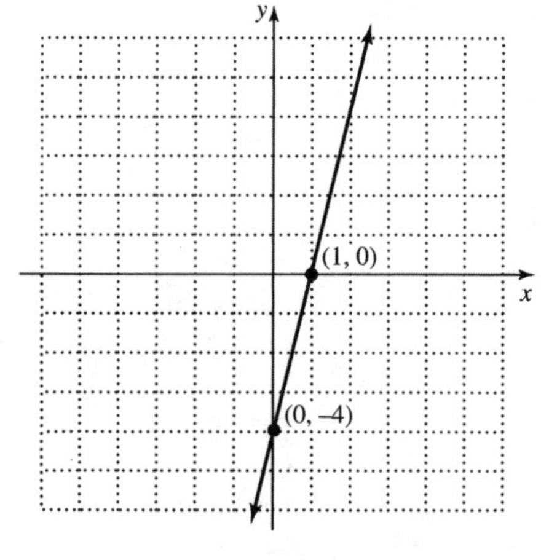

____

Name: Date:
Instructor: Section:

We use the two points shown on the line. The vertical change is the difference in the y-values, or –1 – 1 = –2, and the horizontal change is the difference in the x-values or 5 – (–5) = 10. Thus, the line has

$$\text{slope} = \frac{-2}{10}, \text{ or } -\frac{1}{5}.$$

2. Find the slope of the line.

The line passing through (–5, 4) and (2,–6)

Apply the slope formula.

$(x_1, y_1) = (-5, 4)$ and $(x_2, y_2) = (2, -6)$

$$\text{slope } m = \frac{y_2 - y_1}{x_2 - x_1} = \frac{-6-4}{2-(-5)}$$
$$= \frac{-10}{7}, \text{or } -\frac{10}{7}$$

2. Find the slope of the line.

The line passing through (–6, 7) and (3, –9)

____

3. Find the slope of the line passing through (–9, 3) and (4, 3).

$(x_1, y_1) = (-9, 3)$ and $(x_2, y_2) = (4, 3)$

$$m = \frac{y_2 - y_1}{x_2 - x_1} = \frac{3-3}{4-(-9)} = \frac{0}{13} = 0$$

3. Find the slope of the line passing through (8,–5) and (–7,–5).

____

4. Find the slope of the line passing through (–5, 3) and (–5, 8).

$(x_1, y_1) = (-5, 3)$ and $(x_2, y_2) = (-5, 8)$

$$m = \frac{y_2 - y_1}{x_2 - x_1} = \frac{8-3}{-5-(-5)} = \frac{5}{0} \text{ undefined slope}$$

4. Find the slope of the line passing through (9, 11) and (9,–7).

____

Objective 1 Practice Exercises

For extra help, see Examples 1–4 on pages 225–228 of your text.

Find the slope of the line through the given points.

1. (4, 3) and (3, 5) **1.** ______________

2. (–4, 6) and (–4, –1) **2.** ______________

Name: Date:
Instructor: Section:

3. (−3, 3) and (6, 3) **3.** ________________

Objective 2 Find the slope from the equation of a line.

Video Examples

Review this example for Objective 2:

5. Find the slope of the line.

$4x-3y=7$

Step 1 Solve the equation for y.

$$4x-3y=7$$
$$-3y=-4x+7$$
$$y=\frac{4}{3}x-\frac{7}{3}$$

Step 2 The slope is given by the coefficient of x, so the slope is $\frac{4}{3}$.

Now Try:

5. Find the slope of the line.

$7x-4y=8$

____

Objective 2 Practice Exercises

For extra help, see Example 5 on pages 229–230 of your text.

Find the slope of each line.

4. $7y-4x=11$ **4.** ________________

5. $3y=2x-1$ **5.** ________________

6. $y=-\frac{2}{5}x-4$ **6.** ________________

Name: Date:
Instructor: Section:

Objective 3 Use slope to determine whether two lines are parallel, perpendicular, or neither.

Video Examples

Review these examples for Objective 3:

6. Decide whether each pair of lines is parallel, perpendicular, or neither.

a. $5x - y = 3$
$15x - 3y = 12$

Solve each equation for y.
$y = 5x - 3$
$y = 5x - 4$

Both lines have slope 5, so the lines are parallel.

b. $x + 3y = 8$ and $-3x + y = 5$

Find the slope of each line by first solving each equation for y.

$3y = -x + 8$	$y = 3x + 5$
$y = -\frac{1}{3}x + \frac{8}{3}$	
The slope is $-\frac{1}{3}$.	The slope is 3.

Because the slopes are not equal, the lines are not parallel.

Check the product of the slopes: $-\frac{1}{3}(3) = -1$.

The two lines are perpendicular because the product of their slopes is –1.

Now Try:

6. Decide whether each pair of lines is parallel, perpendicular, or neither.

a. $2x - 4y = 7$
$3x - 6y = 8$

____

b. $9x - y = 7$ and $x + 9y = 11$

____

Objective 3 Practice Exercises

For extra help, see Example 6 on pages 231–232 of your text.

In each pair of equations, give the slope of each line, and then determine whether the two lines are **parallel**, **perpendicular**, *or* **neither**.

7. $-x + y = -7$
$x - y = -3$

7. ____________

Name: Date:
Instructor: Section:

8. $4x+2y=8$
$x+4y=-3$

8. ________________

9. $9x+3y=2$
$x-3y=5$

9. ________________

Name: Date:
Instructor: Section:

Chapter 3 LINEAR EQUATIONS AND INEQUALITIES IN TWO VARIABLES; FUNCTIONS

3.4 Slope-Intercept Form of a Linear Equation

Learning Objectives

1. Use the slope-intercept form of the equation of a line.
2. Graph a line by using its slope and a point on the line.
3. Write an equation of a line by using its slope and any point on the line.
4. Graph and write equations of horizontal and vertical lines.

Key Terms

Use the vocabulary terms listed below to complete each statement in exercises 1–3.

slope-intercept form **point-slope form** **standard form**

1. A linear equation in the form $y - y_1 = m(x - x_1)$ is written in

   ______________________.

2. A linear equation in the form $Ax + By = C$ is written in

   ______________________.

3. A linear equation in the form $y = mx + b$ is written in

   ______________________.

Objective 1 Use the slope-intercept form of the equation of a line.

Video Examples

Review these examples for Objective 1:

1. Identify the slope and y-intercept of the line with each equation.

   a. $y = -8x + 7$

   The slope is –8, and the y-intercept is (0, 7).

   b. $y = \frac{x}{9} - \frac{5}{4}$

   The equation can be written as

   $y = \frac{1}{9}x + \left(-\frac{5}{4}\right)$.

   The slope is $\frac{1}{9}$, and the y-intercept is $\left(0, -\frac{5}{4}\right)$.

Now Try:

1. Identify the slope and y-intercept of the line with each equation.

   a. $y = -12x + 6$

   ____

   b. $y = -\frac{x}{7} - \frac{7}{5}$

   ____

Name: Date:
Instructor: Section:

Objective 1 Practice Exercises

For extra help, see Example 1 on page 238 of your text.

Identify the slope and y-intercept of the line with each equation.

1. $y=\frac{3}{2}x-\frac{2}{3}$ **1.** ________________

____

2. $y=-4x$ **2.** ________________

____

Objective 2 Graph a line by using its slope and a point on the line.

Video Examples

Review this example for Objective 2:

2. Graph the equation by using the slope and y-intercept.

$2x-3y=6$

Step 1 Solve for y to write the equation in slope-intercept form.

$$2x-3y=6$$
$$-3y=-2x+6$$
$$y=\frac{2}{3}x-2$$

Step 2 The y-intercept is (0, –2). Graph this point.

Step 3 The slope is $\frac{2}{3}$. By definition,

$$\text{slope } m=\frac{\text{change in } y \text{ (rise)}}{\text{change in } x \text{ (run)}}=\frac{2}{3}$$

From the y-intercept, count up 2 units and to the right 3 units to obtain the point (3, 0).

Step 4 Draw the line through the points (0, –2) and (3, 0) to obtain the graph.

Now Try:

2. Graph the equation by using the slope and y-intercept.

$y=\frac{2}{3}x$

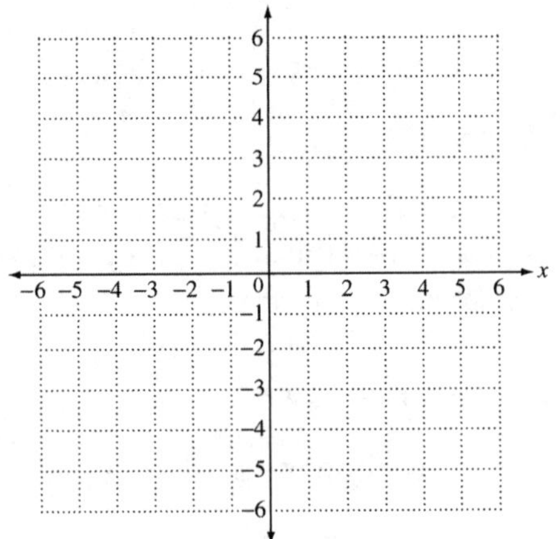

Name: Date:
Instructor: Section:

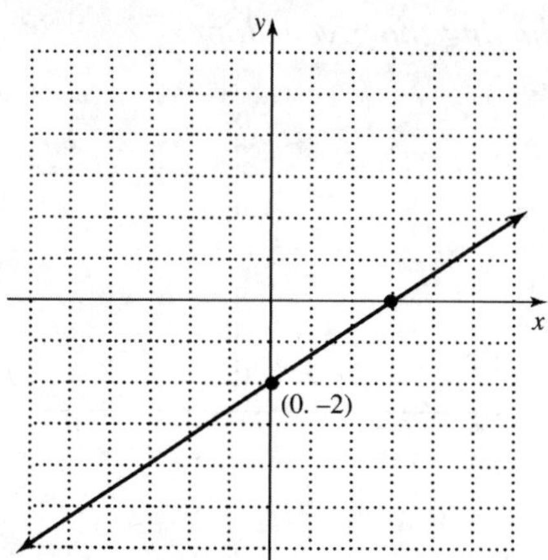

Objective 2 Practice Exercises

For extra help, see Examples 2–3 on pages 239–240 of your text.

Graph each equation by using the slope and y-intercept.

3. $4x - y = 4$

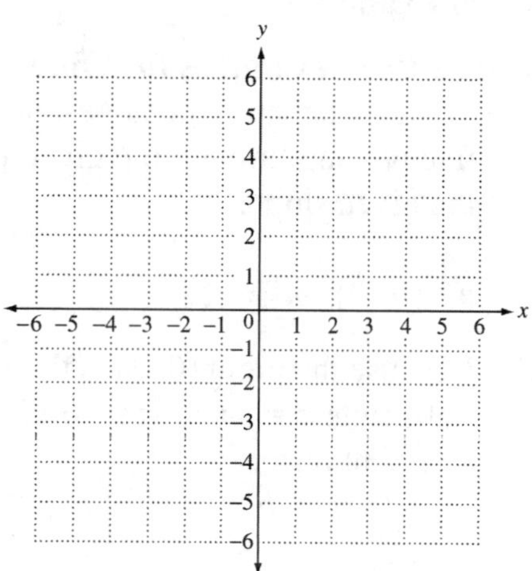

4. $y = -3x + 6$

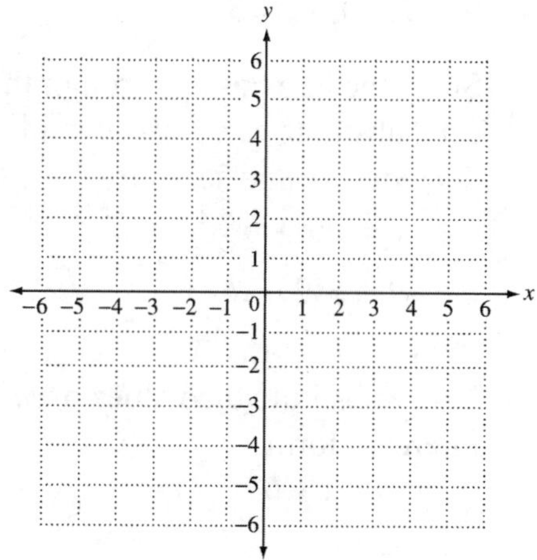

Name: Date:
Instructor: Section:

Graph the line passing through the given point and having the given slope.

5. $(4,-2);\ m=-1$ **5.**

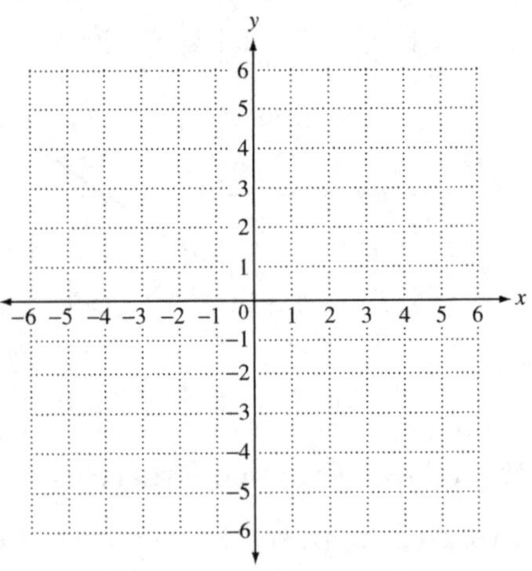

Objective 3 Write an equation of a line using its slope and any point on the line.

Video Examples

Review these examples for Objective 3:

4. Write an equation in slope-intercept form of the line passing through the given point and having the given slope.

a. $(0,\ 2),\ m=-3$

Because the point (0, 2) is the y-intercept, $b=2$. Substitute $b=2$ and $m=-3$ directly in the slope-intercept form.

$$y=mx+b$$
$$y=-3x+2$$

b. $(2,\ 9),\ m=5$

Since the line passes through the point (2, 9), we can substitute $x=2$, $y=9$, and slope $m=5$ into $y=mx+b$ and solve for b.

$$y=mx+b$$
$$9=5(2)+b$$
$$-1=b$$

Now substitute the values of m and b into slope-intercept form.

$$y=mx+b$$
$$y=5x-1$$

Now Try:

4. Write an equation in slope-intercept form of the line passing through the given point and having the given slope.

a. $(0,-5),\ m=\dfrac{3}{4}$

____

b. $(-1,\ 4),\ m=6$

____

Name: Date:
Instructor: Section:

Objective 3 Practice Exercises

For extra help, see Example 4 on pages 240–241 of your text.

Write an equation in slope-intercept form of the line passing through the given point and having the given slope.

6. $(0,-4),\ m=\frac{2}{3}$ **6.** ______________

7. $(3,\ 6),\ m=-2$ **7.** ______________

8. $(-2,\ 0),\ m=1$ **8.** ______________

Objective 4 Graph and write equations of horizontal and vertical lines.

Video Examples

Review these examples for Objective 4:

5. Graph each line passing through the given point and having the given slope.

a. $(-5,-2),\ m=0$

Horizontal lines have slope 0. Plot the point (–5,–2) and draw a horizontal line through it.

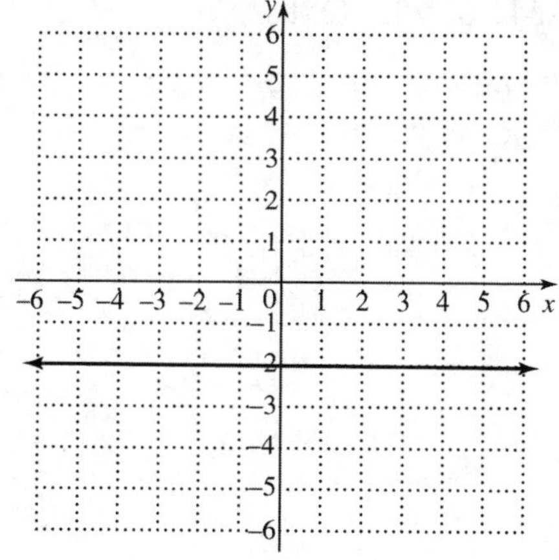

Now Try:

5. Graph each line passing through the given point and having the given slope.

a. $(-2,-2),\ m=0$

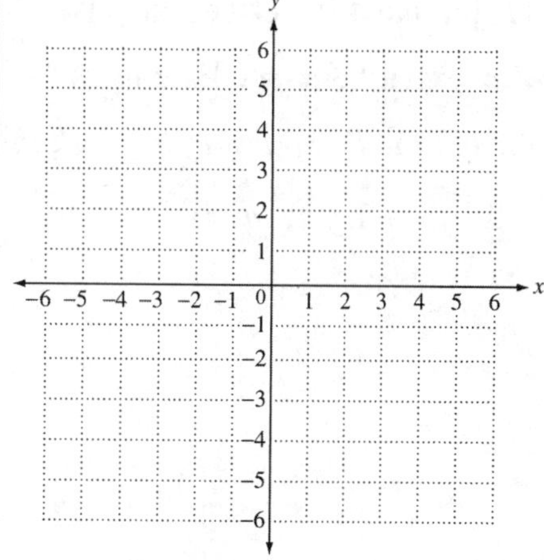

Name: Date:
Instructor: Section:

b. $(-4, 3)$, undefined slope

Vertical lines have undefined slope. Plot the point (–4, 3) and draw a vertical line through it.

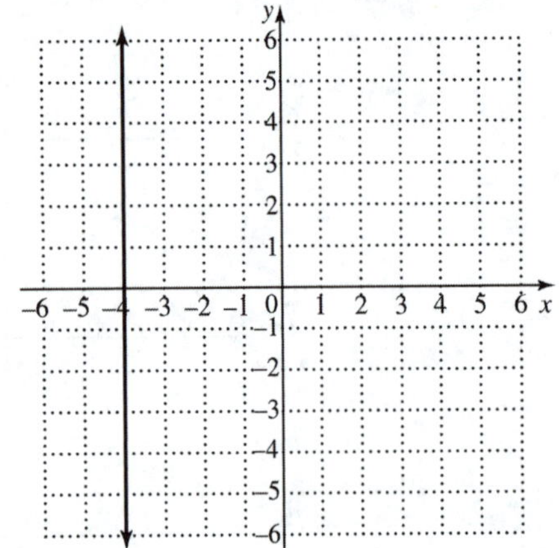

b. $(-3, -1)$, undefined slope

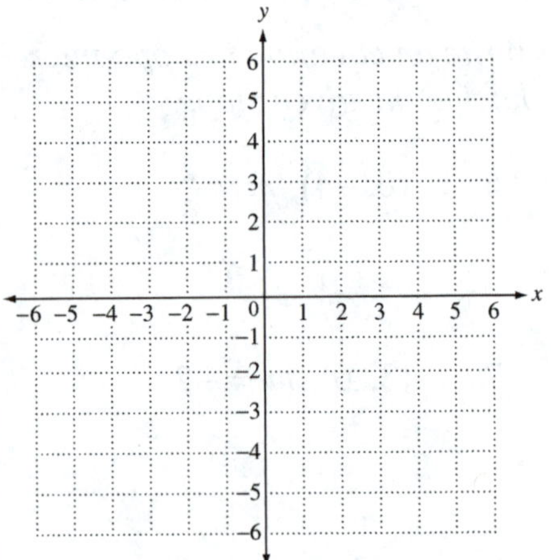

6. Write an equation of the line passing through the point (2,–2) that satisfies the given condition.

a. The line has slope 0.

Since the slope is 0, this is a horizontal line.
$y = -2.$

b. The line has undefined slope.

This is a vertical line, since the slope is undefined.
$x = 2$

6. Write an equation of the line passing through the point (–5, 5) that satisfies the given condition.

a. The line has slope 0.

____

b. The line has undefined slope.

____

Objective 4 Practice Exercises

For extra help, see Examples 5–6 on pages 241–242 of your text.

Graph the line passing through the given point and having the given slope.

9. $(-1, 4); m = 0$

9.

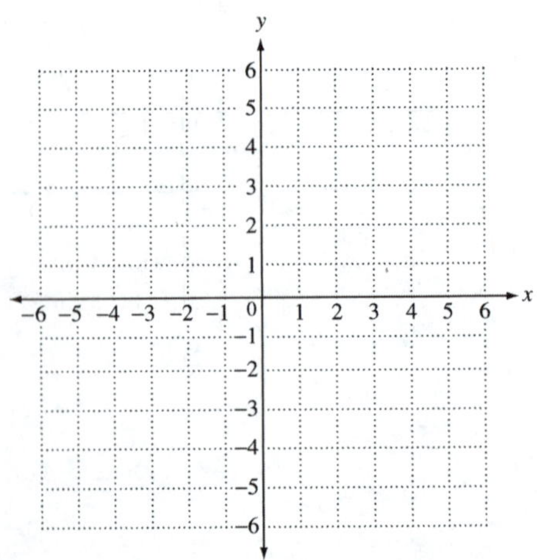

Name: Date:
Instructor: Section:

Write an equation of the line passing through (–3, 3) and having the given slope.

10. Slope 0 **10.** ________________

11. Undefined slope **11.** ________________

Name: Date:
Instructor: Section:

Chapter 3 LINEAR EQUATIONS AND INEQUALITIES IN TWO VARIABLES; FUNCTIONS

3.5 Point-Slope Form of a Linear Equation and Modeling

Learning Objectives

1. Use point-slope form to write an equation of a line.
2. Write an equation of a line using two points on the line.
3. Write an equation of a line that fits a data set.

Key Terms

Use the vocabulary terms listed below to complete each statement in exercises 1–3.

slope-intercept form **point-slope form** **standard form**

1. A linear equation in the form $Ax + By = C$ is written in

   ______________________________.

2. A linear equation in the form $y = mx + b$ is written in

   ______________________________.

3. A linear equation in the form $y - y_1 = m(x - x_1)$ is written in

   ______________________________.

Objective 1 Use point-slope form to write an equation of a line.

Video Examples

Review this example for Objective 1:

1. Write an equation of each line. Give the final answer in slope-intercept form.

   The line passing through (5, –3) with slope $\frac{5}{4}$.

$$y - y_1 = m(x - x_1)$$
$$y - (-3) = \frac{5}{4}(x - 5)$$
$$y + 3 = \frac{5}{4}x - \frac{25}{4}$$
$$y = \frac{5}{4}x - \frac{37}{4}$$

Now Try:

1. Write an equation of each line. Give the final answer in slope-intercept form.
   The line passing through (6, 11) with slope $-\frac{2}{3}$.

   ____

Name: Date:
Instructor: Section:

Objective 1 Practice Exercises

For extra help, see Example 1 on page 246 of your text.

Write an equation for the line passing through the given point and having the given slope. Write the equations in slope-intercept form, if possible.

1. $(-3, 4)$; $m = -\frac{3}{5}$ **1.** ______________

2. $(-4, -3)$; $m = -2$ **2.** ______________

3. $(2, 2)$; $m = -\frac{3}{2}$ **3.** ______________

Objective 2 Write an equation of a line by using two points on the line.

Video Examples

Review this example for Objective 2:

2. Write the equation of the line passing through the points (6, 8) and (–3, 5). Give the final answer in slope-intercept form and then in standard form.

First, find the slope of the line.
$(x_1, y_1) = (6, 8)$ and $(x_2, y_2) = (-3, 5)$

$$\text{slope } m = \frac{y_2 - y_1}{x_2 - x_1} = \frac{5-8}{-3-6} = \frac{-3}{-9} = \frac{1}{3}$$

Now use (x_1, y_1), here (6, 8) and point-slope form.

Now Try:

2. Write the equation of the line passing through the points (7, 15) and (15, 9). Give the final answer in slope-intercept form and then in standard form.

____

Name: Date:
Instructor: Section:

$$y - y_1 = m(x - x_1)$$
$$y - 8 = \frac{1}{3}(x - 6)$$
$$y - 8 = \frac{1}{3}x - 2$$
$$y = \frac{1}{3}x + 6 \quad \text{Slope-intercept form}$$
$$3y = x + 18$$
$$-x + 3y = 18$$
$$x - 3y = -18 \quad \text{Standard form}$$

Objective 2 Practice Exercises

For extra help, see Example 2 on page 247 of your text.

Write an equation for the line passing through each pair of points. Write the equations in standard form.

4. $(-2, 1)$ and $(3, 11)$ **4.** ____________

5. $(2, 3)$ and $(-2, -3)$ **5.** ____________

6. $(3, -4)$ and $(2, 7)$. **6.** ____________

Name: Date:
Instructor: Section:

Objective 3 Write an equation of a line that fits a data set.

Video Examples

Review this example for Objective 3:

3. The table shows the number of internet users in the world from 1998 to 2005, where year 0 represents 1998.

Year	Number of Internet Users (millions)
0	147
2	361
4	587
6	817
8	1093

Plot the data and find an equation that approximates it.

Letting y represent the number of internet users in year x, we plot the data.

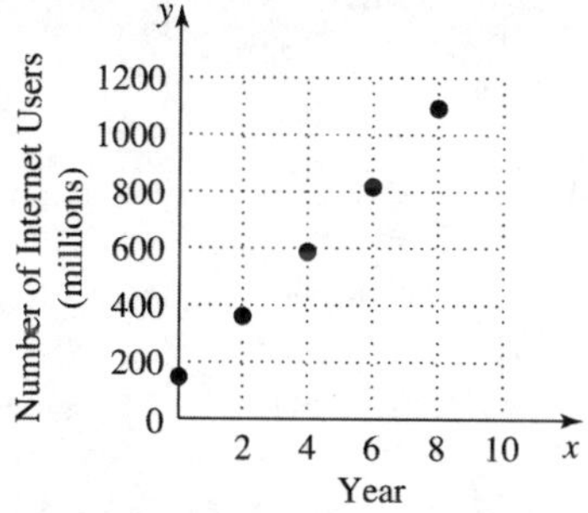

The points appear to lie approximately in a straight line. To find an equation of the line, we choose the ordered pairs (0, 147) and (8, 1093) from the table and find the slope of the line through these points.

$$(x_1, y_1) = (0, 147) \text{ and } (x_2, y_2) = (8, 1093)$$

$$\text{slope } m = \frac{y_2 - y_1}{x_2 - x_1} = \frac{1093 - 147}{8 - 0} = \frac{946}{8}$$
$$= 118.25$$

Use the slope, 118.25, and the point (0, 147) in slope-intercept form.

$$y = mx + b$$
$$147 = 118.25(0) + b$$
$$147 = b$$

Thus, $m = 118.25$ and $b = 147$, so the equation of the line is $y = 118.25x + 147$.

Now Try:

3. The table shows the average annual telephone expenditures for residential and pay telephones from 2001 to 2006, where year 0 represents 2001.

Year	Annual Telephone Expenditures
0	\$686
2	\$620
3	\$592
4	\$570
5	\$542

Plot the data and find an equation that approximates it.

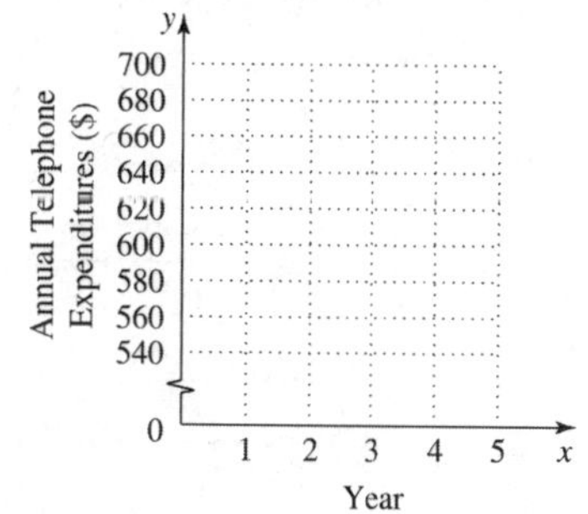

____

Name: Date:
Instructor: Section:

Objective 3 Practice Exercises

For extra help, see Example 3 on pages 248–249 of your text.

Plot the data and find an equation that approximates it.

7. The table shows the U.S. municipal solid waste recycling percents since 1985, where year 0 represents 1985.

Year	Recycling Percent
0	10.1
5	16.2
10	26.0
15	29.1
20	32.5

7.

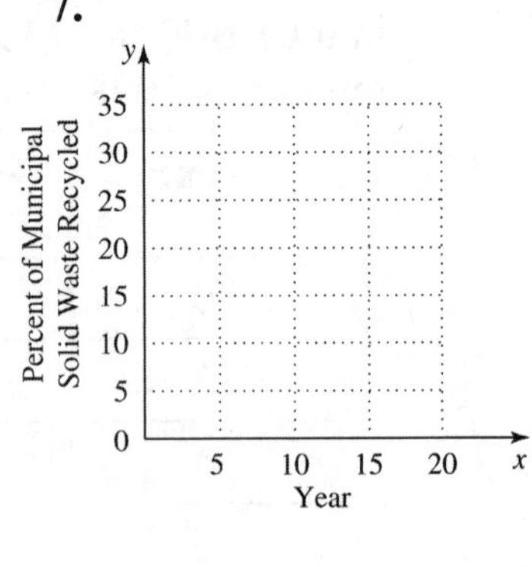

____

8. The table shows the approximate consumer expenditures for food in the U.S. in billions of dollars for selected years, where year 0 represents 1985.

Year	Food Expenditures (billions of dollars)
0	233
5	298
10	343
15	417
20	515

8.

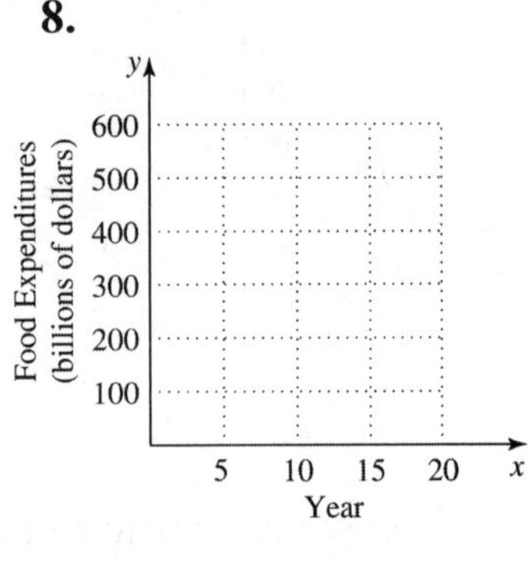

____

Name: Date:
Instructor: Section:

Chapter 3 LINEAR EQUATIONS AND INEQUALITIES IN TWO VARIABLES; FUNCTIONS

3.6 Graphing Linear Inequalities in Two Variables

Learning Objectives	
1	Graph linear inequalities in two variables.
2	Graph an inequality with a boundary line through the origin.

Key Terms

Use the vocabulary terms listed below to complete each statement in exercises 1–2.

linear inequality in two variables **boundary line**

1. In the graph of a linear inequality, the ____________________________ separates the region that satisfies the inequality from the region that does not satisfy the inequality.

2. An inequality that can be written in the form $Ax + By < C$, $Ax + By > C$, $Ax + By \leq C$, or $Ax + By \geq C$ is called a ________________________________.

Objective 1 Graph linear inequalities in two variables.

Video Examples

Review these examples for Objective 1:

1. Graph $3x - 2y \leq 6$.

The inequality $3x - 2y \leq 6$ means that
$3x - 2y < 6$ or $3x - 2y = 6$.
We begin by graphing the line $3x - 2y = 6$ with intercepts (0, –3) and (2, 0). This boundary line divides the plane into two regions, one of which satisfies the inequality. We use the test point (0, 0) to see whether the resulting statement is true or false, thereby determining whether the point is in the shaded region or not.

$$3x - 2y \leq 6$$
$$3(0) - 2(0) \overset{?}{\leq} 6$$
$$0 - 0 \overset{?}{\leq} 6$$
$$0 \leq 6 \text{ True}$$

Since the last statement is true, we shade the region that includes the test point (0, 0). The shaded region, along with the boundary line, is the desired graph.

Now Try:

1. Graph $2x + 5y \leq -8$.

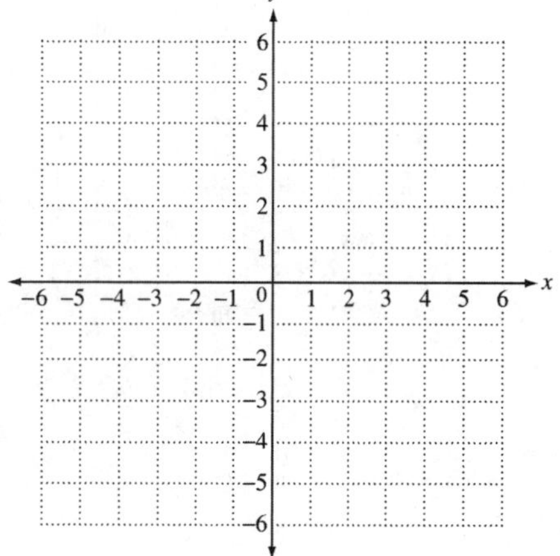

Name: Date:
Instructor: Section:

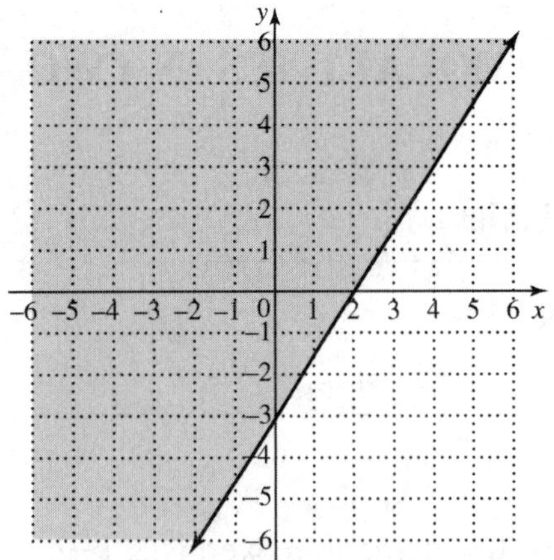

3. Graph $x - 4 \leq -1$.

First, solve the inequality for x.

$x \leq 3$

Now graph the line $x = 3$, a vertical line through the point (3, 0). Use a solid line, and choose (0, 0) as a test point.

$0 \leq 3$ True

Because $0 \leq 3$ is true, we shade the region containing (0, 0).

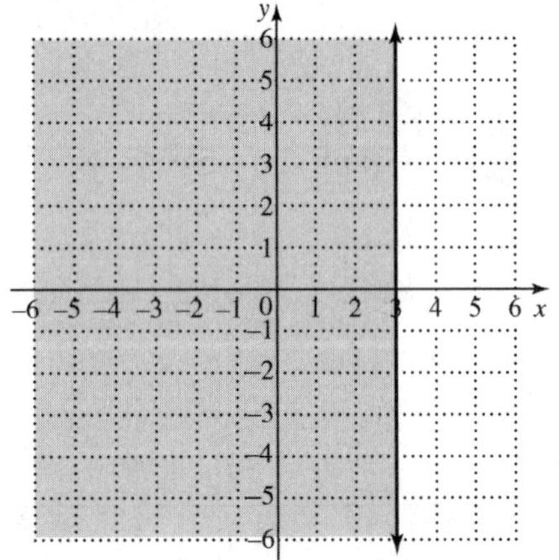

3. Graph $y \geq -1$.

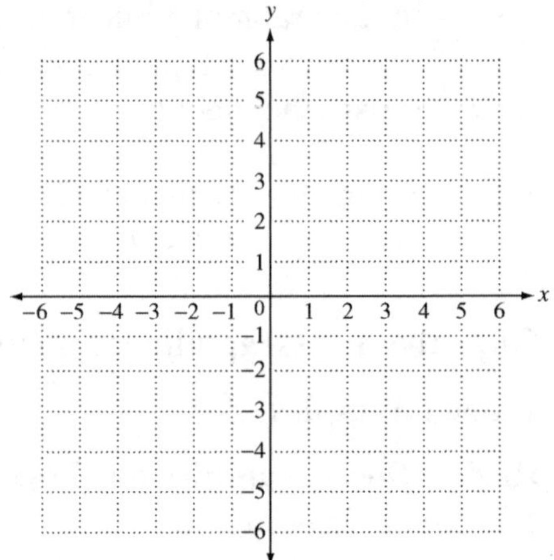

Name: Date:
Instructor: Section:

Objective 1 Practice Exercises

For extra help, see Examples 1–3 on pages 254–257 of your text.

Graph each linear inequality.

1. $y \geq x - 1$

1.

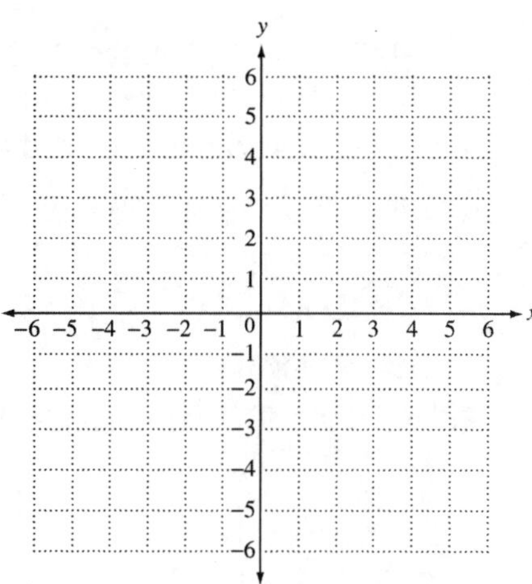

2. $y > -x + 2$

2.

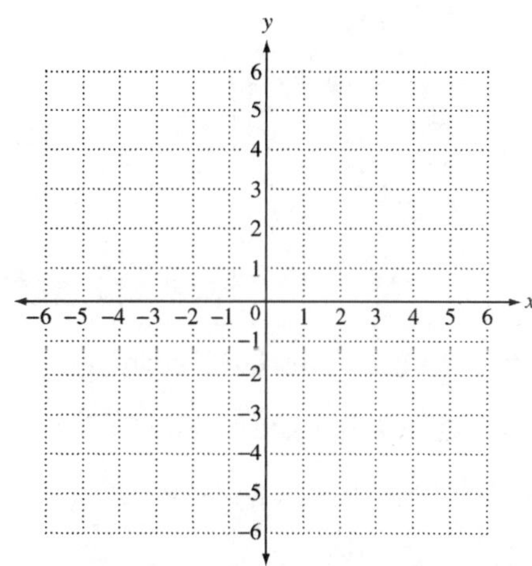

3. $3x - 4y - 12 > 0$

3.

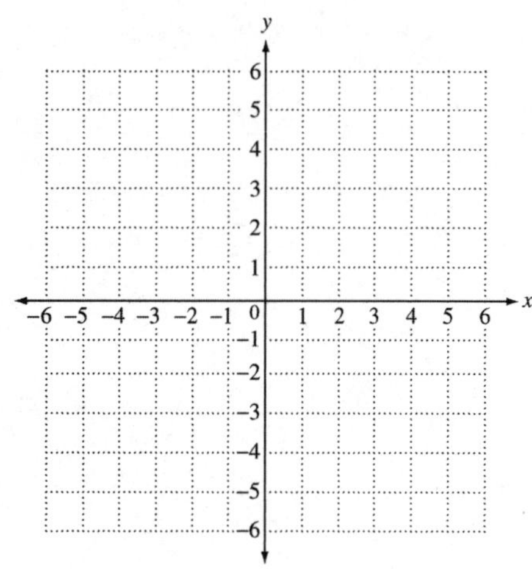

Name: Date:
Instructor: Section:

Objective 2 Graph an inequality with a boundary line through the origin.

Video Examples

Review this example for Objective 2:

4. Graph $y \geq 3x$.

We graph $y = 3x$ using a solid line through (0, 0), (1, 3) and (2, 6). Because (0, 0) is on the line $y \geq 3x$, it cannot be used as a test point. Instead we choose a test point off the line, say (3, 0).

$0 \overset{?}{\geq} 3(3)$

$0 \geq 9$ False

Because $0 \geq 9$ is false, shade the other region.

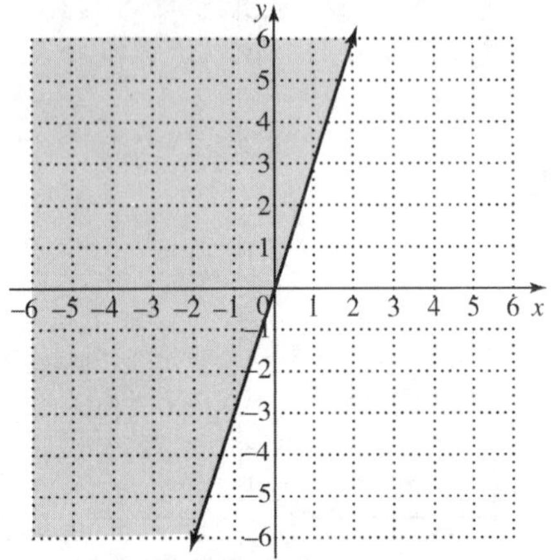

Now Try:

4. Graph $y \geq x$.

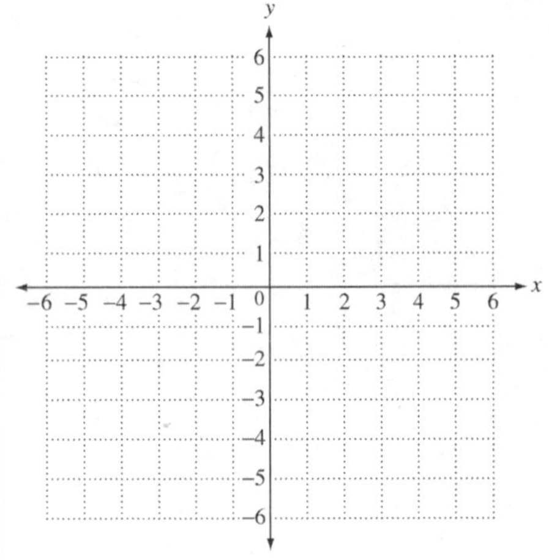

Objective 2 Practice Exercises

For extra help, see Example 4 on page 257 of your text.

Graph each linear inequality.

4. $y \leq \frac{2}{5}x$

4.

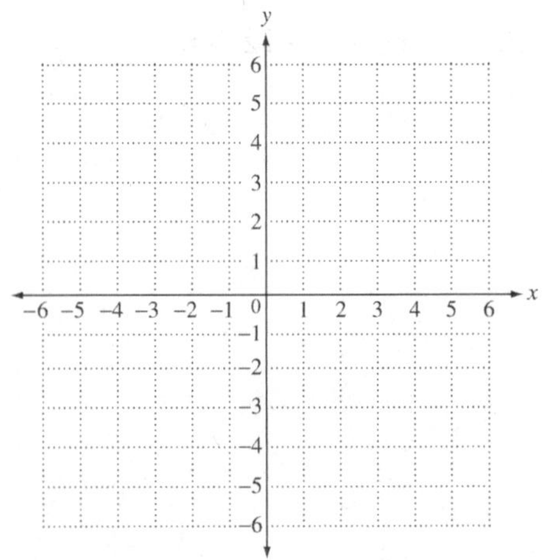

Name: Date:
Instructor: Section:

5. $y \geq \frac{1}{3}x$

5.

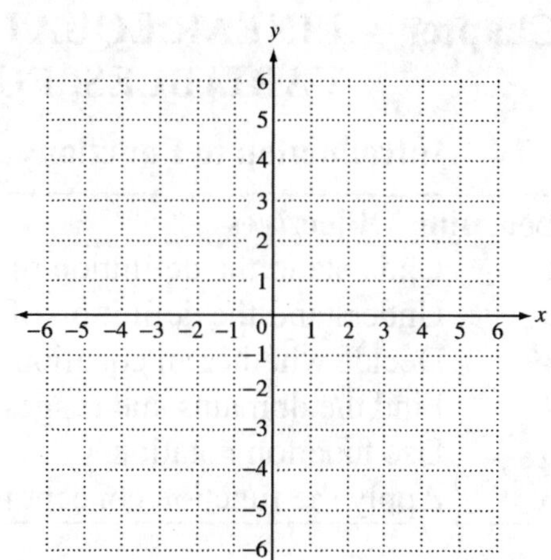

6. $x < -2y$

6.

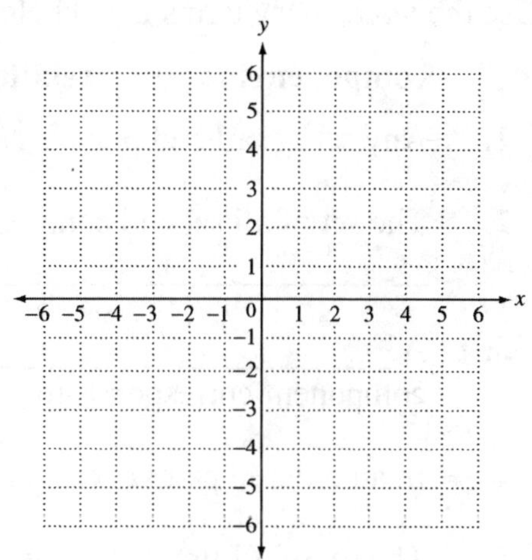

Name: Date:
Instructor: Section:

Chapter 3 LINEAR EQUATIONS AND INEQUALITIES IN TWO VARIABLES; FUNCTIONS

3.7 Introduction to Functions

Learning Objectives

1. Understand the definition of a relation.
2. Understand the definition of a function.
3. Decide whether an equation defines a function.
4. Find the domains and ranges.
5. Use function notation.
6. Apply the function concept in an application.

Key Terms

Use the vocabulary terms listed below to complete each statement in exercises 1−5.

components **relation** **domain** **range** **function**

1. Any set of ordered pairs is called a ________________________.

2. The set of all second components in the ordered pairs of a relation is the ____________________________ of the relation.

3. A ________________________ is a set of ordered pairs in which each first component corresponds to exactly one second component.

4. In an ordered pair (x, y), x and y are the __________________________.

5. The set of all first components in the ordered pairs of a relation is the ______________ of the relation.

Objective 1 Understand the definition of a relation.

Video Examples

Review this example for Objective 1:

1. Identify the domain and range of the relation.

   $\{(1, 0), (2, 7), (5, 9), (6, 2)\}$

   This relation has
   domain: $\{1, 2, 5, 6\}$
   and
   range: $\{0, 7, 9, 2\}$

Now Try:

1. Identify the domain and range of the relation.
   $\{(6, 7), (8, 9), (10, 11), (12, 13)\}$

   ____

Name: Date:
Instructor: Section:

Objective 1 Practice Exercises

For extra help, see Example 1 on page 260 of your text.

Identify the domain and range of each relation.

1. $\{(2,7),(5,-4),(-3,-1),(0,-8),(5,2)\}$

1. ________________

domain:__________

range: __________

2. $\{(3,5),(3,8),(3,-4),(3,1),(3,0)\}$

2. ________________

domain:__________

range: __________

3. $\{(-3,5),(-2,5),(-1,0),(0,-5),(1,5)\}$

3. ________________

domain:__________

range: __________

Objective 2 Understand the definition of a function.

Video Examples

Review these examples for Objective 2:

2. Determine whether each relation is a function.

a. {(7, 4), (–7, 3), (7, 2)}

The first component 7 appears in two ordered pairs and corresponds to two different second components. Therefore, this relation is not a function.

b. Domain Range

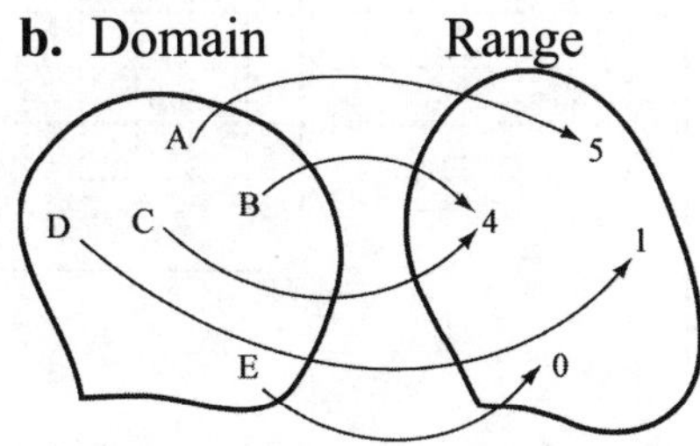

The points here include (A, 5), (B, 4), (C, 4), (D, 1), and (E, 0).
Each first component appears once and only once. The relation is a function.

Now Try:

2. Determine whether each relation is a function.

a. {(10, 0), (10, 5), (10, 20)}

____

b. Domain Range

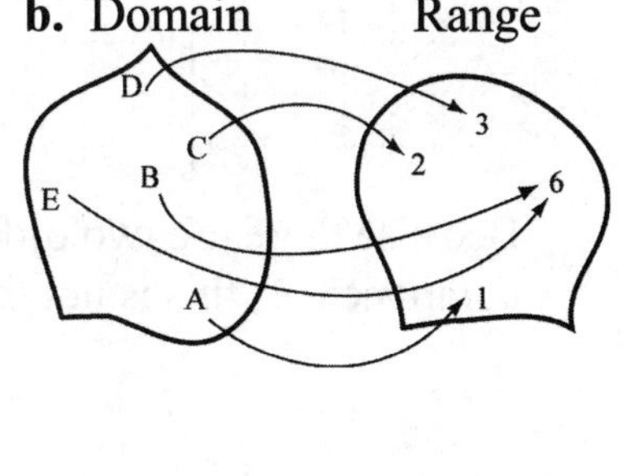

____

Name: Date:
Instructor: Section:

Objective 2 Practice Exercises

For extra help, see Example 2 on page 261 of your text.

Determine whether each relation is a function.

4. $\{(1,3),(5,7),(11,9),(8,-2),(6,-7),(-4,-3)\}$ **4.** ______________

5. $\{(-1.2,4),(1.8,-2.5),(3.7,-3.8),(3.7,3.8)\}$ **5.** ______________

6. $\{(-3,5),(-2,5),(-1,0),(0,-5),(1,5)\}$ **6.** ______________

Objective 3 Decide whether an equation defines a function.

Video Examples

Review these examples for Objective 3:

3. Determine whether each relation represented by a graph or an equation is a function.

a.

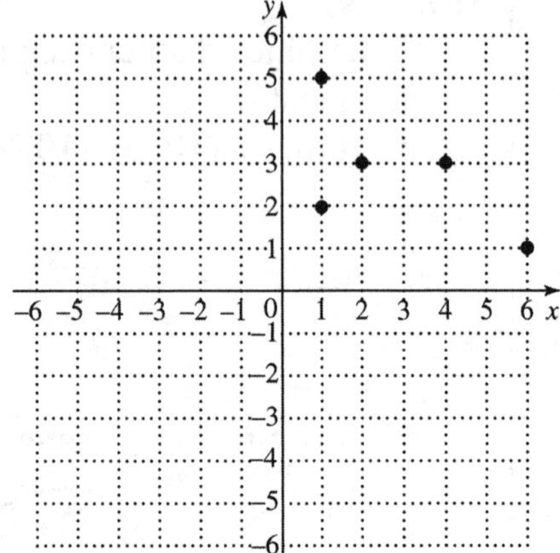

Because there are two ordered pairs with first component 1, this is not the graph of a function.

Now Try:

3. Determine whether each relation represented by a graph or an equation is a function.

a.

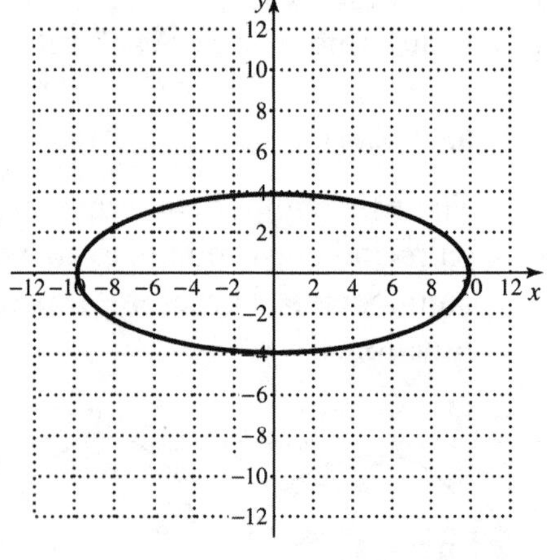

____

Name: Date:
Instructor: Section:

b.

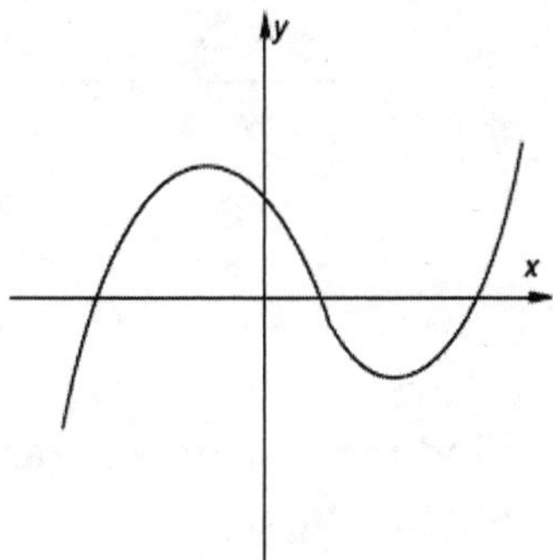

Use the vertical line test. Any vertical line intersects the graph just once, so this is the graph of a function.

c. $y = 4x - 2$

This linear equation is in the form $y = mx + b$. Since the graph of this equation is a line that is not vertical, the equation defines a function.

b.

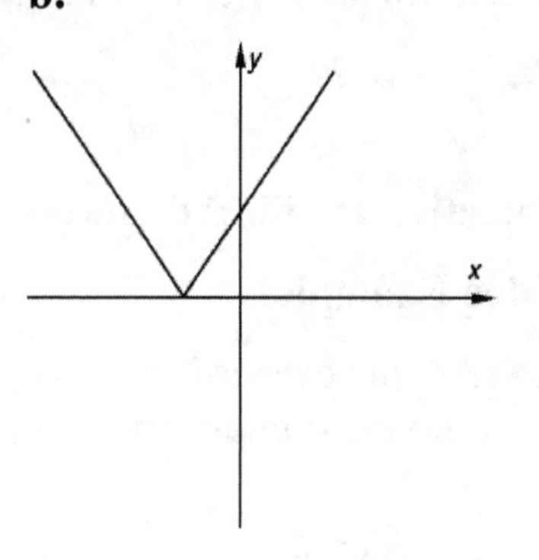

____

c. $y = -3x + 5$

____

Objective 3 Practice Exercises

For extra help, see Example 3 on pages 262–263 of your text.

Use the vertical line test to determine whether each relation graphed is a function.

7.

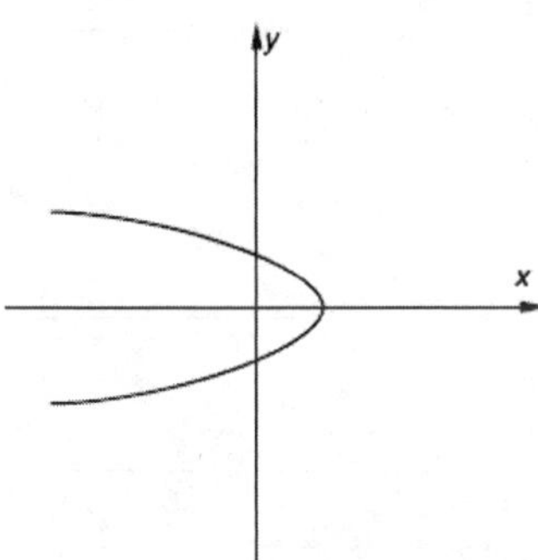

7. ______________

8.

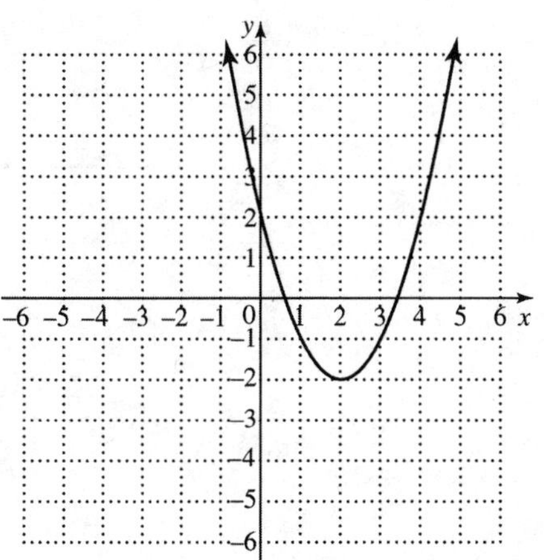

8. ______________

Name: Date:
Instructor: Section:

Decide whether the equation defines y as a function of x.

9. $y = 7$ **9.** ________________

Objective 4 Find domains and ranges.

Video Examples

Review this example for Objective 4:

4. Find the domain and range of the function.

$y = 4x - 5$

Any number may be input for x, so the domain is the set of all real numbers, or $(-\infty, \infty)$.
Any number may be the output for y, so the range is also the set of all real numbers, or $(-\infty, \infty)$.

Now Try:

4. Find the domain and range of the function.
$y = -5x + 6$

____

____

Objective 4 Practice Exercises

For extra help, see Example 4 on page 263 of your text.

Find the domain and range of each function.

10. $y = -2x + 3$ **10.** ________________

____

11. $y = -x - 1$ **11.** ________________

____

12. $y = 2x^2$ **12.** ________________

____

Name: Date:
Instructor: Section:

Objective 5 Use function notation.

Video Examples

Review these examples for Objective 5:

5. For the function $f(x) = x^2 - 5$, find each function value.

 a. $f(3)$

 Substitute 3 for x.

$$f(x) = x^2 - 5$$
$$f(3) = 3^2 - 5$$
$$f(3) = 9 - 5$$
$$f(3) = 4$$

 b. $f(0)$

$$f(0) = 0^2 - 5$$
$$f(0) = 0 - 5$$
$$f(0) = -5$$

 c. $f(-4)$

$$f(-4) = (-4)^2 - 5$$
$$f(-4) = 16 - 5$$
$$f(-4) = 11$$

Now Try:

5. For the function $f(x) = 5x - 10$, find each function value.

 a. $f(3)$ ____________

 b. $f(0)$ ____________

 c. $f(-1)$ ____________

Objective 5 Practice Exercises

For extra help, see Example 5 on page 264 of your text.

For each function f, find (a) $f(-2)$, (b) $f(0)$, *and* (c) $f(4)$.

13. $f(x) = 3x - 7$

13. a.____________

b.____________

c.____________

14. $f(x) = x^2 + 2$

14. a.____________

b.____________

c.____________

Name: Date:
Instructor: Section:

15. $f(x)=9$

15. a. ______________

b. ______________

c. ______________

Objective 6 Apply the function concept in an application.

Video Examples

Review this example for Objective 6:

6. Write the information in the graph as a set of ordered pairs. Does this set define a function?

Year	Profit (millions of dollars)
2006	34
2008	40
2010	46
2012	48

{(2006, 34), (2008, 40), (2010, 46), (2012, 48)}
Yes, this set defines a function.

Now Try:

6. Write the information in the graph as a set of ordered pairs. Does this set define a function?

Year	Worldwide Internet Users (in millions)
2003	719
2004	817
2005	1018
2006	1093
2007	1262

____

Objective 6 Practice Exercises

For extra help, see Example 6 on page 265 of your text.

The yearly revenue for a small business is shown in the table below. Use the table to answer problems 16–18.

Year	Revenue (thousands of dollars)
2008	596
2009	625
2010	872
2011	795
2012	625

16. Write the information in the table as a set of ordered pairs. Does this set define a function?

16. ______________

Name: Date:
Instructor: Section:

17. Suppose that r is the name given to this relation. Give the domain and range of r.

17. domain __________

range____________

18. Find $r(2009)$ and $r(2011)$.

18. _________________

Name: Date:
Instructor: Section:

Chapter 4 SYSTEMS OF LINEAR EQUATIONS AND INEQUALITIES

4.1 Solving Systems of Linear Equations by Graphing

Learning Objectives
1. Decide whether a given ordered pair is a solution of a system.
2. Solve linear systems by graphing.
3. Solve special systems by graphing.
4. Identify special systems without graphing.

Key Terms

Use the vocabulary terms listed below to complete each statement in exercises 1−7.

system of linear equations	**solution of the system**
solution set of the system	**consistent system**
inconsistent system	**independent equations**
dependent equations	

1. Equations of a system that have different graphs are called ____________________.

2. A system of equations with at least one solution is a ____________________.

3. The set of all ordered pairs that are solutions of a system is the ______________________________.

4. The __________________________ of linear equations is an ordered pair that makes all the equations of the system true at the same time.

5. Equations of a system that have the same graph (because they are different forms of the same equation) are called ____________________.

6. A system with no solution is called a(n) ____________________.

7. A(n) ________________________ consists of two or more linear equations with the same variables.

Name: Date:
Instructor: Section:

Objective 1 Decide whether a given ordered pair is a solution of a system.

Video Examples

Review this example for Objective 1:

1. Determine whether the ordered pair (5, –2) is a solution of the system.

$$4x+5y=10$$
$$3x+8y=6$$

Again, substitute 5 for x and –2 for y in each equation.

$4x+5y=10$	$3x+8y=6$
$4(5)+5(-2)\overset{?}{=}10$	$3(5)+8(-2)\overset{?}{=}6$
$20-10\overset{?}{=}10$	$15-16\overset{?}{=}6$
$10=10$ True	False $-1=6$

The ordered pair (5, –2) is not a solution of this system because it does not satisfy the second equation.

Now Try:

1. Determine whether the ordered pair (6, 5) is a solution of the system.

$$5x-6y=0$$
$$6x+5y=50$$

____

Objective 1 Practice Exercises

For extra help, see Example 1 on page 278 of your text.

Decide whether the given ordered pair is a solution of the given system.

1. $(2,-4)$

$$2x+3y=6$$
$$3x-2y=14$$

1. ______________

2. $(-3,-1)$

$$5x-3y=-12$$
$$2x+3y=-9$$

2. ______________

3. $(4,\ 0)$

$$4x+3y=16$$
$$x-4y=-4$$

3. ______________

Name: Date:
Instructor: Section:

Objective 2 Solve linear systems by graphing.

Video Examples

Review this example for Objective 2:

2. Solve the system of equations by graphing both equations on the same axes.

$$6x - 5y = 4$$
$$2x - 5y = 8$$

Graph these equations by plotting several points for each line. To find the x-intercept, let $y = 0$. To find the y-intercept, let $x = 0$.
The tables show the intercepts and a check point for each graph.

$6x - 5y = 4$

x	y
0	$-\frac{4}{5}$
$\frac{2}{3}$	0
4	4

$2x - 5y = 8$

x	y
0	$\frac{8}{5}$
4	0
-6	-4

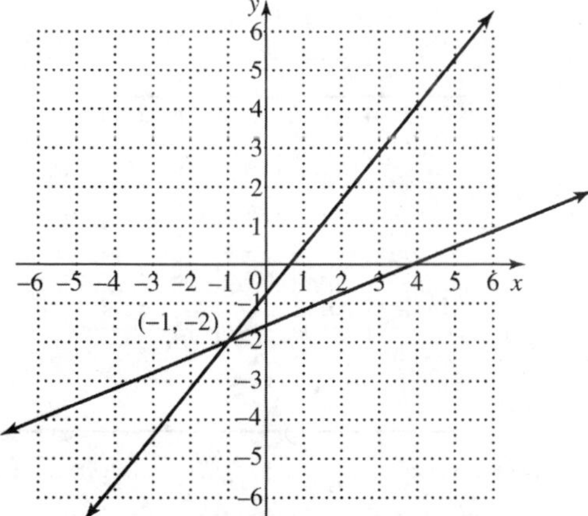

The lines suggest that the graphs intersect at the point (–1, –2). We check by substituting –1 for x and –2 for y in both equations.

$$6x - 5y = 4 \qquad 2x - 5y = 8$$
$$6(-1) - 5(-2) \stackrel{?}{=} 4 \qquad 2(-1) - 5(-2) \stackrel{?}{=} 8$$
$$-6 + 10 \stackrel{?}{=} 4 \qquad -2 + 10 \stackrel{?}{=} 8$$
$$4 = 4 \text{ True} \qquad \text{True } 8 = 8$$

Because (–1, –2) satisfies both equations, the solution set of this system is {(–1, –2)}.

Now Try:

2. Solve the system of equations by graphing both equations on the same axes.

$$3x - y = -7$$
$$2x + y = -3$$

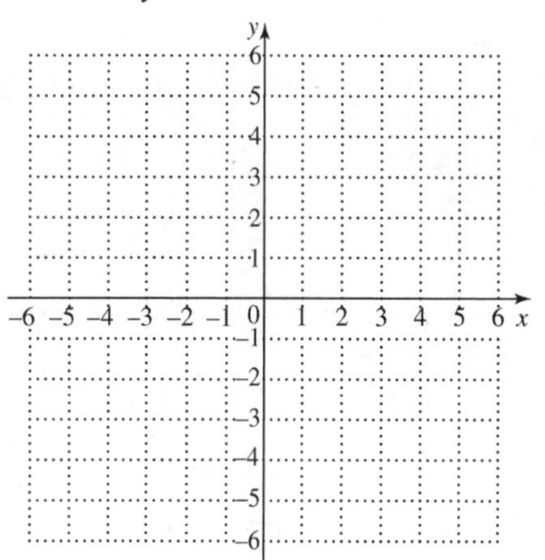

Name: Date:
Instructor: Section:

Objective 2 Practice Exercises

For extra help, see Example 2 on page 279 of your text.

Solve each system by graphing both equations on the same axes.

4. $x-2y=6$
$2x+y=2$

4. ____________________

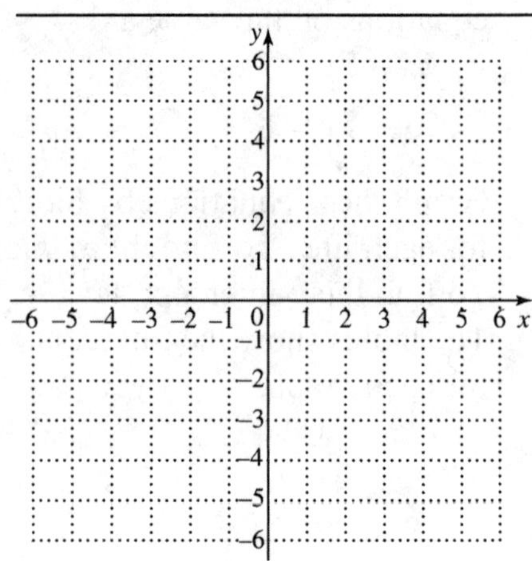

5. $2x=y$
$5x+3y=0$

5. ____________________

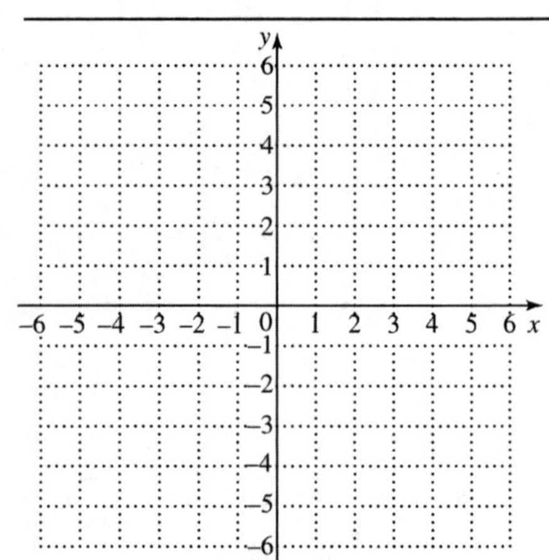

6. $3x+2=y$
$2x-y=0$

6. ____________________

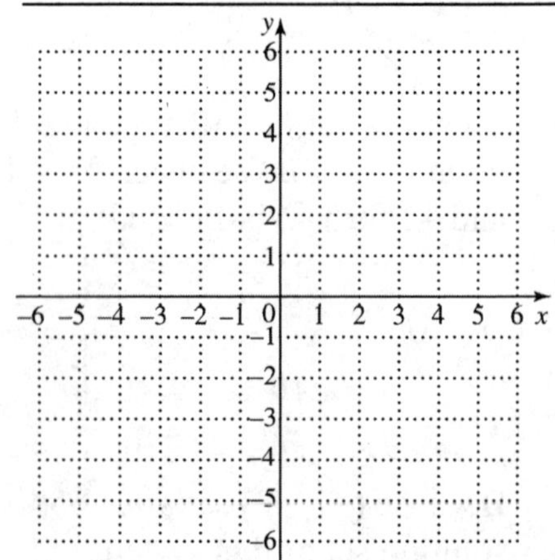

Name: Date:
Instructor: Section:

Objective 3 Solve special systems by graphing.

Video Examples

Review these examples for Objective 3:

3. Solve each system by graphing.

a. $x - y = 1$

$x - y = -1$

The graphs of these two equations are parallel and have no points in common. There is no solution for this system. It is inconsistent. The solution set is $\varnothing$.

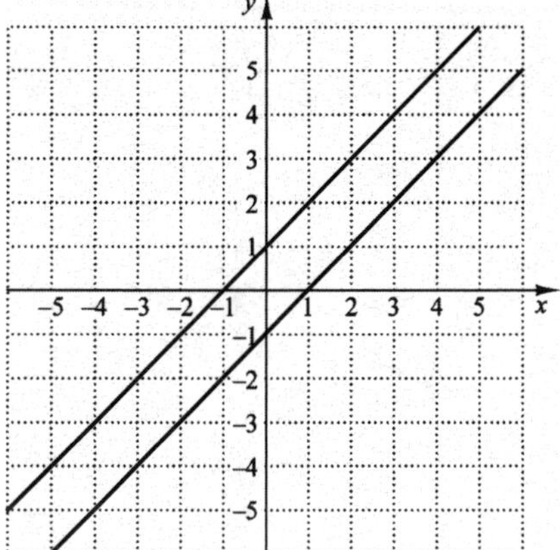

b. $3x - y = 0$

$2y = 6x$

The graphs of these two equations are the same line.

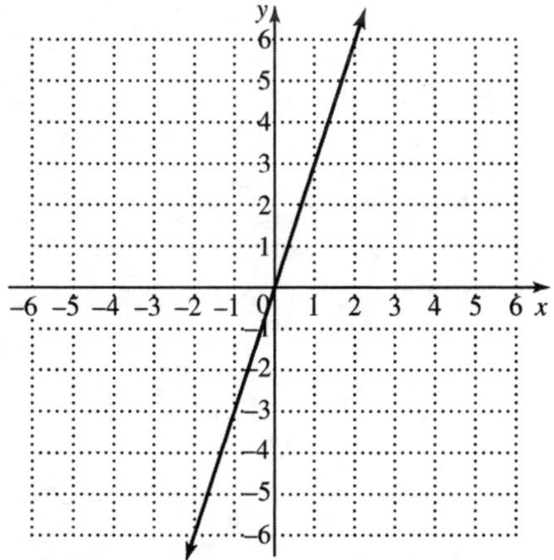

In this case, every point on the line is a solution of the system, and the solution set contains an infinite number of ordered pairs, each of which satisfies both equations of the system. We write the solution set as

$\{(x, y) | \, 3x - y = 0\}$.

Now Try:

3. Solve each system by graphing.

a. $x - 3y = 6$

$x - 3y = 4$

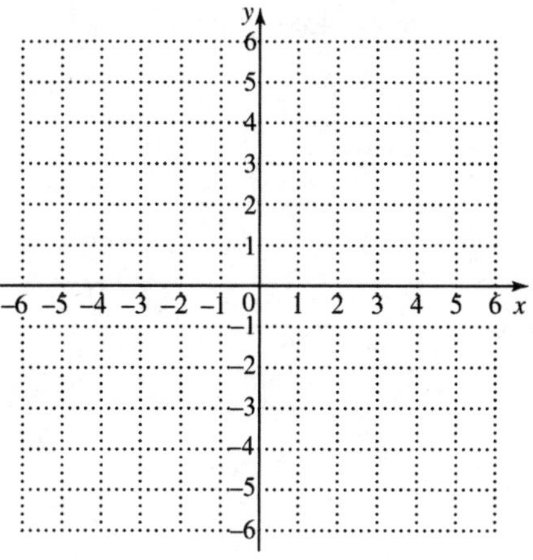

____

b. $4x - 2y = 8$

$6x - 3y = 12$

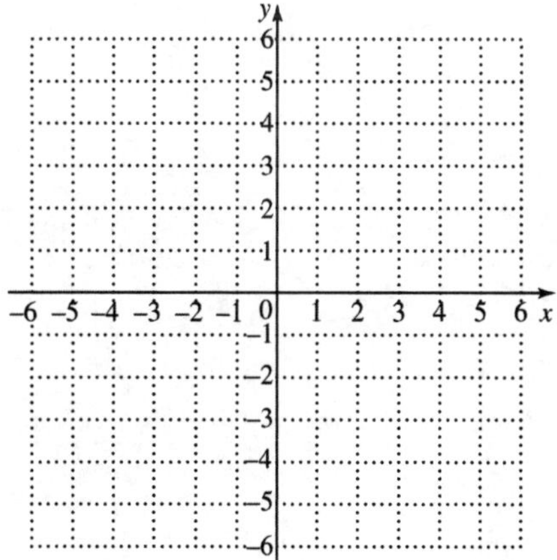

____

Name: Date:
Instructor: Section:

Objective 3 Practice Exercises

For extra help, see Example 3 on page 280 of your text.

Solve each system of equations by graphing both equations on the same axes. If the two equations produce parallel lines, write **no solution***. If the two equations produce the same line, write* ***infinite number of solutions****.*

7. $8x+4y=-1$
$4x+2y=3$

7. ______________________

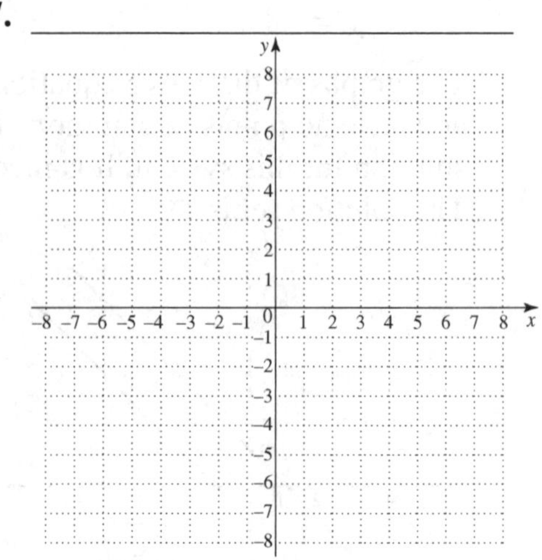

8. $-3x+2y=6$
$-6x+4y=12$

8. ______________________

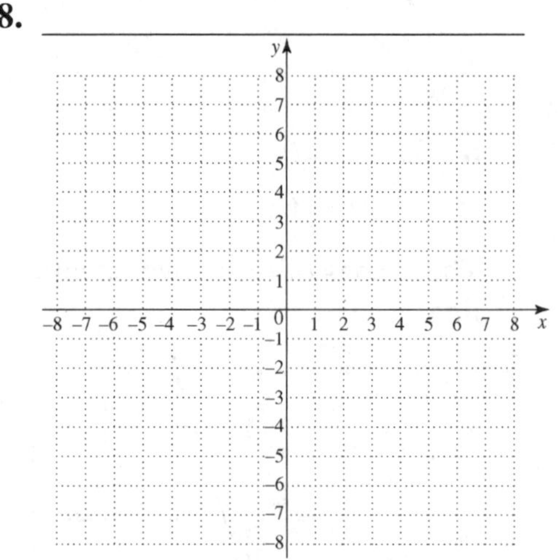

Name: Date:
Instructor: Section:

Objective 4 Identify special systems without graphing.

Video Examples

Review these examples for Objective 4:

4. Given the linear system,

$$3x-4y=12$$
$$2x-3y=6$$

answer the following questions without graphing.

a. Is the system inconsistent, are the equations dependent, or neither?

Write each equation in slope-intercept form.

$3x-4y=12$	$2x-3y=6$
$-4y=-3x+12$	$-3y=-2x+6$
$y=\frac{3}{4}x-3$	$y=\frac{2}{3}x-2$

The slopes are different. This system is not inconsistent nor is it dependent. The system is neither.

b. Is the graph a pair of intersecting lines, a pair of parallel lines, or one line?

From part (a), we have written the system in slope-intercept form.

$$y=\frac{3}{4}x-3 \qquad y=\frac{2}{3}x-2$$

The graphs are neither parallel nor the same line, since the slopes are different. The system is a pair of intersecting lines.

c. Does the system have one solution, no solution, or an infinite number of solutions?

From part (a), we have written the system in slope-intercept form.

$$y=\frac{3}{4}x-3 \qquad y=\frac{2}{3}x-2$$

Since the slopes are different, the graph is neither parallel nor the same line, and therefore cannot have no solution nor an infinite number of solutions. This system has exactly one solution.

Now Try:

4. Given the linear system,

$$x-6y=4$$
$$6x-y=9$$

answer the following questions without graphing.

a. Is the system inconsistent, are the equations dependent, or neither?

____

b. Is the graph a pair of intersecting lines, a pair of parallel lines, or one line?

____

c. Does the system have one solution, no solution, or an infinite number of solutions?

____

Name: Date:
Instructor: Section:

Objective 4 Practice Exercises

For extra help, see Example 4 on page 282 of your text.

Without graphing, answer the following equations for each linear system.
(a) Is the system inconsistent, are the equations dependent, or neither?
(b) Is the graph a pair of intersecting lines, a pair of parallel lines, or one line?
(c) Does the system have one solution, no solution, or an infinite number of solutions?

9. $y = 2x + 1$
$3x - y = 7$

9. (a)____________
(b)____________
(c)____________

10. $-2x + y = 4$
$-4x + 2y = -2$

10. (a)____________
(b)____________
(c)____________

11. $4x + 3y = 12$
$-12x = -36 + 9y$

11. (a)____________
(b)____________
(c)____________

Name: Date:
Instructor: Section:

Chapter 4 SYSTEMS OF LINEAR EQUATIONS AND INEQUALITIES

4.2 Solving Systems of Linear Equations by Substitution

Learning Objectives
1. Solve linear systems by substitution.
2. Solve special systems by substitution.
3. Solve linear systems with fractions and decimals.

Key Terms

Use the vocabulary terms listed below to complete each statement in exercises 1–4.

substitution **ordered pair** **inconsistent system** **dependent system**

1. The solution of a linear system of equations is written as a(n) ________________.
2. When one expression is replaced by another, ____________________ is being used.
3. A system of equations in which all solutions of the first equation are also solutions of the second equation is a(n) ________________________________.
4. A system of equations that has no common solution is called a(n) ___________________________.

Objective 1 Solve linear systems by substitution.

Video Examples

Review these examples for Objective 1:

1. Solve the system by the substitution method.

$$2x+5y=22 \quad (1)$$
$$y=4x \quad (2)$$

Equation (2) is already solved for y. We substitute $4x$ for y in equation (1).

$$2x+5y=22$$
$$2x+5(4x)=22$$
$$2x+20x=22$$
$$22x=22$$
$$x=1$$

Find the value of y by substituting 1 for x in either equation. We use equation (2).

$$y=4x$$
$$y=4(1)=4$$

Now Try:

1. Solve the system by the substitution method.

$$x+y=7$$
$$y=6x$$

____

Name: Date:
Instructor: Section:

We check the solution (1, 4) by substituting 1 for x and 4 for y in both equations.

$2x+5y=22$	$y=4x$
$2(1)+5(4)\stackrel{?}{=}22$	$4\stackrel{?}{=}4(1)$
$2+20\stackrel{?}{=}22$	True $4=4$
$22=22$ True	

Since (1, 4) satisfies both equations, the solution set of the system is $\{(1, 4)\}$.

2. Solve the system by the substitution method.

$$4x+5y=13 \quad (1)$$
$$x=-y+2 \quad (2)$$

Equation (2) gives x in terms of y. We substitute $-y+2$ for x in equation (1).

$$4x+5y=13$$
$$4(-y+2)+5y=13$$
$$-4y+8+5y=13$$
$$y+8=13$$
$$y=5$$

Find the value of x by substituting 5 for y in either equation. We use equation (2).

$$x=-y+2$$
$$x=-5+2=-3$$

We check the solution (–3, 5) by substituting –3 for x and 5 for y in both equations.

$4x+5y=13$	$x=-y+2$
$4(-3)+5(5)\stackrel{?}{=}13$	$-3\stackrel{?}{=}-5+2$
$-12+25\stackrel{?}{=}13$	True $-3=-3$
$13=13$ True	

Both results are true, so the solution set of the system is $\{(-3, 5)\}$.

3. Solve the system by the substitution method.

$$4x=5-y \quad (1)$$
$$7x+3y=15 \quad (2)$$

Step 1 Solve one of the equations for x or y. Solve equation (1) for y to avoid fractions.

$$4x=5-y$$
$$y+4x=5$$
$$y=-4x+5$$

2. Solve the system by the substitution method.

$$2x+3y=6$$
$$x=5-y$$

____

3. Solve the system by the substitution method.

$$2x+7y=2$$
$$3y=2-x$$

____

Step 2 Now substitute $-4x+5$ for y in equation (2).

$$7x+3y=15$$
$$7x+3(-4x+5)=15$$

Step 3 Solve the equation from Step 2.

$$7x-12x+15=15$$
$$-5x+15=15$$
$$-5x=0$$
$$x=0$$

Step 4 Equation (1) solved for y is $y=-4x+5$. Substitute 0 for x.

$$y=-4(0)+5=5$$

Step 5 Check that (0, 5) is the solution.

$4x=5-y$	$7x+3y=15$
$4(0)\stackrel{?}{=}5-5$	$7(0)+3(5)\stackrel{?}{=}15$
$0=0$ True	True $15=15$

Since both results are true, the solution set of the system is $\{(0, 5)\}$.

Objective 1 Practice Exercises

For extra help, see Examples 1–3 on pages 286–289 of your text.

Solve each system by the substitution method. Check each solution.

1. $3x+2y=14$
$y=x+2$

1. ________________

2. $x+y=9$
$5x-2y=-4$

2. ________________

3. $3x-21=y$
$y+2x=-1$

3. ________________

Name: Date:
Instructor: Section:

Objective 2 Solve special systems by substitution.

Video Examples

Review these examples for Objective 2:

4. Use substitution to solve the system.

$$x = 7 - 3y \quad (1)$$
$$5x + 15y = 1 \quad (2)$$

Because equation (1) is already solved for x, we substitute $7-3y$ for x in equation (2).

$$5x + 15y = 1$$
$$5(7-3y) + 15y = 1$$
$$35 - 15y + 15y = 1$$
$$35 = 1 \quad \text{False}$$

A false result, here 35 = 1, means that the equations in the system have graphs that are parallel lines. The system is inconsistent and has no solution, so the solution set is $\varnothing$.

5. Use substitution to solve the system.

$$14x - 7y = 21 \quad (1)$$
$$-2x + y = -3 \quad (2)$$

Begin by solving equation (2) for y to get $y = 2x - 3$. Substitute $2x-3$ for y in equation (1).

$$14x - 7y = 21$$
$$14x - 7(2x-3) = 21$$
$$14x - 14x + 21 = 21$$
$$0 = 0$$

This true result means that every solution of one equation is also a solution of the other, so the system has an infinite number of solutions. The solution set is $\{(x, y) | 14x - 7y = 21\}$.

Now Try:

4. Use substitution to solve the system.

$$5x - 10y = 8$$
$$x = 2y + 5$$

____

5. Use substitution to solve the system.

$$5x + 4y = 20$$
$$-10x + 40 = 8y$$

____

Objective 2 Practice Exercises

For extra help, see Examples 4–5 on pages 289–290 of your text.

Solve each system by the substitution method. Use set-builder notation for dependent equations.

4. $y = -\frac{1}{3}x + 5$

$3y + x = -9$

4. ______________

Name: Date:
Instructor: Section:

5. $\frac{1}{2}x+3=y$

$6=-x+2y$

5. ______________

6. $4x+3y=2$

$8x+6y=6$

6. ______________

Objective 3 Solve linear systems with fractions and decimals.

Video Examples

Review these examples for Objective 3:

6. Solve the system by the substitution method.

$$\frac{1}{2}x-y=3 \quad (1)$$

$$\frac{1}{5}x+\frac{1}{2}y=\frac{3}{10} \quad (2)$$

Clear equation (1) of fractions by multiplying each side by 2.

$$2\left(\frac{1}{2}x-y\right)=2(3)$$

$$2\left(\frac{1}{2}x\right)-2y=2(3)$$

$$x-2y=6$$

Clear equation (2) of fractions by multiplying each side by 10.

$$10\left(\frac{1}{5}x+\frac{1}{2}y\right)=10\left(\frac{3}{10}\right)$$

$$10\left(\frac{1}{5}x\right)+10\left(\frac{1}{2}y\right)=10\left(\frac{3}{10}\right)$$

$$2x+5y=3$$

The given system of equations has been simplified to an equivalent system.

$$x-2y=6 \quad (3)$$

$$2x+5y=3 \quad (4)$$

To solve the system by substitution, solve equation (3) for x.

$$x-2y=6$$

$$x=2y+6$$

Now substitute the result for x in equation (4).

Now Try:

6. Solve the system by the substitution method.

$$x+\frac{1}{2}y=\frac{1}{2}$$

$$\frac{1}{2}x+\frac{1}{5}y=0$$

____

$$2x+5y=3$$
$$2(2y+6)+5y=3$$
$$4y+12+5y=3$$
$$9y+12=3$$
$$9y=-9$$
$$y=-1$$

Substitute –1 for y in $x=2y+6$ (equation (3) solved for x).

$x=2(-1)+6=4$

Check (4, –1) in both of the original equations. The solution set is $\{(4, -1)\}$.

7. Solve the system by the substitution method.

$0.4x+2.5y=8$ (1)

$-0.1x+3.5y=14.5$ (2)

Clear each equation of decimals, by multiplying by 10.

$$0.4x+2.5y=8$$
$$10(0.4x+2.5y)=10(8)$$
$$10(0.4x)+10(2.5y)=10(8)$$
$$4x+25y=80$$

$$-0.1x+3.5y=14.5$$
$$10(-0.1x+3.5y)=10(14.5)$$
$$10(-0.1x)+10(3.5y)=10(14.5)$$
$$-x+35y=145$$

Now solve the equivalent system of equations by substitution.

$4x+25y=80$ (3)

$-x+35y=145$ (4)

Equation (4) can be solved for x.

$x=35y-145$

Substitute this result for x in equation (3).

$$4x+25y=80$$
$$4(35y-145)+25y=80$$
$$140y-580+25y=80$$
$$165y-580=80$$
$$165y=660$$
$$y=4$$

7. Solve the system by the substitution method.

$0.6x+0.8y=-2.2$

$0.7x-0.1y=2.6$

____

Since equation (4) solved for x is $x = 35y - 145$, substitute 4 for y.

$x = 35(4) - 145 = -5$

Check (–5, 4) in both of the original equations.
The solution set is {(–5, 4)}.

Objective 3 Practice Exercises

For extra help, see Examples 6–7 on pages 290–292 of your text.

Solve each system by the substitution method. Check each solution.

7. $\frac{5}{4}x - y = -\frac{1}{4}$

$-\frac{7}{8}x + \frac{5}{8}y = 1$

7. ______________

8. $\frac{1}{4}x + \frac{3}{8}y = -3$

$\frac{5}{6}x - \frac{3}{7}y = -10$

8. ______________

9. $0.6x + 0.8y = 1$

$0.4y = 0.5 - 0.3x$

9. ______________

Name: Date:
Instructor: Section:

Chapter 4 SYSTEMS OF LINEAR EQUATIONS AND INEQUALITIES

4.3 Solving Systems of Linear Equations by Elimination

Learning Objectives

1 Solve linear systems by elimination.
2 Multiply when using the elimination method.
3 Use an alternative method to find the second value in a solution.
4 Solve special systems by elimination.

Key Terms

Use the vocabulary terms listed below to complete each statement in exercises 1–3.

addition property of equality **elimination method** **substitution**

1. Using the addition property to solve a system of equations is called the ____________________.

2. The ____________________ states that the same added quantity to each side of an equation results in equal sums.

3. ____________________ is being used when one expression is replaced by another.

Objective 1 Solve linear systems by elimination.

Video Examples

Review this example for Objective 1:

1. Use the elimination method to solve the system.

$$x+y=6 \quad (1)$$
$$-x+y=4 \quad (2)$$

Add the equations vertically.

$$\begin{array}{r} x+y=6 \quad (1) \\ -x+y=4 \quad (2) \\ \hline 2y=10 \\ y=5 \end{array}$$

To find the x-value, substitute 5 for y in either of the two equations of the system. We choose equation (1).

$$x+y=6$$
$$x+5=6$$
$$x=1$$

Now Try:

1. Use the elimination method to solve the system.

$$x+y=11$$
$$x-y=5$$

____

Name: Date:
Instructor: Section:

Check the solution (1, 5), by substituting 1 for x and 5 for y in both equations of the given system.

$x+y=6$	$-x+y=4$
$1+5\stackrel{?}{=}6$	$-1+5\stackrel{?}{=}4$
$6=6$ True	True $4=4$

Since both results are true, the solution set of the system is $\{(1, 5)\}$.

Objective 1 Practice Exercises

For extra help, see Examples 1–2 on pages 294–295 of your text.

Solve each system by the elimination method. Check your answers.

1. $x-4y=-4$
$-x+y=-5$

1. ________________

2. $2x-y=10$
$3x+y=10$

2. ________________

3. $x-3y=5$
$-x+4y=-5$

3. ________________

Name: Date:
Instructor: Section:

Objective 2 Multiply when using the elimination method.

Video Examples

Review this example for Objective 2:

3. Solve the system.

$$3x+8y=-2 \quad (1)$$
$$2x+7y=2 \quad (2)$$

To eliminate x, multiply equation (1) by 2 and multiply equation (2) by –3. Then add.

$$\begin{aligned} 6x+16y&=-4 \\ -6x-21y&=-6 \\ \hline -5y&=-10 \\ y&=2 \end{aligned}$$

Find the value of x by substituting 2 for y in either equation (1) or (2).

$$\begin{aligned} 2x+7y&=2 \\ 2x+7(2)&=2 \\ 2x+14&=2 \\ 2x&=-12 \\ x&=-6 \end{aligned}$$

Check that the solution set of the system is $\{(-6, 2)\}$.

Now Try:

3. Solve the system.

$$3x+4y=24$$
$$4x+3y=11$$

____

Objective 2 Practice Exercises

For extra help, see Example 3 on page 296 of your text.

Solve each system by the elimination method. Check your answers.

4. $6x+7y=10$
$2x-3y=14$

4. ____________

5. $8x+6y=10$
$4x-y=1$

5. ____________

6. $6x+y=1$
$3x-4y=23$

6. ____________

Name: Date:
Instructor: Section:

Objective 3 Use an alternative method to find the second value in a solution.

Video Examples

Review this example for Objective 3:

4. Solve the system.

$$6x = 7 - 3y \quad (1)$$
$$8x - 5y = 5 \quad (2)$$

Write equation (1) in standard form.

$$6x + 3y = 7 \quad (3)$$
$$8x - 5y = 5 \quad (4)$$

One way to proceed is to eliminate y by multiplying each side of equation (3) by 5 and each side of equation (4) by 3, and then adding.

$$\begin{array}{r} 30x + 15y = 35 \\ 24x - 15y = 15 \\ \hline 54x \qquad = 50 \end{array}$$

$$x = \frac{50}{54} \text{ or } \frac{25}{27}$$

Substituting $\frac{25}{27}$ for x in one of the given equations would give y, but the arithmetic would be complicated. Instead, solve for y by starting again with the original equations in standard form, and eliminating x.

Multiply equation (3) by 4 and equation (4) by –3.

$$\begin{array}{r} 24x + 12y = 28 \\ -24x + 15y = -15 \\ \hline 27y = 13 \end{array}$$

$$y = \frac{13}{27}$$

The solution set is $\left\{\left(\frac{25}{27}, \frac{13}{27}\right)\right\}$.

Now Try:

4. Solve the system.

$$8x = 5y + 1$$
$$6x - 8y = -2$$

____

Objective 3 Practice Exercises

For extra help, see Example 4 on page 297 of your text.

Solve each system by the elimination method. Check your answers.

7. $4x - 3y - 20 = 0$
$6x + 5y + 8 = 0$

7. ______________

Name: Date:
Instructor: Section:

8. $6x = 16 - 7y$
$4x = 3y + 26$

8. ______________

9. $2x = 14 + 4y$
$6y = -5x + 3$

9. ______________

Objective 4 Solve special systems by elimination.

Video Examples

Review these examples for Objective 4:

5. Solve each system by the elimination method.

a. $7x + y = 9$ (1)
$-14x - 2y = -18$ (2)

Multiply each side of equation (1) by 2.

$$\begin{aligned} 14x + 2y &= 18 \\ -14x - 2y &= -18 \\ \hline 0 &= 0 \quad \text{True} \end{aligned}$$

A true statement occurs when the equations are equivalent. This indicates that every solution of one equation is also a solution of the other. The solution set is $\{(x, y) \mid 7x + y = 9\}$.

b. $5x + 10y = 9$ (1)
$3x + 6y = 8$ (2)

Multiply each side of equation (1) by 3 and each side of equation (2) by –5.

$$\begin{aligned} 15x + 30y &= 27 \\ -15x - 30y &= -40 \\ \hline 0 &= -13 \quad \text{False} \end{aligned}$$

The false statement 0 = –13 indicates that the system has solution set ∅.

Now Try:

5. Solve each system by the elimination method.

a. $9x - 7y = 5$
$18x = 14y + 10$

____

b. $2x + 6y = 5$
$5x + 15y = 8$

____

Name: Date:
Instructor: Section:

Objective 4 Practice Exercises

For extra help, see Example 5 on pages 297–298 of your text.

Solve each system by the elimination method. Use set-builder notation for dependent equations. Check your answers.

10. $12x - 8y = 3$
$6x - 4y = 6$

10. ________________

11. $2x + 4y = -6$
$-x - 2y = 3$

11. ________________

12. $15x + 6y = 9$
$10x + 4y = 18$

12. ________________

Name: Date:
Instructor: Section:

Chapter 4 SYSTEMS OF LINEAR EQUATIONS AND INEQUALITIES

4.4 Applications of Linear Systems

Learning Objectives
1 Solve problems about unknown numbers.
2 Solve problems about quantities and their costs.
3 Solve problems about mixtures.
4 Solve problems about distance, rate (or speed), and time.

Key Terms

Use the vocabulary terms listed below to complete each statement in exercises 1–2.

system of linear equations $d = rt$

1. The formula that relates distance, rate, and time is ______________________.

2. A ______________________ consists of at least two linear equations with different variables.

Objective 1 Solve problems about unknown numbers.

Video Examples

Review this example for Objective 1:

1. Two towns have a combined population of 9045. There are 2249 more people living in one than in the other. Find the population in each town.

   Step1 Read the problem carefully. We are to find the population of each town.

   Step 2 Assign variables. Let x = the population of the larger town, and y = the population of smaller town.

   Step 3 Write two equations. There are 2249 more people living in one town. The total population is 9045.

   $x = y + 2249 \quad (1)$

   $x + y = 9045 \quad (2)$

   Step 4 Solve the system. We use substitution. Substitute $y + 2249$ for x in equation (2).

Now Try:

1. A rope 82 centimeters long is cut into two pieces with one piece four more than twice as long as the other. Find the length of each piece.

   ____

$$x + y = 9045$$
$$(y + 2249) + y = 9045$$
$$2249 + 2y = 9045$$
$$2y = 6796$$
$$y = 3398$$

To find x, substitute 3398 into either original equation. We use equation (1).
$x = 3398 + 2249 = 5647$

Step 5 State the answer. The population of the larger town is 5647. The population of the smaller town is 3398.

Step 6 Check the answer in the original problem. $3398 + 2249 = 5647$ and $5647 + 3398 = 9045$ The answers check.

Objective 1 Practice Exercises

For extra help, see Example 1 on pages 301–302 of your text.

Write a system of equations for each problem, then solve the problem.

1. The difference between two numbers is 14. If two times the smaller is added to one-half the larger, the result is 52. Find the numbers.

1.
larger number __________
smaller number __________

2. There are a total of 49 students in the two second grade classes at Jefferson School. If Carla has 7 more students in her class than Linda, find the number of students in each class.

2.
Carla's class __________
Linda's class __________

Name: Date:
Instructor: Section:

3. The perimeter of a rectangular room is 50 feet. The length is three feet greater than the width. Find the dimensions of the rectangle.

3.
length ______________

width ______________

Objective 2 Solve problems about quantities and their costs.

Video Examples

Review this example for Objective 2:

2. The total receipts for a basketball game were \$4690.50. There were 723 tickets sold, some for children and some for adults. If the adult tickets cost \$9.50 and the children's tickets cost \$4, how many of each type were there?

Step1 Read the problem.

Step 2 Assign variables. Let x = the number of adult tickets and y = the number of children's tickets.

Step 3 Write two equations. The total number of tickets is 723. The total value of the tickets is \$4690.50.

$$x + y = 723 \quad (1)$$
$$9.50x + 4y = 4690.5 \quad (2)$$

Step 4 Solve the system. Solve using elimination. Multiply equation (1) by –4.

$$\begin{aligned} -4x - 4y &= -2892 \\ 9.50x + 4y &= 4690.5 \\ \hline 5.50x \quad &= 1798.50 \\ x &= 327 \end{aligned}$$

Substitute 327 for x in equation (1) to find y.

$$\begin{aligned} x + y &= 723 \\ 327 + y &= 723 \\ y &= 396 \end{aligned}$$

Step 5 State the answer. The number of adult tickets is 327 and the number of children's tickets is 396.

Now Try:

2. Twice as many general admission tickets to a basketball game were sold as reserved seat tickets. General admission tickets cost \$10 and reserved seat tickets cost \$15. If the total value of both kinds of tickets was \$26,250, how many tickets of each kind were sold?

____

Step 6 Check. The sum of the tickets is 723. The value of the tickets is

$$9.50(327) + 4(396) = 4690.50$$

This checks.

Objective 2 Practice Exercises

For extra help, see Example 2 on pages 302–303 of your text.

Write a system of equations for each problem, then solve the problem.

4. There were 411 tickets sold for a soccer game, some for students and some for nonstudents. Student tickets cost \$4.25 and nonstudent tickets cost \$8.50 each. The total receipts were \$3021.75. How many of each type were sold?

4.
student tix ____________
nonstudent tix ____________

5. A cashier has some \$5 bills and some \$10 bills. The total value of the money is \$750. If the number of tens is equal to twice the number of fives, how many of each type are there?

5.
\$5 bills ____________
\$10 bills ____________

6. Luke plans to buy 10 ties with exactly \$162. If some ties cost \$14, and the others cost \$25, how many ties of each price should he buy?

6.
\$14 ties ____________
\$25 ties ____________

Name: Date:
Instructor: Section:

Objective 3 Solve problems about mixtures.

Video Examples

Review this example for Objective 3:

3. A mixture of 75% solution should be mixed with a 55% solution to get 70 liters of 63% solution. Determine the number of liters required of the 55% and 75% solutions.

Step1 Read the problem carefully.

Step 2 Assign variables. Let x = the number of liters of the 75% liquid and y = the number of liters of the 55% liquid.

Step 3 Write two equations. The total amount of liquid of the final mixture is 70 liters. The amount of 75% solution mixed with 55% solution will equal the 70 liters of 63% solution.

$$x+y=70 \quad (1)$$
$$0.75x+0.55y=0.63(70) \quad (2)$$

Step 4 Solve the system. Solve by substitution. Solving equation (1) for x results in $70-y$.

$$0.75x+0.55y=0.63(70)$$
$$0.75(70-y)+0.55y=44.1$$
$$52.5-0.75y+0.55y=44.1$$
$$-0.20y+52.5=44.1$$
$$-0.20y=-8.4$$
$$y=42$$

Substitute 42 for y in equation (1).

$$x=70-42=28$$

Step 5 State the answer. There should be 28 liters of 75% solution and 42 liters of 55% solution.

Step 6 Check the answer in the original problems. 28 + 42 = 70 and $0.75(28)+0.55(42)=44.1$ The answer checks.

Now Try:

3. A pharmacist wants to add water to a solution that contains 80% medicine. She wants to obtain 12 oz. of a solution that is 20% medicine. How much water and how much of the 80% solution should she use?

____

Name: Date:
Instructor: Section:

Objective 3 Practice Exercises

For extra help, see Example 3 on pages 303–304 of your text.

Write a system of equations for each problem, then solve the problem.

7. Jorge wishes to make 150 pounds of coffee blend that can be sold for $8 per pound. The blend will be a mixture of coffee worth $6 per pound and coffee worth $12 per pound. How many pounds of each kind of coffee should be used in the mixture?

7.
$6 coffee__________
$12 coffee__________

8. How many liters of water should be added to 25% antifreeze solution to get 30 liters of a 20% solution? How many liters of 25% solution are needed?

8.
water__________
25% solution __________

9. Ben wishes to blend candy selling for $1.60 a pound with candy selling for $2.50 a pound to get a mixture that will be sold for $1.90 a pound. How many pounds of the $1.60 and the $2.50 candy should be used to get 30 pounds of the mixture?

9.
$1.60 candy __________
$2.50 candy __________

Name: Date:
Instructor: Section:

Objective 4 Solve problems about distance, rate (or speed), and time.

Video Examples

Review this example for Objective 4:

4. Bill and Hillary start in Washington and fly in opposite directions. At the end of 4 hours, they are 4896 kilometers apart. If Bill flies 60 kilometers per hour faster than Hillary, what are their speeds?

Step1 Read the problem carefully.

Step 2 Assign variables. Let x = Bill's rate of speed and y = Hillary's rate of speed.

Step 3 Write two equations.

$x = 60 + y$ (1)

$4x + 4y = 4896$ (2)

Step 4 Solve the system. . Solve by substitution.

$$\begin{aligned} 4(60+y)+4y &= 4896 \\ 240+4y+4y &= 4896 \\ 240+8y &= 4896 \\ 8y &= 4656 \\ y &= 582 \end{aligned}$$

Substitute 582 for y in equation (1).

$x = 60 + 582 = 642$

Step 5 State the answer. Bill's rate is 642 kmh and Hillary's rate is 582 kmh.

Step 6 Check. Since $4(642)+4(582)=4896$ and $642 = 582 + 60$ the answers check.

Now Try:

4. Enid leaves Cherry Hill, driving by car toward New York, which is 90 miles away. At the same time, Jerry, riding his bicycle, leaves New York cycling toward Cherry Hill. Enid is traveling 28 miles per hour faster than Jerry. They pass each other $1\frac{1}{2}$ hours later. What are their speeds?

____

Objective 4 Practice Exercises

For extra help, see Examples 4–5 on pages 305–306 of your text.

Write a system of equations for each problem, and then solve the problem.

10. It takes Carla's boat $\frac{1}{2}$ hour to go 8 miles downstream and 1 hour to make the return trip upstream. Find the speed of the current and the speed of Carla's boat in still water.

10.
boat speed____________
current speed __________

Name: Date:
Instructor: Section:

11. Two planes left Philadelphia traveling in opposite directions. Plane A left 15 minutes before plane B. After plane B had been flying for 1 hour, the planes were 860 miles apart. What were the speeds of the two planes if plane A was flying 40 miles per hour faster than plane B?

11.
plane A ____________
plane B ____________

12. At the beginning of a fund-raising walk, Steve and Vic are 30 miles apart. If they leave at the same time and walk in the same direction, Steve would overtake Vic in 15 hours. If they walked toward each other, they would meet in 3 hours. What are their speeds?

12.
Steve ____________
Vic ____________

Name: Date:
Instructor: Section:

Chapter 4 SYSTEMS OF LINEAR EQUATIONS AND INEQUALITIES

4.5 Solving Systems of Linear Inequalities

Learning Objectives
1 Solve systems of linear inequalities by graphing.

Key Terms

Use the vocabulary terms listed below to complete each statement in exercises 1–2.

system of linear inequalities

solution set of a system of linear inequalities

1. All ordered pairs that make all inequalities of the system true at the same time is called the ________________________________.

2. A ________________________________ contains two or more linear inequalities (and no other kinds of inequalities).

Objective 1 Solve systems of linear inequalities by graphing.

Video Examples

Review these examples for Objective 1:

1. Graph the solution set of the system.

$$x + y \le 3$$
$$5x - y \ge 5$$

To graph $x + y \le 3$, graph the solid boundary line $x + y = 3$ using the intercepts (0, 3) and (3, 0). Determine the region to shade using (0, 0) as a test point.

$$x + y \le 3$$
$$0 + 0 \overset{?}{\le} 3$$
$$0 \le 3 \quad \text{True}$$

Shade the region containing (0, 0).

To graph $5x - y \ge 5$, graph the solid boundary line $5x - y = 5$ using the intercepts (0, –5) and (1, 0). Determine the region to shade using (0, 0) as a test point.

$$5x - y \ge 5$$
$$5(0) - 0 \overset{?}{\ge} 5$$
$$0 \ge 5 \quad \text{False}$$

Shade the region that does not contain (0, 0).

Now Try:

1. Graph the solution set of the system.

$$3x - y \le 3$$
$$x + y \le 0$$

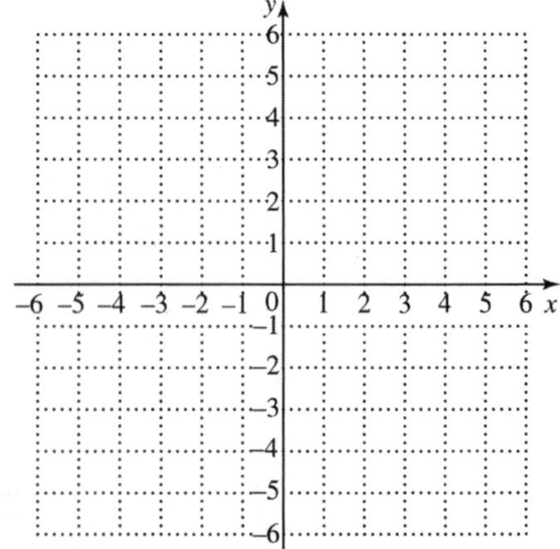

Name: Date:
Instructor: Section:

The solution set of this system includes all points in the intersection (overlap) of the graph of the two inequalities. This intersection is the gray shaded region and portions of the two boundary lines that surround it.

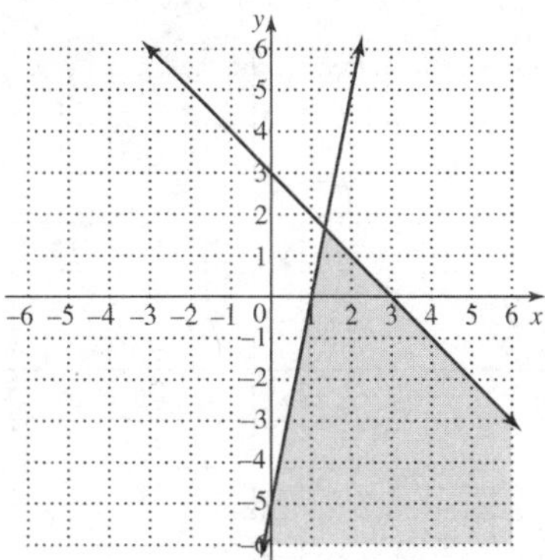

2. Graph the solution set of the system.

$$4x - y > 2$$
$$x + y > -2$$

Graph $4x - y > 2$ using a dashed line for $4x - y = 2$ and intercepts (0, –2) and $\left(\frac{1}{2}, 0\right)$.

Using the test point (0, 0), we shade the region that does not contain the point (0, 0).

Graph $x + y > -2$ using a dashed line for $x + y = -2$ and intercepts (–2, 0) and (0, –2). Using the test point (0, 0), we shade the region that does contain the point (0, 0).

The solution set is marked in gray. The solution set does not include either boundary line.

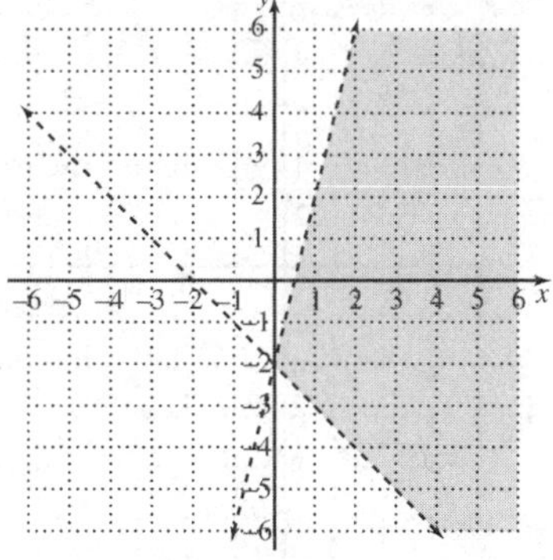

2. Graph the solution set of the system.

$$6x - y > 6$$
$$2x + 5y < 10$$

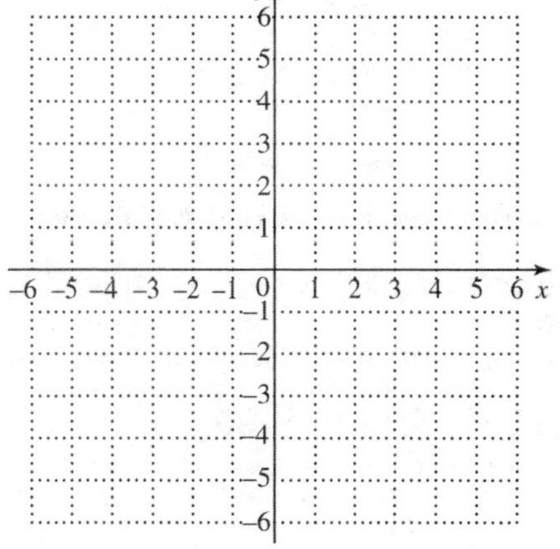

Name: Date:
Instructor: Section:

3. Graph the solution set of the system.

$$y \geq -1$$
$$2x - y > -1$$

Recall that $y = -1$ is a horizontal line through the point (0, –1). Use a solid line, and shade the region with the point (0, 0).

The graph of $2x - y > -1$ is created using a dashed line through the intercepts (0, 1) and $\left(-\frac{1}{2}, 0\right)$. Shade the region with the point (0, 0).

The solution set is marked in gray. The solution set includes the boundary line $y \geq -1$.

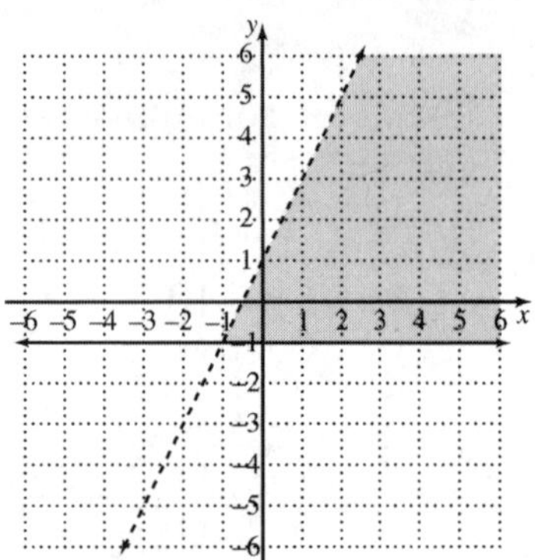

3. Graph the solution set of the system.

$$x - 3y \leq -7$$
$$x < 2$$

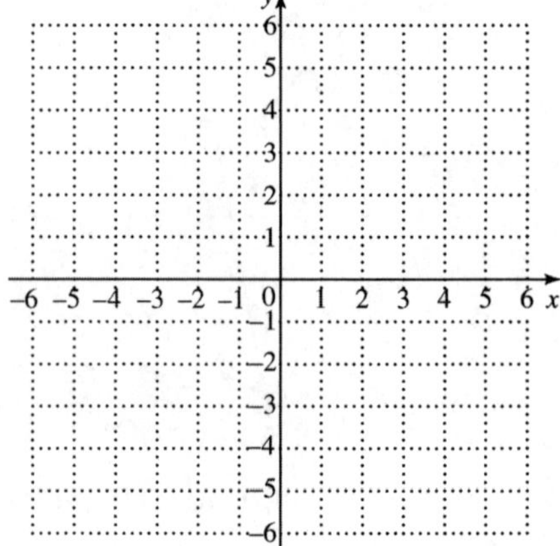

Objective 1 Practice Exercises

For extra help, see Examples 1–3 on pages 313–315 of your text.

Graph the solution of each system of linear inequalities.

1. $4x + 5y \leq 20$

$y \leq x + 3$

1.

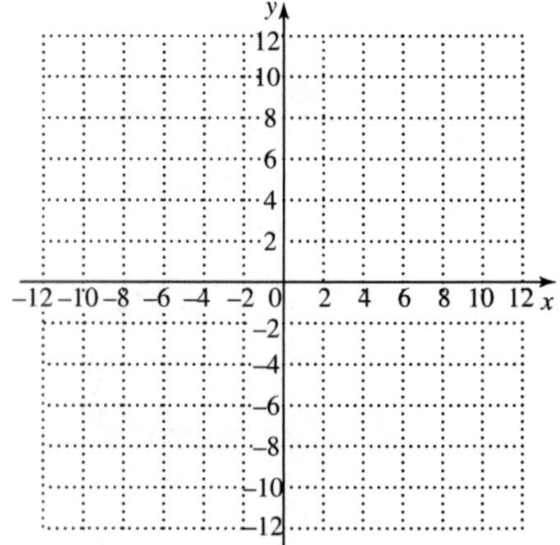

Name: Date:
Instructor: Section:

2. $x < 2y + 3$

$0 < x + y$

2.

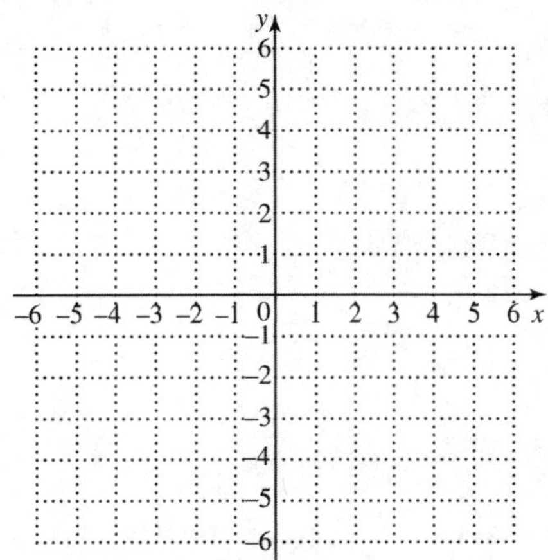

3. $y < 4$

$x \geq -3$

3.

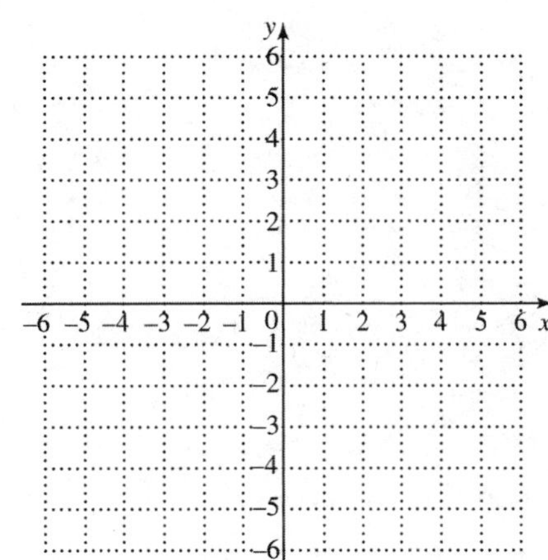

Name: Date:
Instructor: Section:

Chapter 5 EXPONENTS AND POLYNOMIALS

5.1 The Product Rule and Power Rules for Exponents

Learning Objectives

1 Use exponents.
2 Use the product rule for exponents.
3 Use the rule $(a^m)^n = a^{mn}$.
4 Use the rule $(ab)^m = a^m b^m$.
5 Use the rule $\left(\frac{a}{b}\right)^m = \frac{a^m}{b^m}$.
6 Use combinations of the rules for exponents.
7 Use the rules for exponents in a geometry application.

Key Terms

Use the vocabulary terms listed below to complete each statement in exercises 1−3.

exponential expression **base** **power**

1. 2^5 is read "2 to the fifth ____________________".

2. A number written with an exponent is called a(n) ________________________.

3. The ______________ is the number being multiplied repeatedly.

Objective 1 Use exponents.

Video Examples

Review these examples for Objective 1:

1. Write $5 \cdot 5 \cdot 5$ in exponential form.

   Since 5 occurs as a factor three times, the base is 5 and the exponent is 3.

   $5 \cdot 5 \cdot 5 = 5^3$

2. Name the base and exponent of each expression. Then evaluate.

   a. 3^4

   Base: 3
   Exponent: 4
   Value: $3^4 = 3 \cdot 3 \cdot 3 \cdot 3 = 81$

Now Try:

1. Write $4 \cdot 4 \cdot 4 \cdot 4 \cdot 4$ in exponential form.

   ____

2. Name the base and exponent of each expression. Then evaluate.

   a. 2^6

   ____

b. $(-3)^4$

Base: –3
Exponent: 4
Value: $(-3)^4 = (-3)(-3)(-3)(-3) = 81$

b. $(-2)^6$

______

Objective 1 Practice Exercises

For extra help, see Examples 1–2 on page 326 of your text.

Write the expression in exponential form and evaluate, if possible.

1. $\left(\frac{1}{3}\right)\left(\frac{1}{3}\right)\left(\frac{1}{3}\right)\left(\frac{1}{3}\right)\left(\frac{1}{3}\right)$

1. ______

Evaluate each exponential expression. Name the base and the exponent.

2. $(-4)^4$

2. ______
base______
exponent______

3. -3^8

3. ______
base______
exponent______

Objective 2 Use the product rule for exponents.

Video Examples

Review these examples for Objective 2:

3. Use the product rule for exponents to simplify, if possible.

a. $8^4 \cdot 8^5$

$8^4 \cdot 8^5 = 8^{4+5}$
$= 8^9$

b. $m^7 m^8 m^9$

$m^7 m^8 m^9 = m^{7+8+9}$
$= m^{24}$

Now Try:

3. Use the product rule for exponents to simplify, if possible.

a. $9^6 \cdot 9^7$

______

b. $m^{11} m^9 m^7$

______

Name: Date:
Instructor: Section:

c. $(5x^4)(6x^9)$

$$5x^4 \cdot 6x^9 = (5\cdot 6)\cdot(x^4 \cdot x^9)$$
$$= 30x^{4+9}$$
$$= 30x^{13}$$

d. $5^2 + 5^3$

$$5^2 + 5^3 = 25 + 125$$
$$= 150$$

c. $(6x^5)(3x^6)$

____

d. $3^4 + 3^3$

____

Objective 2 Practice Exercises

For extra help, see Example 3 on page 327 of your text.

Use the product rule to simplify each expression, if possible. Write each answer in exponential form.

4. $7^4 \cdot 7^3$ **4.** ____________

5. $(-2c^7)(-4c^8)$ **5.** ____________

6. $(3k^7)(-8k^2)(-2k^9)$ **6.** ____________

Objective 3 Use the rule $(a^m)^n = a^{mn}$.

Video Examples

Review these examples for Objective 3:

4. Use power rule (a) for exponents to simplify.

a. $(5^6)^3$

$$(5^6)^3 = 5^{6\cdot 3}$$
$$= 5^{18}$$

b. $(x^3)^4$

$$(x^3)^4 = x^{3\cdot 4}$$
$$= x^{12}$$

Now Try:

4. Use power rule (a) for exponents to simplify.

a. $(7^2)^4$

____

b. $(x^5)^6$

____

Name: Date:
Instructor: Section:

Objective 3 Practice Exercises

For extra help, see Example 4 on page 328 of your text.

Simplify each expression. Write all answers in exponential form.

7. $(7^3)^4$ **7.** ______________

8. $-(v^4)^9$ **8.** ______________

9. $[(-3)^3]^7$ **9.** ______________

Objective 4 Use the rule $(ab)^m = a^m b^m$.

Video Examples

Review this example for Objective 4:

5. Use power rule (b) for exponents to simplify.

$(5xy)^3$

$$(5xy)^3 = 5^3x^3y^3$$
$$= 125x^3y^3$$

Now Try:

5. Use power rule (b) for exponents to simplify.

$(4ab)^3$

____

Objective 4 Practice Exercises

For extra help, see Example 5 on page 329 of your text.

Simplify each expression.

10. $(5r^3t^2)^4$ **10.** ______________

11. $(-0.2a^4b)^3$ **11.** ______________

12. $(-2w^3z^7)^4$ **12.** ______________

Name: Date:
Instructor: Section:

Objective 5 Use the rule $\left(\frac{a}{b}\right)^m = \frac{a^m}{b^m}$.

Video Examples

Review this example for Objective 5:

6. Use power rule (c) for exponents to simplify.

$$\left(\frac{1}{8}\right)^3$$

$$\left(\frac{1}{8}\right)^3 = \frac{1^3}{8^3} = \frac{1}{512}$$

Now Try:

6. Use power rule (c) for exponents to simplify.

$$\left(\frac{1}{4}\right)^5$$

____

Objective 5 Practice Exercises

For extra help, see Example 6 on page 329 of your text.

Simplify each expression.

13. $\left(-\frac{2x}{5}\right)^3$ **13.** __________

14. $\left(\frac{xy}{z^2}\right)^4$ **14.** __________

15. $\left(\frac{-2a}{b^2}\right)^7$ **15.** __________

Objective 6 Use combinations of the rules for exponents.

Video Examples

Review these examples for Objective 6:

7. Simplify each expression.

a. $\left(\frac{3}{4}\right)^3 \cdot 3^2$

$$\left(\frac{3}{4}\right)^3 \cdot 3^2 = \frac{3^3}{4^3} \cdot \frac{3^2}{1}$$
$$= \frac{3^3 \cdot 3^2}{4^3 \cdot 1}$$
$$= \frac{3^{3+2}}{4^3}$$
$$= \frac{3^5}{4^3}, \text{ or } \frac{243}{64}$$

Now Try:

7. Simplify each expression.

a. $\left(\frac{5}{2}\right)^3 \cdot 5^2$

____

Name: Date:
Instructor: Section:

b. $\left(-x^5y\right)^4\left(-x^6y^5\right)^3$

$\left(-x^5y\right)^4\left(-x^6y^5\right)^3$

$=\left(-1x^5y\right)^4\left(-1x^6y^5\right)^3$

$=(-1)^4\left(x^5\right)^4\left(y^4\right)\cdot(-1)^3\left(x^6\right)^3\left(y^5\right)^3$

$=(-1)^4\left(x^{20}\right)\left(y^4\right)\cdot(-1)^3\left(x^{18}\right)\left(y^{15}\right)$

$=(-1)^7\,x^{20+18}y^{4+15}$

$=-1x^{38}y^{19}$

$=-x^{38}y^{19}$

b. $\left(-x^5y\right)^3\left(-x^6y^5\right)^2$

____

Objective 6 Practice Exercises

For extra help, see Example 7 on page 330–331 of your text.

Simplify. Write all answers in exponential form.

16. $\left(-x^3\right)^2\left(-x^5\right)^4$

16. ____________

17. $\left(2ab^2c\right)^5(ab)^4$

17. ____________

18. $\left(5x^2y^3\right)^7\left(5xy^4\right)^4$

18. ____________

Name: Date:
Instructor: Section:

Objective 7 Use the rules for exponents in a geometry application.

Video Examples

Review this example for Objective 7:

8. Find the area of the figure.

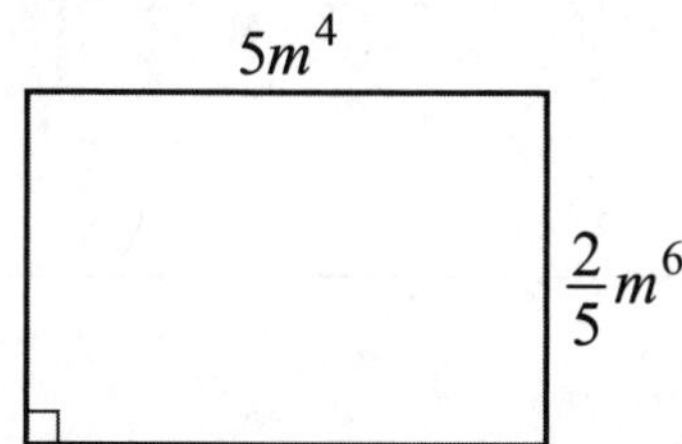

Use the formula for the area of a rectangle.

$$A = LW$$

$$A = (5m^4)\left(\frac{2}{5}m^6\right)$$

$$A = 5 \cdot \frac{2}{5} \cdot m^{4+6}$$

$$A = 2m^{10}$$

Now Try:

8. Find the area of the figure.

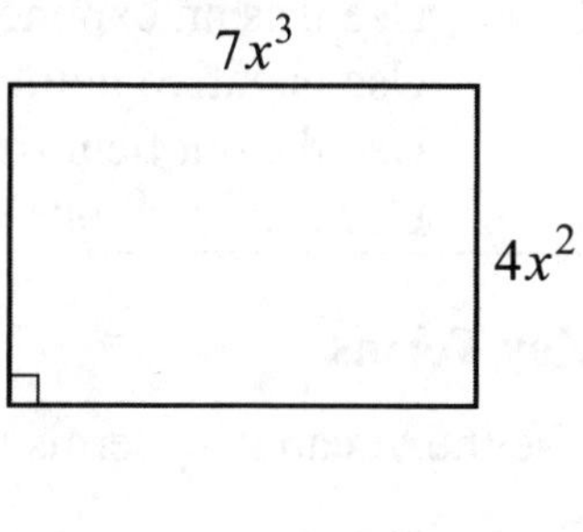

Objective 7 Practice Exercises

For extra help, see Example 8 on page 331 of your text.

Find a polynomial that represents the area of each figure.

19.

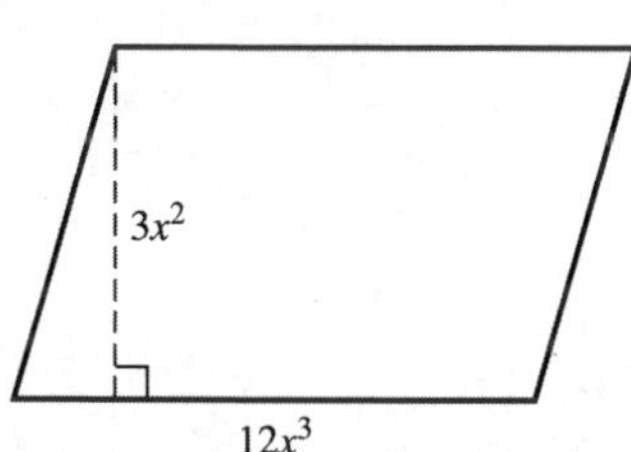

19. ________

20.

$3p^8$

$2q^6$

20. ________

21.

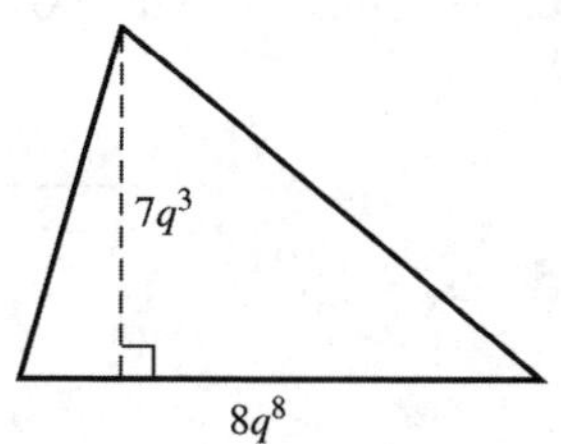

21. ________

Name: Date:
Instructor: Section:

Chapter 5 EXPONENTS AND POLYNOMIALS

5.2 Integer Exponents and the Quotient Rule

Learning Objectives
1 Use 0 as an exponent.
2 Use negative numbers as exponents.
3 Use the quotient rule for exponents.
4 Use combinations of the rules for exponents.

Key Terms

Use the vocabulary terms listed below to complete each statement in exercises 1–3.

exponent **base** **product rule for exponents**

power rule for exponents

1. The statement "If *m* and *n* are any integers, then $(a^m)^n = a^{mn}$" is an example of the ______________________________.

2. In the expression a^m, *a* is the ______________ and *m* is the ______________.

3. The statement "If *m* and *n* are any integers, then $a^m \cdot a^n = a^{m+n}$ is an example of the ______________________________.

Objective 1 Use 0 as an exponent.

Video Examples

Review these examples for Objective 1:

1. Evaluate.

a. $75^0 = 1$

b. $-75^0 = -(1)$ or -1

c. $(-9x)^0 = 1 \quad (x \neq 0)$

d. $3^0 + 12^0 = 1 + 1 = 2$

Now Try:

1. Evaluate.

a. 88^0 ______________

b. -88^0 ______________

c. $(-88a)^0 \quad (a \neq 0)$ ______________

d. $5^0 - 16^0$ ______________

Name: Date:
Instructor: Section:

Objective 1 Practice Exercises

For extra help, see Example 1 on page 335 of your text.

Evaluate each expression.

1. -12^0

1. ______________

2. $-15^0-(-15)^0$

2. ______________

3. $\frac{0^8}{8^0}$

3. ______________

Objective 2 Use negative numbers as exponents.

Video Examples

Review these examples for Objective 2:

2. Simplify by writing with positive exponents. Assume that all variables represent nonzero real numbers.

a. 4^{-3}

$4^{-3}=\frac{1}{4^3}$, or $\frac{1}{64}$

b. $\left(\frac{1}{3}\right)^{-3}$

$\left(\frac{1}{3}\right)^{-3}=3^3$, or 27

c. $\left(\frac{2}{3}\right)^{-5}$

$$\left(\frac{2}{3}\right)^{-5}=\left(\frac{3}{2}\right)^5 = \frac{3^5}{2^5} = \frac{243}{32}$$

Now Try:

2. Simplify by writing with positive exponents. Assume that all variables represent nonzero real numbers.

a. 3^{-3}

____

b. $\left(\frac{1}{5}\right)^{-2}$

____

c. $\left(\frac{3}{2}\right)^{-3}$

____

Name: Date:
Instructor: Section:

d. $5^{-1} - 3^{-1}$

$$5^{-1} - 3^{-1} = \frac{1}{5} - \frac{1}{3}$$
$$= \frac{3}{15} - \frac{5}{15}$$
$$= -\frac{2}{15}$$

e. $q^{-3} \quad (q \neq 0)$

$$q^{-3} = \frac{1}{q^3}$$

3. Simplify. Assume that all variables represent nonzero real numbers.

a. $\frac{3^{-4}}{7^{-2}} = \frac{7^2}{3^4}$, or $\frac{49}{81}$

b. $a^{-6}b^4 = \frac{b^4}{a^6}$

c. $\frac{x^{-3}y}{4z^{-4}} = \frac{yz^4}{4x^3}$

d. $4^{-1} - 8^{-1}$ __________

e. $p^{-5} \quad (p \neq 0)$ __________

3. Simplify. Assume that all variables represent nonzero real numbers.

a. $\frac{6^{-2}}{5^{-3}}$ __________

b. $x^{-7}y^2$ __________

c. $\frac{p^{-3}q}{4r^{-5}}$ __________

Objective 2 Practice Exercises

For extra help, see Examples 2–3 on pages 336–337 of your text.

Evaluate or simplify each expression, and write it using only positive exponents. Assume that all variables represent nonzero real numbers.

4. $-2k^{-4}$ **4.** __________

5. $(m^2n)^{-9}$ **5.** __________

6. $\frac{2x^{-4}}{3y^{-7}}$ **6.** __________

Name: Date:
Instructor: Section:

Objective 3 Use the quotient rule for exponents.

Video Examples

Review these examples for Objective 3:

4. Simplify. Assume that all variables represent nonzero real numbers.

a. $\frac{4^9}{4^6} = 4^{9-6} = 4^3 = 64$

b. $\frac{p^6}{p^{-4}} = p^{6-(-4)} = p^{10}$

c. $\frac{(x+7)^{-3}}{(x+7)^{-5}} \quad x \neq -7$

$$\frac{(x+7)^{-3}}{(x+7)^{-5}} = (x+7)^{-3-(-5)}$$
$$= (x+7)^{-3+5}$$
$$= (x+7)^2$$

d. $\frac{8x^{-4}y^3}{5^{-1}x^3y^{-4}}$

$$\frac{8x^{-4}y^3}{5^{-1}x^3y^{-4}} = \frac{8 \cdot 5y^3y^4}{x^3x^4}$$
$$= \frac{40y^7}{x^7}$$

Now Try:

4. Simplify. Assume that all variables represent nonzero real numbers.

a. $\frac{3^{18}}{3^{16}}$ __________

b. $\frac{z^6}{z^{-4}}$ __________

c. $\frac{(a-b)^{-8}}{(a-b)^{-10}} \quad a \neq b$ __________

d. $\frac{9a^{-5}b^3}{4^{-1}a^3b^{-4}}$ __________

Objective 3 Practice Exercises

For extra help, see Example 4 on pages 338–339 of your text.

Use the quotient rule to simplify each expression, and write it using only positive exponents. Assume that all variables represent nonzero real numbers.

7. $\frac{4k^7m^{10}}{8k^3m^5}$ **7.** __________

8. $\dfrac{a^4b^3}{a^{-2}b^{-3}}$

8. ________________

9. $\dfrac{3^{-1}m^{-4}p^6}{3^4m^{-1}p^{-2}}$

9. ________________

Objective 4 Use combinations of the rules for exponents.

Video Examples

Review these examples for Objective 4:

5. Simplify each expression. Assume that all variables represent nonzero real numbers.

a. $\dfrac{(5^4)^2}{5^6}$

$$\frac{(5^4)^2}{5^6} = \frac{5^8}{5^6} = 5^{8-6} = 5^2 = 25$$

b. $(3a)^4(3a)^2$

$$(3a)^4(3a)^2 = (3a)^6 = 3^6a^6 = 729a^6$$

c. $\left(\dfrac{3x^4}{4}\right)^{-5}$

$$\left(\frac{3x^4}{4}\right)^{-5} = \left(\frac{4}{3x^4}\right)^5 = \frac{4^5}{3^5x^{20}} = \frac{1024}{243x^{20}}$$

Now Try:

5. Simplify each expression. Assume that all variables represent nonzero real numbers.

a. $\dfrac{(6^3)^2}{6^5}$

____

b. $(5b)^3(5b)^2$

____

c. $\left(\dfrac{2p^4}{3}\right)^{-5}$

____

Name: Date:
Instructor: Section:

d. $\dfrac{(k^2m^{-3}n)^{-5}}{(3km^2n^{-4})^{-6}}$

$$\frac{(k^2m^{-3}n)^{-5}}{(3km^2n^{-4})^{-6}}=\frac{(k^2)^{-5}(m^{-3})^{-5}n^{-5}}{3^{-6}k^{-6}(m^2)^{-6}(n^{-4})^{-6}}$$
$$=\frac{k^{-10}m^{15}n^{-5}}{3^{-6}k^{-6}m^{-12}n^{24}}$$
$$=\frac{3^6m^{15+12}}{k^{-6+10}n^{24+5}}$$
$$=\frac{729m^{27}}{k^4n^{29}}$$

d. $\dfrac{(7xy^{-2}z^3)^{-3}}{(x^{-4}yz^{-2})^4}$

Objective 4 Practice Exercises

For extra help, see Example 5 on page 340 of your text.

Simplify each expression, and write it using only positive exponents. Assume that all variables represent nonzero real numbers.

10. $(9xy)^7(9xy)^{-8}$

10. ______________

11. $\dfrac{(a^{-1}b^{-2})^{-4}(ab^2)^6}{(a^3b)^{-2}}$

11. ______________

12. $\left(\dfrac{k^3t^4}{k^2t^{-1}}\right)^{-4}$

12. ______________

Name: Date:
Instructor: Section:

Chapter 5 EXPONENTS AND POLYNOMIALS

5.3 Scientific Notation

Learning Objectives
1 Express numbers in scientific notation.
2 Convert numbers in scientific notation standard notation.
3 Use scientific notation in calculations.

Key Terms

Use the vocabulary terms listed below to complete each statement in exercises 1–3.

scientific notation **quotient rule** **power rule**

1. A number written as $a \times 10^n$, where $1 \le |a| < 10$ and n is an integer, is written in ____________________.

2. The statement "If m and n are any integers and $b \neq 0$, then $\left(\frac{a}{b}\right)^m = \frac{a^m}{b^m}$" is an example of the ____________________.

3. The statement "If m and n are any integers and $b \neq 0$, then $\frac{a^m}{a^n} = a^{m-n}$" is an example of the ____________________.

Objective 1 Express numbers in scientific notation.

Video Examples

Review these examples for Objective 1:

1. Write each number in scientific notation.

a. 84,300,000,000

Move the decimal point 10 places to the left.

$84{,}300{,}000{,}000 = 8.43 \times 10^{10}$

b. 0.00573

The first nonzero digit is 5. Count the places.
Move the decimal point 3 places to the right.

$0.00573 = 5.73 \times 10^{-3}$

Now Try:

1. Write each number in scientific notation.

a. 47,710,000,000

____

b. 0.0463

____

Name: Date:
Instructor: Section:

Objective 1 Practice Exercises

For extra help, see Example 1 on page 345 of your text.

Write each number in scientific notation.

1. 23,651 **1.** ______________

2. −429,600,000,000 **2.** ______________

3. −0.0002208 **3.** ______________

Objective 2 Convert numbers in scientific notation standard notation.

Video Examples

Review these examples for Objective 2:

2. Write each number without exponents.

a. 3.57×10^{6}

Move the decimal point 6 places to the right, and attach four zeros.

$3.57\times10^{6}=3,570,000$

b. 8.98×10^{-3}

Move the decimal point 3 places to the left.

$8.98\times10^{-3}=0.00898$

Now Try:

2. Write each number without exponents.

a. 2.796×10^{7}

____

b. 1.64×10^{-4}

____

Objective 2 Practice Exercises

For extra help, see Example 2 on page 346 of your text.

Write each number in standard notation.

4. -2.45×10^{6} **4.** ______________

5. 6.4×10^{-3} **5.** ______________

6. -4.02×10^{4} **6.** ______________

Name: Date:
Instructor: Section:

Objective 3 Use scientific notation in calculations.

Video Examples

Review these examples for Objective 3:

3. Perform each calculation. Write answers in scientific notation and also without exponents.

a. $(8\times10^4)(7\times10^3)$

$$\begin{aligned}(8\times10^4)(7\times10^3) &= (8\times7)(10^4\times10^3)\\ &= 56\times10^7\\ &= (5.6\times10^1)\times10^7\\ &= 5.6\times10^8\\ &= 560{,}000{,}000\end{aligned}$$

b. $\dfrac{6\times10^{-4}}{3\times10^2}$

$$\begin{aligned}\frac{6\times10^{-4}}{3\times10^2} &= \frac{6}{3}\times\frac{10^{-4}}{10^2}\\ &= 2\times10^{-6}\\ &= 0.000002\end{aligned}$$

4. The Sahara desert covers approximately 3.5×10^6 square miles. Its sand is, on average, 12 feet deep. Find the volume, in cubic feet, of sand in the Sahara. $(\text{Hint: } 1 \text{ mi}^2 = 5280^2 \text{ ft}^2)$ Round your answer to two decimal places.

$$\begin{aligned}&(3.5\times10^6)(5280^2)(12)\\ &= 97574400(12)\times10^6\\ &= 1170892800\times10^6\\ &\approx 1.17\times10^{15} \text{ ft}^3\end{aligned}$$

The volume is 1.17×10^{15} cubic feet.

Now Try:

3. Perform each calculation. Write answers in scientific notation and also without exponents.

a. $(9\times10^5)(3\times10^2)$

____

____

b. $\dfrac{39\times10^{-3}}{13\times10^5}$

____

____

4. The Sahara desert covers approximately 3.5×10^6 square miles. Its sand is, on average, 12 feet deep. The volume of a single grain of sand is approximately 1.3×10^{-9} cubic feet. About how many grains of sand are in the Sahara?

____

Name: Date:
Instructor: Section:

Objective 3 Practice Exercises

For extra help, see Examples 3–5 on pages 346–347 of your text.

Perform the indicated operations, and write the answers in scientific notation.

7. $(2.3\times10^{4})\times(1.1\times10^{-2})$ **7.** ________________

8. $\dfrac{9.39\times10^{1}}{3\times10^{3}}$ **8.** ________________

Work the problem. Give answer in scientific notation.

9. There are about 6×10^{23} atoms in a mole of atoms. How many atoms are there in 8.1×10^{-5} mole? **9.** ________________

Name: Date:
Instructor: Section:

Chapter 5 EXPONENTS AND POLYNOMIALS

5.4 Adding, Subtracting, and Graphing Polynomials

Learning Objectives
1 Identify terms and coefficients.
2 Combine like terms.
3 Know the vocabulary for polynomials.
4 Evaluate polynomials.
5 Add and subtract polynomials.
6 Graph equations defined by polynomials of degree 2.

Key Terms

Use the vocabulary terms listed below to complete each statement in exercises 1–9.

term	**like terms**	**polynomial**
descending powers		**degree of a term**
degree of a polynomial		**monomial**
binomial		**trinomial**

parabola **vertex** **axis** **line of symmetry**

1. The ______________________ is the sum of the exponents on the variables in that term.

2. A polynomial in x is written in __________________________ if the exponents on x in its terms are decreasing order.

3. A _______________ is a number, a variable, or a product or quotient of a number and one or more variables raised to powers.

4. A polynomial with exactly three terms is called a _______________________.

5. A __________________ is a term, or the sum of a finite number of terms with whole number exponents.

6. A polynomial with exactly one term is called a _____________________.

7. The __________________________ is the greatest degree of any term of the polynomial.

8. A __________________ is a polynomial with exactly two terms.

9. Terms with exactly the same variables (including the same exponents) are called _________________________.

Name: Date:
Instructor: Section:

10. If a graph is folded on its____________________, the two sides coincide.

11. The ____________________ of a parabola that opens upward or downward is the lowest or highest point on the graph.

12. The ____________________ of a parabola that opens upward or downward is a vertical line through the vertex.

13. The graph of the quadratic equation $y = ax^2 + bx + c$ is called a ______________.

Objective 1 Identify terms and coefficients.

Video Examples

Review this example for Objective 1:

1. For each expression, determine the number of terms and name the coefficients of the terms.

$6 - 3x^4 - x^2$

Rewrite the expression as $6x^0 - 3x^4 - 1x^2$.
There are three terms: $6, -3x^4$, and $-x^2$.
The coefficients are 6, –3, and –1.

Now Try:

1. For each expression, determine the number of terms and name the coefficients of the terms.

$x^2 + 7 - 2x$

____

Objective 1 Practice Exercises

For extra help, see Example 1 on page 352 of your text.

For each expression, determine the number of terms and name the coefficients of the terms.

1. $3x^2 - 2 + x$ **1.** ____________

2. $5 + 6z^3$ **2.** ____________

3. $8y - y^3 - 1$ **3.** ____________

Name: Date:
Instructor: Section:

Objective 2 Combine like terms.

Video Examples

Review this example for Objective 2:

2. Simplify the expression by combining like terms.

$19m^3 + 6m + 5m^3$

$$19m^3 + 6m + 5m^3 = (19+5)m^3 + 6m$$
$$= 24m^3 + 6m$$

Now Try:

2. Simplify the expression by combining like terms.

$22m^2 + 15m^3 + 7m^2$

____

Objective 2 Practice Exercises

For extra help, see Example 2 on pages 352–353 of your text.

In each polynomial, combine like terms whenever possible. Write the result with descending powers.

4. $7z^3 - 4z^3 + 5z^3 - 11z^3$ **4.** ____________

5. $-1.3z^7 + 0.4z^7 + 2.6z^8$ **5.** ____________

6. $6c^3 - 9c^2 - 2c^2 + 14 + 3c^2 - 6c - 8 + 2c^3$ **6.** ____________

Objective 3 Know the vocabulary for polynomials.

Video Examples

Review these examples for Objective 3:

3. Simplify each polynomial, if possible, and write in descending powers of the variable. Then give the degree and tell whether the polynomial is a monomial, a binomial, a trinomial, or none of these.

a. $5x + 6x^3 - 7x - 2x^2 + 4x$

$5x + 6x^3 - 7x - 2x^2 + 4x = 6x^3 - 2x^2 + 2x$
The degree is 3. The simplified polynomial is a trinomial.

Now Try:

3. Simplify each polynomial, if possible, and write in descending powers of the variable. Then give the degree and tell whether the polynomial is a monomial, a binomial, a trinomial, or none of these.

a. $2x - 7x^2 - 6x + 3x^3 + 8x$

____

Name: Date:
Instructor: Section:

b. $9y^3 - 7y^5 + 3y^3 + 2y^5$

$9y^3 - 7y^5 + 3y^3 + 2y^5 = -5y^5 + 12y^3$
The degree is 5. The simplified polynomial is a binomial.

b. $w^5 - 4w^2 + 3w^5 + w^2$

____

Objective 3 Practice Exercises

For extra help, see Example 3 on page 354 of your text.

For each polynomial, first simplify, if possible, and write the resulting polynomial in descending powers of the variable. Then give the degree of this polynomial, and tell whether it is a monomial, *a* binomial, *a* trinomial, *or* none of these.

7. $3n^8 - n^2 - 2n^8$

7. ____________

degree: ________

type: __________

8. $-d^2 + 3.2d^3 - 5.7d^8 - 1.1d^5$

8. ____________

degree: ________

type: __________

9. $-6c^4 - 6c^2 + 9c^4 - 4c^2 + 5c^5$

9. ____________

degree: ________

type: __________

Objective 4 Evaluate polynomials.

Video Examples

Review this example for Objective 4:

4. Find the value of $4x^3 + 6x^2 - 5x - 5$ for

$x = 2$

$$\begin{aligned} &4x^3 + 6x^2 - 5x - 5 \\ &\quad = 4(2)^3 + 6(2)^2 - 5(2) - 5 \\ &\quad = 4(8) + 6(4) - 5(2) - 5 \\ &\quad = 32 + 24 - 10 - 5 \\ &\quad = 41 \end{aligned}$$

Now Try:

4. Find the value of $5x^4 + 3x^2 - 9x - 7$ for $x = 4$

____

Name: Date:
Instructor: Section:

Objective 4 Practice Exercises

For extra help, see Example 4 on page 354 of your text.

Find the value of each polynomial (a) *when* $x = -2$ *and* (b) *when* $x = 3$.

10. $3x^3 + 4x - 19$

10. a.______________

b.______________

11. $-4x^3 + 10x^2 - 1$

11. a.______________

b.______________

12. $x^4 - 3x^2 - 8x + 9$

12. a.______________

b.______________

Objective 5 Add and subtract polynomials.

Video Examples

Review these examples for Objective 5:

6. Find each sum.

a. Add $5x^4 - 7x^3 + 9$ and $-3x^4 + 8x^3 - 7$

$$(5x^4 - 7x^3 + 9) + (-3x^4 + 8x^3 - 7)$$
$$= 5x^4 - 3x^4 - 7x^3 + 8x^3 + 9 - 7$$
$$= 2x^4 + x^3 + 2$$

b. $(5x^4 - 7x^2 + 6x) + (-3x^3 + 4x^2 - 7)$

$$(5x^4 - 7x^2 + 6x) + (-3x^3 + 4x^2 - 7)$$
$$= 5x^4 - 3x^3 - 7x^2 + 4x^2 + 6x - 7$$
$$= 5x^4 - 3x^3 - 3x^2 + 6x - 7$$

Now Try:

6. Find each sum.

a. Add $15x^3 - 5x + 3$ and $-11x^3 + 6x + 9$

____

b. $(8x^2 - 6x + 4) + (7x^3 - 8x - 5)$

____

Name: Date:
Instructor: Section:

7. Perform the subtraction.

Subtract $8x^3 - 5x^2 + 8$ from $9x^3 + 6x^2 - 7$.

$$\left(9x^3 + 6x^2 - 7\right) - \left(8x^3 - 5x^2 + 8\right)$$
$$= \left(9x^3 + 6x^2 - 7\right) + \left(-8x^3 + 5x^2 - 8\right)$$
$$= x^3 + 11x^2 - 15$$

7. Perform the subtraction.

Subtract $18x^3 + 4x - 6$ from $7x^3 - 3x - 5$.

____

9. Perform the indicated operations to simplify the expression

$$\left(5 - 2x + 9x^2\right) - \left(7 - 5x + 8x^2\right) + \left(6 + 3x - 5x^2\right)$$

Rewrite, changing the subtraction to adding the opposite.

$$\left(5-2x+9x^2\right)-\left(7-5x+8x^2\right)+\left(6+3x-5x^2\right)$$
$$=\left(5-2x+9x^2\right)+\left(-7+5x-8x^2\right)+\left(6+3x-5x^2\right)$$
$$=\left(-2+3x+x^2\right)+\left(6+3x-5x^2\right)$$
$$=4+6x-4x^2$$

9. Perform the indicated operations to simplify the expression

$$\left(10-7x+6x^2\right)-\left(5-11x+3x^2\right)$$
$$+\left(2+4x-7x^2\right)$$

____

10. Add or subtract as indicated.

$$\left(3x^2y + 5xy + y^2\right) - \left(4x^2y + xy - 3y^2\right)$$

Change the signs of the terms in the parentheses and add like terms vertically.

$$\begin{array}{r} 3x^2y + 5xy + y^2 \\ -4x^2y - xy + 3y^2 \\ \hline -x^2y + 4xy + 4y^2 \end{array}$$

10. Add or subtract as indicated.

$$\left(7x^2y + 3xy + 4y^2\right)$$
$$-\left(6x^2y - xy + 4y^2\right)$$

____

Objective 5 Practice Exercises

For extra help, see Examples 5–10 on pages 355–357 of your text.

Add or subtract as indicated.

13. $\left(3r^3 + 5r^2 - 6\right) + \left(2r^2 - 5r + 4\right)$

13. ____________

Name: Date:
Instructor: Section:

14. $\left(-8w^3+11w^2-12\right)-\left(-10w^2+3\right)$ **14.** ________________

15. $\left(2x^2y+2xy-4xy^2\right)+\left(6xy+9xy^2\right)-\left(9x^2y+5xy\right)$ **15.** ________________

Objective 6 Graph equations defined by polynomials of degree 2.

Video Examples

Review this example for Objective 6:

11. Graph the equation.

$y=x^2-3$

Find several ordered pairs. Let $x = 0$ to find the y-intercept.

$y=x^2-3=0^2-3=-3$

This gives the ordered pair (0, –3). Select several values for x and find the corresponding values for y. Plot the ordered pairs and join them with a smooth curve.

x	y
2	1
1	−2
0	−3
−1	−2
−2	1

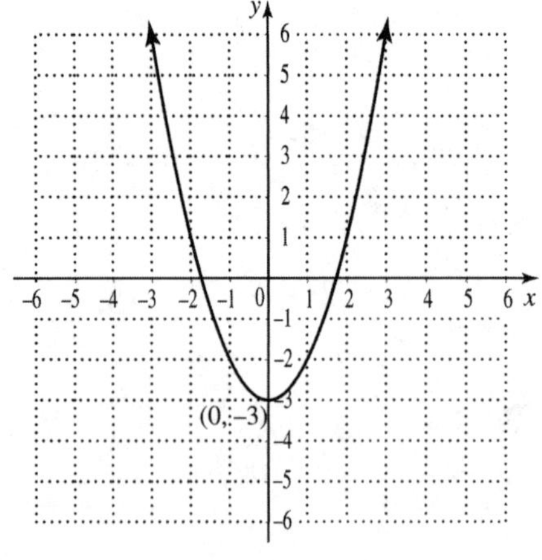

Now Try:

11. Graph the equation.

$y=9-x^2$

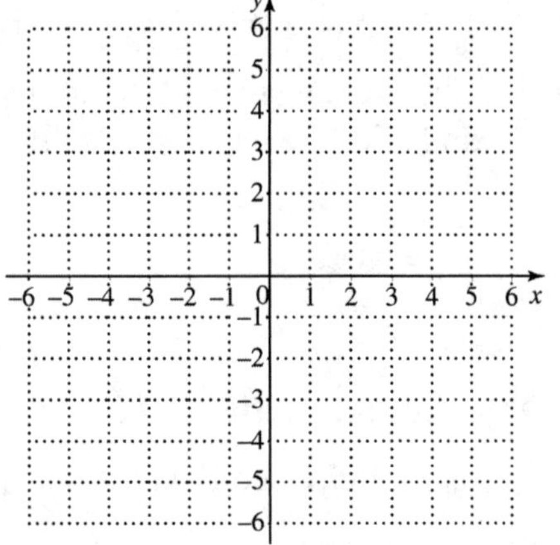

Name: Date:
Instructor: Section:

Objective 6 Practice Exercises

For extra help, see Example 11 on pages 357–358 of your text.

Graph each equation.

16. $y = -x^2 - 1$

16.

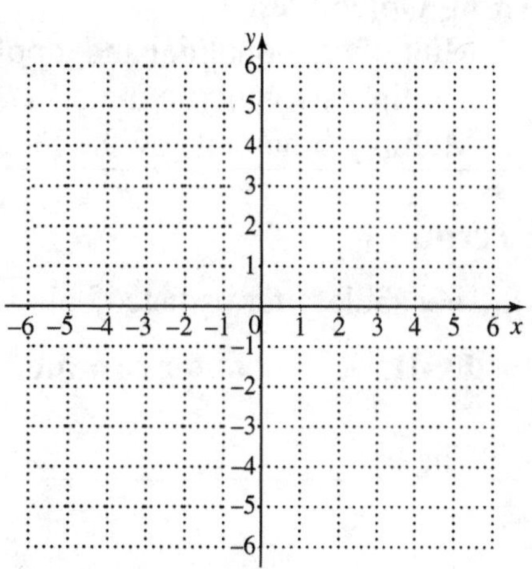

vertex: ______________________

17. $y = x^2 + 2$

17.

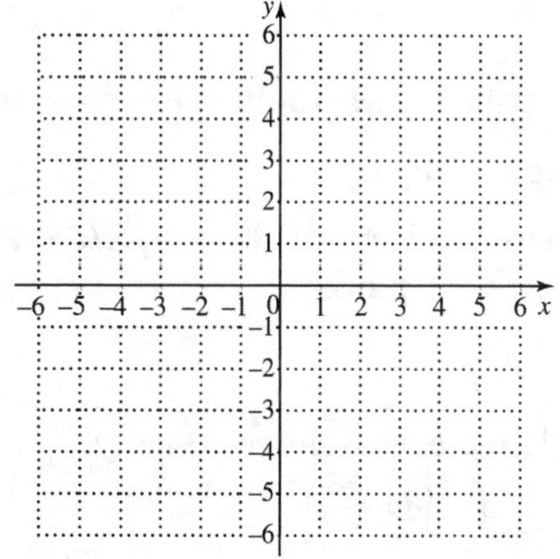

vertex: ______________________

Name: Date:
Instructor: Section:

Chapter 5 EXPONENTS AND POLYNOMIALS

5.5 Multiplying Polynomials

Learning Objectives
1 Multiply a monomial and a polynomial.
2 Multiply two polynomials.
3 Multiply binomials by the FOIL method.

Key Terms

Use the vocabulary terms listed below to complete each statement in exercises 1–3.

FOIL **outer product** **inner product**

1. The ______________________________ of $(2y-5)(y+8)$ is $-5y$.

2. ____________________ is a shortcut method for finding the product of two binomials.

3. The ______________________________ of $(2y-5)(y+8)$ is $16y$.

Objective 1 Multiply a monomial and a polynomial.

Video Examples

Review this example for Objective 1:

1. Find the product.

$5x^2(7x+3)$

Use the distributive property.

$$5x^2(7x+3) = 5x^2(7x) + 5x^2(3)$$
$$= 35x^3 + 15x^2$$

Now Try:

1. Find the product.

$8x^3(4x+8)$

____

Objective 1 Practice Exercises

For extra help, see Example 1 on page 362 of your text.

Find each product.

1. $7z(5z^3+2)$ 1. ______________

2. $2m(3+7m^2+3m^3)$ 2. ______________

3. $-3y^2(2y^3+3y^2-4y+11)$ 3. ______________

Name: Date:
Instructor: Section:

Objective 2 Multiply two polynomials.

Video Examples

Review these examples for Objective 2:

2. Multiply $(x^2+6)(5x^3-4x^2+3x)$.

Multiply each term of the second polynomial by each term of the first.

$$(x^2+6)(5x^3-4x^2+3x)$$
$$=x^2(5x^3)+x^2(-4x^2)+x^2(3x)$$
$$+6(5x^3)+6(-4x^2)+6(3x)$$
$$=5x^5-4x^4+3x^3+30x^3-24x^2+18x$$
$$=5x^5-4x^4+33x^3-24x^2+18x$$

3. Multiply $(2x^3+7x^2+5x-1)(4x+6)$ vertically.

Write the polynomials vertically.

$$\begin{array}{r} 2x^3+7x^2+5x-1 \\ \underline{4x+6} \end{array}$$

Begin by multiplying each term in the top row by 6.

$$\begin{array}{rrrr} 2x^3 & +7x^2 & +5x & -1 \\ & & 4x & +6 \\ \hline 12x^3 & +42x^2 & +30x & -6 \end{array}$$

Now multiply each term in the top row by $4x$. Then add like terms.

$$\begin{array}{rrrrr} & 2x^3 & +7x^2 & +5x & -1 \\ & & & 4x & +6 \\ \hline & 12x^3 & +42x^2 & +30x & -6 \\ 8x^4 & +28x^3 & +20x^2 & -4x & \\ \hline 8x^4 & +40x^3 & +62x^2 & +26x & -6 \end{array}$$

The product is $8x^4+40x^3+62x^2+26x-6$.

Now Try:

2. Multiply $(x^3+9)(4x^4-2x^2+x)$

____

3. Multiply $(4x^3-3x^2+6x+5)(7x-3)$ vertically.

____

Name: Date:
Instructor: Section:

4. Find the product of $-16m^3+12m^2+4$ and $\frac{1}{4}m^2+\frac{3}{4}$.

Multiply each term of the second polynomial by each term of the first.

$$\left(-16m^3+12m^2+4\right)\left(\frac{1}{4}m^2+\frac{3}{4}\right)$$
$$=-16m^3\left(\frac{1}{4}m^2\right)-16m^3\left(\frac{3}{4}\right)+12m^2\left(\frac{1}{4}m^2\right)$$
$$+12m^2\left(\frac{3}{4}\right)+4\left(\frac{1}{4}m^2\right)+4\left(\frac{3}{4}\right)$$
$$=-12m^3+9m^2+3-5m^5+3m^4+m^2$$
$$=-4m^5+3m^4-12m^3+10m^2+3$$

The product is $-4m^5+3m^4-12m^3+10m^2+3$.

4. Find the product of $12x^3-36x^2+6$ and $\frac{1}{6}x^2+\frac{5}{6}$.

____

Objective 2 Practice Exercises

For extra help, see Examples 2–4 on page 363 of your text.

Find each product.

4. $(x+3)(x^2-3x+9)$ **4.** ____________

5. $(2m^2+1)(3m^3+2m^2-4m)$ **5.** ____________

6. $(3x^2+x)(2x^2+3x-4)$ **6.** ____________

Objective 3 Multiply binomials by the FOIL method.

Video Examples

Review these examples for Objective 3:

5. Use the FOIL method to find the product $(x+7)(x-5)$.

Step 1 F Multiply the first terms: $x(x)=x^2$.

Now Try:

5. Use the FOIL method to find the product $(x+9)(x-6)$.

____

Name: Date:
Instructor: Section:

Step 2 O Find the outer product: $x(-5) = -5x$.

Step 3 I Find the inner product: $7(x) = 7x$.
Add the outer and inner products mentally:
$$-5x + 7x = 2x$$
Step 4 L Multiply the last terms: $7(-5) = -35$.

The product $(x+7)(x-5)$ is $x^2 + 2x - 35$.

6. Multiply $(7x-3)(4y+5)$.

First $7x(4y) = 28xy$

Outer $7x(5) = 35x$

Inner $-3(4y) = -12y$

Last $-3(5) = -15$

The product $(7x-3)(4y+5)$ is
$28xy + 35x - 12y - 15$.

6. Multiply $(8y-7)(2x+9)$.

____

7. Find the product.

$(3k+7m)(2k+9m)$

$(3k+7m)(2k+9m)$
$= 3k(2k) + 3k(9m) + 7m(2k) + 7m(9m)$
$= 6k^2 + 27km + 14km + 63m^2$
$= 6k^2 + 41km + 63m^2$

7. Find the product.

$(5k+8n)(3k+4n)$

____

Objective 3 Practice Exercises

For extra help, see Examples 5–7 on page 365 of your text.

Find each product.

7. $(5a-b)(4a+3b)$ **7.** ______________

8. $(3+4a)(1+2a)$ **8.** ______________

9. $(2m+3n)(-3m+4n)$ **9.** ______________

Name: Date:
Instructor: Section:

Chapter 5 EXPONENTS AND POLYNOMIALS

5.6 Special Products

Learning Objectives
1 Square binomials.
2 Find the product of the sum and difference of two terms.
3 Find greater powers of binomials.

Key Terms

Use the vocabulary terms listed below to complete each statement in exercises 1–2.

conjugate **binomial**

1. A polynomial with two terms is called a ______________________.

2. The ________________ of $a + b$ is $a - b$.

Objective 1 Square binomials.

Video Examples

Review these examples for Objective 1:

2. Square each binomial.

a. $(6x+3y)^2$

$$(6x+3y)^2 = (6x)^2 + 2(6x)(3y) + (3y)^2$$
$$= 36x^2 + 36xy + 9y^2$$

b. $\left(6n+\frac{1}{4}\right)^2$

$$\left(6n+\frac{1}{4}\right)^2 = (6n)^2 + 2(6n)\left(\frac{1}{4}\right) + \left(\frac{1}{4}\right)^2$$
$$= 36n^2 + 3n + \frac{1}{16}$$

Now Try:

2. Square each binomial.

a. $(2a+9k)^2$

____

b. $\left(3p+\frac{1}{6}\right)^2$

____

Objective 1 Practice Exercises

For extra help, see Examples 1–2 on pages 369–370 of your text.

Find each square by using the pattern for the square of a binomial.

1. $(7+x)^2$ 1. ______________

Name: Date:
Instructor: Section:

2. $(2m-3p)^2$

2. ______________

3. $(4y-0.7)^2$

3. ______________

Objective 2 Find the product of the sum and difference of two terms.

Video Examples

Review these examples for Objective 2:

3. Find each product.

a. $(x+5)(x-5)$

Use the rule for the product of the sum and difference of two terms.

$$(x+5)(x-5) = x^2 - 5^2$$
$$= x^2 - 25$$

b. $\left(\frac{3}{4}-y\right)\left(\frac{3}{4}+y\right)$

$$\left(\frac{3}{4}-y\right)\left(\frac{3}{4}+y\right) = \left(\frac{3}{4}\right)^2 - y^2$$
$$= \frac{9}{16} - y^2$$

4. Find each product.

a. $(6x+w)(6x-w)$

$$(6x+w)(6x-w) = (6x)^2 - w^2$$
$$= 36x^2 - w^2$$

b. $3q(q^2+4)(q^2-4)$

First, multiply the conjugates.

$$3q(q^2+4)(q^2-4) = 3q(q^4-16)$$
$$= 3q^5 - 48q$$

Now Try:

3. Find each product.

a. $(x+9)(x-9)$

____

b. $\left(\frac{5}{6}+a\right)\left(\frac{5}{6}-a\right)$

____

4. Find each product.

a. $(11x-y)(11x+y)$

____

b. $4p(p^2+6)(p^2-6)$

____

Name: Date:
Instructor: Section:

Objective 2 Practice Exercises

For extra help, see Examples 3–4 on page 371 of your text.

Find each product by using the pattern for the sum and difference of two terms.

4. $(12+x)(12-x)$ **4.** ______________

5. $(8k+5p)(8k-5p)$ **5.** ______________

6. $\left(\frac{4}{7}t+2u\right)\left(\frac{4}{7}t-2u\right)$ **6.** ______________

Objective 3 Find greater powers of binomials.

Video Examples

Review these examples for Objective 3:

5. Find each product.

a. $(x+4)^3$

$$(x+4)^3$$
$$=(x+4)^2(x+4)$$
$$=(x^2+8x+16)(x+4)$$
$$=x^3+8x^2+16x+4x^2+32x+64$$
$$=x^3+12x^2+48x+64$$

b. $(5y-4)^4$

$$(5y-4)^4$$
$$=(5y-4)^2(5y-4)^2$$
$$=(25y^2-40y+16)(25y^2-40y+16)$$
$$=625y^4-1000y^3+400y^2$$
$$-1000y^3+1600y^2-640y$$
$$+400y^2-640y+256$$
$$=625y^4-2000y^3+2400y^2-1280y+256$$

Now Try:

5. Find each product.

a. $(x+6)^3$

____

b. $(3x-5)^4$

____

Name: Date:
Instructor: Section:

Objective 3 Practice Exercises

For extra help, see Example 5 on page 372 of your text.

Find each product.

7. $(a-3)^3$ **7.** ________________

8. $(j+3)^4$ **8.** ________________

9. $(4s+3t)^4$ **9.** ________________

Name: Date:
Instructor: Section:

Chapter 5 EXPONENTS AND POLYNOMIALS

5.7 Dividing Polynomials

Learning Objectives
1 Divide a polynomial by a monomial.
2 Divide a polynomial by a polynomial.
3 Use division in a geometry problem.

Key Terms

Use the vocabulary terms listed below to complete each statement in exercises 1–3.

quotient **dividend** **divisor**

1. In the division $\frac{5x^5-10x^3}{5x^2}=x^3-2x$, the expression $5x^5-10x^3$ is the __________________.

2. In the division $\frac{5x^5-10x^3}{5x^2}=x^3-2x$, the expression x^3-2x is the __________.

3. In the division $\frac{5x^5-10x^3}{5x^2}=x^3-2x$, the expression $5x^2$ is the _____________.

Objective 1 Divide a polynomial by a monomial.

Video Examples

Review these examples for Objective 1:

1. Divide $6x^4-18x^3$ by $6x^2$.

$$\frac{6x^4-18x^3}{6x^2}=\frac{6x^4}{6x^2}-\frac{18x^3}{6x^2}$$
$$=x^2-3x$$

Check Multiply. $6x^2(x^2-3x)=6x^4-18x^3$

2. Divide. $\frac{25a^6-15a^4+10a^2}{5a^3}$

Divide each term by $5a^3$.

$$\frac{25a^6-15a^4+10a^2}{5a^3}=\frac{25a^6}{5a^3}-\frac{15a^4}{5a^3}+\frac{10a^2}{5a^3}$$
$$=5a^3-3a+\frac{2}{a}$$

Now Try:

1. Divide $20x^4-10x^2$ by $2x$.

____

2. Divide. $\frac{27n^5-36n^4-18n^2}{9n^3}$

____

Name: Date:
Instructor: Section:

3. Divide $-12x^4+15x^5-5x$ by $-5x$.

Write the polynomial in descending powers before dividing.

$$\frac{15x^5-12x^4-5x}{-5x}=\frac{15x^5}{-5x}-\frac{12x^4}{-5x}-\frac{5x}{-5x}$$
$$=-3x^4+\frac{12}{5}x^3+1$$

Check $-5x\left(-3x^4+\frac{12}{5}x^3+1\right)$
$$=-5x\left(-3x^4\right)-5x\left(\frac{12}{5}x^3\right)-5x(1)$$
$$=15x^5-12x^4-5x$$

4. Divide
$225x^5y^9-150x^3y^7+110x^2y^5-80xy^3+75y^2$
by $-25xy^2$.

$$\frac{225x^5y^9-150x^3y^7+110x^2y^5-80xy^3+75y^2}{-25xy^2}$$
$$=\frac{225x^5y^9}{-25xy^2}-\frac{150x^3y^7}{-25xy^2}+\frac{110x^2y^5}{-25xy^2}-\frac{80xy^3}{-25xy^2}+\frac{75y^2}{-25xy^2}$$
$$=-9x^4y^7+6x^2y^5-\frac{22xy^3}{5}+\frac{16y}{5}-\frac{3}{x}$$

Check by multiplying the quotient by the divisor.

3. Divide $-8z^5+7z^6-10z-6$ by $2z^2$.

____

4. Divide $80a^5b^3+160a^4b^2-120a^2b$ by $-40a^2b$.

____

Objective 1 Practice Exercises

For extra help, see Examples 1–4 on pages 375–376 of your text.

Perform each division.

1. $\frac{16a^5-24a^3}{8a^2}$

1. ____________

2. $\frac{12z^5+28z^4-8z^3+3z}{4z^3}$

2. ____________

3. $\frac{39m^4-12m^3+15}{-3m^2}$

3. ____________

Name: Date:
Instructor: Section:

Objective 2 Divide a polynomial by a polynomial.

Video Examples

Review these examples for Objective 2:

5. Divide $\frac{2x^2-11x+15}{x-3}$.

Step 1 $2x^2$ divided by x is $2x$
$2x(x-3)=2x^2-6x$

Step 2 Subtract. Bring down the next term.

Step 3 $-5x$ divided by x is -5.
$-5(x-3)=-5x+15$

Step 4 Subtract. The remainder is 0.

$$\begin{array}{r} 2x\ -5 \\ x-3\overline{)2x^2-11x+15} \\ \underline{2x^2-\ 6x} \\ -5x+15 \\ \underline{-5x+15} \\ 0 \end{array}$$

$$\frac{2x^2-11x+15}{x-3}=2x-5$$

Check $(x-3)(2x-5)=2x^2-5x-6x+15$
$=2x^2-11x+15$

6. Divide $\frac{8x+9x^3-7-9x^2}{3x-1}$.

Write the dividend in descending powers as $9x^3-9x^2+8x-7$.

Step 1 $9x^3$ divided by $3x$ is $3x^2$.
$3x^2(3x-1)=9x^3-3x^2$

Step 2 Subtract. Bring down the next term.

Step 3 $-6x^2$ divided by $3x$ is $-2x$.
$-2x(3x-1)=-6x^2+2x$

Step 4 Subtract. Bring down the next term.

Step 5 $6x$ divided by $3x$ is 2.
$2(3x-1)=6x-2$

Now Try:

5. Divide $\frac{4x^2-5x-6}{x-2}$.

____

6. Divide $\frac{-12x^2+10x^3-3-8x}{5x-1}$.

____

$$\begin{array}{r}
3x^2-2x+2 \\
3x-1\overline{)9x^3-9x^2+8x-7} \\
\underline{9x^3-3x^2} \qquad\quad \\
-6x^2+8x \quad\ \\
\underline{-6x^2+2x} \quad\ \\
6x-7 \\
\underline{6x-2} \\
-5
\end{array}$$

$$\frac{9x^3-9x^2+8x-7}{3x-1}=3x^2-2x+2+\frac{-5}{3x-1}$$

Step 7 Multiply to check.

Check $(3x-1)\left(3x^2-2x+2+\frac{-5}{3x-1}\right)$

$$=(3x-1)(3x^2)+(3x-1)(-2x)$$
$$+(3x-1)(2)+(3x-1)\left(\frac{-5}{3x-1}\right)$$
$$=9x^3-3x^2-6x^2+2x+6x-2-5$$
$$=9x^3-9x^2+8x-7$$

7. Divide x^3-64 by $x-4$.

Here the dividend is missing the x^2-term and the x-term. We use 0 as the coefficient for each missing term.

$$\begin{array}{r}
x^2+4x+16 \\
x-4\overline{)x^3+0x^2+0x-64} \\
\underline{x^3-4x^2} \qquad\qquad \\
4x^2\ +0x \qquad \\
\underline{4x^2-16x} \qquad \\
16x-64 \\
\underline{16x-64} \\
0
\end{array}$$

The remainder is 0. The quotient is $x^2+4x+16$.

Check $(x-4)(x^2+4x+16)$

$$=x^3+4x^2+16x-4x^2-16x-64$$
$$=x^3-64$$

7. Divide x^3-1000 by $x-10$.

____

Name: Date:
Instructor: Section:

8. Divide $x^4 - 3x^3 + 7x^2 - 8x + 14$ by $x^2 + 2$.

Since $x^2 + 2$ is missing the x-term, we write it as $x^2 + 0x + 2$.

$$
\begin{array}{r}
x^2 - 3x + 5 \\
x^2 + 0x + 2 \overline{\smash{)}\, x^4 - 3x^3 + 7x^2 - 8x + 14} \\
\underline{x^4 + 0x^3 + 2x^2 } \\
-3x^3 + 5x^2 - 8x \\
\underline{-3x^3 + 0x^2 - 6x } \\
5x^2 - 2x + 14 \\
\underline{5x^2 + 0x + 10} \\
-2x + 4
\end{array}
$$

The quotient is $x^2 - 3x + 5 + \dfrac{-2x + 4}{x^2 + 2}$

The check shows that the quotient multiplied by the divisor gives the original dividend.

8. Divide $3x^4 + 5x^3 - 7x^2 - 12x + 9$ by $x^2 - 4$.

____

Objective 2 Practice Exercises

For extra help, see Examples 5–9 on pages 378–380 of your text.

Perform each division.

4. $\dfrac{-6x^2 + 23x - 20}{2x - 5}$

4. ______________

5. $\dfrac{6x^4 - 12x^3 + 13x^2 - 5x - 1}{2x^2 + 3}$

5. ______________

Name: Date:
Instructor: Section:

6. $\dfrac{2a^4+5a^2+3}{2a^2+3}$

6. ____________

Objective 3 Use division in a geometry problem.

Video Examples

Review this example for Objective 3:

10. The area of a rectangle is given by $12p^3-7p^2+5p-1$ square units, and the width is $4p-1$ units. What is the length of the rectangle?

For a rectangle, $A=LW$. Solving for L gives $L=\dfrac{A}{W}$. Divide the area, $12p^3-7p^2+5p-1$ by the width $4p-1$.

$$\begin{array}{r} 3p^2-p+1 \\ 4p-1\overline{)12p^3-7p^2+5p-1} \\ \underline{12p^3-3p^2} \\ -4p^2+5p \\ \underline{-4p^2+p} \\ 4p-1 \\ \underline{4p-1} \\ 0 \end{array}$$

The length is $3p^2-p+1$ units.

Now Try:

10. The area of a rectangle is given by $6r^3-5r^2+16r-5$ square units, and the width is $3r-1$ units. What is the length of the rectangle?

____

Name: Date:
Instructor: Section:

Objective 3 Practice Exercises

For extra help, see Example 10 on page 380 of your text.

Work each problem.

7. The area of a parallelogram is given by $4y^3 - 44y - 600$ square units, and the height is $y - 6$ units. What is the base of the parallelogram?

7. ________________

8. The area of a parallelogram is given by $3t^3 + 16t^2 - 32t - 64$ square units, and the base is $t^2 + 4t - 16$ units. What is the height of the parallelogram?

8. ________________

Name: Date:
Instructor: Section:

Chapter 6 FACTORING AND APPLICATIONS

6.1 The Greatest Common Factor; Factoring by Grouping

Learning Objectives
1. Find the greatest common factor of a list of terms.
2. Factor out the greatest common factor.
3. Factor by grouping.

Key Terms

Use the vocabulary terms listed below to complete each statement in exercises 1–4.

factor **factored form** **greatest common factor (GCF)**
factoring

1. The process of writing a polynomial as a product is called ________________.

2. An expression is in ________________________ when it is written as a product.

3. The __________________________ is the largest quantity that is a factor of each of a group of quantities.

4. An expression A is a ___________________ of an expression B if B can be divided by A with 0 remainder.

Objective 1 Find the greatest common factor of a list of terms.

Video Examples

Review these examples for Objective 1:

1. Find the greatest common factor for each list of numbers.

   a. 60, 45

   Write the prime factored form of each number.
   $60 = 2 \cdot 2 \cdot 3 \cdot 5$

   $45 = 3 \cdot 3 \cdot 5$

   Use each prime the least number of times it appears in all the factored forms.

   $\text{GCF} = 3^1 \cdot 5^1 = 15$

Now Try:

1. Find the greatest common factor for each list of numbers.

   a. 18, 36, 42

   ____

Name: Date:
Instructor: Section:

b. 90, 36, 108

Write the prime factored form of each number.

$$90 = 2 \cdot 3 \cdot 3 \cdot 5$$
$$36 = 2 \cdot 2 \cdot 3 \cdot 3$$
$$108 = 2 \cdot 2 \cdot 3 \cdot 3 \cdot 3$$

There is one factor of 2 and two factors of 3.

$$\text{GCF} = 2^1 \cdot 3^2 = 18$$

c. 17, 18, 24

Write the prime factored form of each number.

$$17 = 17$$
$$18 = 2 \cdot 3 \cdot 3$$
$$24 = 2 \cdot 2 \cdot 2 \cdot 3$$

There are no primes common to all three numbers, so the GCF is 1.

2. Find the greatest common factor for the list of terms.

$28m^4,\ 35m^6,\ 49m^9,\ 70m^5$

$$28m^4 = 2 \cdot 2 \cdot 7 \cdot m^4$$
$$35m^6 = 5 \cdot 7 \cdot m^6$$
$$49m^9 = 7 \cdot 7 \cdot m^9$$
$$70m^5 = 2 \cdot 5 \cdot 7 \cdot m^5$$

Then, GCF $= 7m^4$.

b. 32, 40, 72

____

c. 26, 27, 28

____

2. Find the greatest common factor for the list of terms.

$54x^5,\ 48x^7,\ 42x^9,\ 30x^4$

____

Objective 1 Practice Exercises

For extra help, see Examples 1–2 on pages 395–396 of your text.

Find the greatest common factor for each list of terms.

1. 84, 280, 112 **1.** __________

2. $6k^2m^4n^5,\ 8k^3m^7n^4,\ k^4m^8n^7$ **2.** __________

3. $9xy^4,\ 72x^4y^7,\ 27xy^2,\ 108x^2y^5$ **3.** __________

Name: Date:
Instructor: Section:

Objective 2 Factor out the greatest common factor.

Video Examples

Review these examples for Objective 2:

3. Write in factored form by factoring out the greatest common factor.

$12x^5 + 27x^4 - 15x^3$

$\text{GCF} = 3x^3$

$$12x^5 + 27x^4 - 15x^3$$
$$= 3x^3(4x^2) + 3x^3(9x) + 3x^3(-5)$$
$$= 3x^3(4x^2 + 9x - 5)$$

Check Multiply the factored form.

$$3x^3(4x^2 + 9x - 5)$$
$$= 3x^3(4x^2) + 3x^3(9x) + 3x^3(-5)$$
$$= 12x^5 + 27x^4 - 15x^3$$

5. Write in factored form by factoring out the greatest common factor.

a. $x(x+9) + 7(x+9)$

Factor out $x+9$.

$$x(x+9) + 7(x+9) = (x+9)(x+7)$$

b. $a^2(a+6) - 7(a+6)$

Factor out $a+6$.

$$a^2(a+6) - 7(a+6) = (a+6)(a^2 - 7)$$

Now Try:

3. Write in factored form by factoring out the greatest common factor.

$20y^4 - 12y^3 + 4y^2$

____

5. Write in factored form by factoring out the greatest common factor.

a. $y(y+8) + 4(y+8)$

____

b. $z^2(z+5) - 11(z+5)$

____

Objective 2 Practice Exercises

For extra help, see Examples 3–5 on pages 396–398 of your text.

Factor out the greatest common factor or a negative common factor if the coefficient of the term of greatest degree is negative.

4. $20x^2 + 40x^2y - 70xy^2$ **4.** ____________

5. $2a(x-2y) + 9b(x-2y)$ **5.** ____________

6. $26x^8 - 13x^{12} + 52x^{10}$

6. ________________

Objective 3 Factor by grouping.

Video Examples

Review these examples for Objective 3:

6. Factor by grouping.

a. $7by + 28y + b + 4$

Group the terms, then factor each group.

$$7by + 28y + b + 4$$
$$= (7by + 28y) + (b + 4)$$
$$= 7y(b + 4) + 1(b + 4)$$
$$= (b + 4)(7y + 1)$$

Check Use the FOIL method.

$$(b + 4)(7y + 1) = 7by + 1b + 4(7y) + 4$$
$$= 7by + 28y + b + 4$$

b. $3x^2 - 18x + 5xy - 30y$

$$3x^2 - 18x + 5xy - 30y$$
$$= (3x^2 - 18x) + (5xy - 30y)$$
$$= 3x(x - 6) + 5y(x - 6)$$
$$= (x - 6)(3x + 5y)$$

Check by multiplying using the FOIL method.

c. $r^3 + 3r^2 - 5r - 15$

$$r^3 + 3r^2 - 5r - 15$$
$$= (r^3 + 3r^2) + (-5r - 15)$$
$$= r^2(r + 3) - 5(r + 3)$$
$$= (r + 3)(r^2 - 5)$$

Check by multiplying using the FOIL method.

Now Try:

6. Factor by grouping.

a. $36x + 4tx + 9 + t$

____

b. $4x^2 - 28x + 5xy - 35y$

____

c. $x^3 + 7x^2 - 2x - 14$

____

Name: Date:
Instructor: Section:

7. Factor by grouping.

$18x^2 - 20y + 24x - 15xy$

Group the terms, then factor each group.

$18x^2 - 20y + 24x - 15xy$
$= 2(9x^2 - 10y) + 3x(8 - 5y)$

This does not lead to a common factor, so we try rearranging the terms.

$18x^2 - 20y + 24x - 15xy$
$= 18x^2 - 15xy + 24x - 20y$
$= (18x^2 - 15xy) + (24x - 20y)$
$= 3x(6x - 5y) + 4(6x - 5y)$
$= (6x - 5y)(3x + 4)$

Check Use the FOIL method.

$(6x - 5y)(3x + 4)$
$= 18x^2 + 24x - 15xy - 20y$
$= 18x^2 - 20y + 24x - 15xy$

7. Factor by grouping.

$56x^2 + 32x - 21xy - 12y$

Objective 3 Practice Exercises

For extra help, see Examples 6–7 on pages 398–400 of your text.

Factor each polynomial by grouping.

7. $15 - 5x - 3y + xy$

7. ________________

8. $2x^2 - 14xy + xy - 7y^2$

8. ________________

9. $3r^3 - 2r^2s + 3s^2r - 2s^3$

9. ________________

Name: Date:
Instructor: Section:

Chapter 6 FACTORING AND APPLICATIONS

6.2 Factoring Trinomials

Learning Objectives
1 Factor trinomials with coefficient 1 for the second-degree term.
2 Factor such trinomials after factoring out the greatest common factor.

Key Terms

Use the vocabulary terms listed below to complete each statement in exercises 1–3.

prime polynomial **factoring** **greatest common factor**

1. ______________________ is the process of writing a polynomial as a product.

2. The ______________________________ of a polynomial is the greatest term that is a factor of all the terms in the polynomial.

3. A ___________________________ is a polynomial that cannot be factored using only integers.

Objective 1 Factor trinomials with coefficient 1 for the second-degree term.

Video Examples

Review these examples for Objective 1:

1. Factor $m^2+8m+15$.

Look for integers whose product is 15 and whose sum is 8. Only positive signs are needed.

Factors of 15	Sums of Factors
15, 1	$15+1=16$
5, 3	$5+3=8$

From the table, 5 and 3 are the required integers.

$m^2+8m+15$ factors as $(m+5)(m+3)$

Check Use the FOIL method.

$$(m+5)(m+3)=m^2+3m+5m+15$$
$$=m^2+8m+15$$

Now Try:

1. Factor $x^2+11x+24$.

____

Name:
Instructor:

Date:
Section:

2. Factor $x^2 - 11x + 28$.

Look for integers whose product is 28 and whose sum is –11. Since the numbers have a positive product and a negative sum, we consider only pairs of negative integers.

Factors of 28	Sums of Factors
$-28, -1$	$-28 + (-1) = -29$
$-14, -2$	$-14 + (-2) = -16$
$-7, -4$	$-7 + (-4) = -11$

The required integers are –7 and –4.

$x^2 - 11x + 28$ factors as $(x-7)(x-4)$

Check Use the FOIL method.

$$(x-7)(x-4) = x^2 - 4x - 7x + 28$$
$$= x^2 - 11x + 28$$

3. Factor $x^2 + 2x - 15$.

Look for integers whose product is –15 and whose sum is 2. To get a negative product, the pairs of integers must have different signs.

Factors of -15	Sums of Factors
$15, -1$	$15 + (-1) = 14$
$-15, 1$	$-15 + 1 = -14$
$5, -3$	$5 + (-3) = 2$

The required integers are 5 and –3.

$x^2 + 2x - 15$ factors as $(x+5)(x-3)$

Check Use the FOIL method.

$$(x+5)(x-3) = x^2 - 3x + 5x - 15$$
$$= x^2 + 2x - 15$$

5. Factor the trinomial.

$x^2 - 7x + 18$

Look for integers whose product is 18 and whose sum is –7. Since the numbers have a positive product and a negative sum, we consider only pairs of negative integers.

2. Factor $y^2 - 12y + 35$.

____

3. Factor $p^2 + 6p - 27$.

____

5. Factor the trinomial.

$m^2 - 7m + 5$

____

Name: Date:
Instructor: Section:

Factors of 18	Sums of Factors
$-18, -1$	$-18+(-1)=-19$
$-9, -2$	$-9+(-2)=-11$
$-6, -3$	$-6+(-3)=-9$

None of the pairs of integers has a sum of –7.

$x^2-7x+18$ cannot be factored.

It is a prime polynomial.

6. Factor $x^2-6xy-7y^2$.

Here, the coefficient of x in the middle term is $-6y$, so we need to find two expressions whose product is $-7y^2$ and whose sum is $-6y$.

Factors of $-7y^2$	Sums of Factors
$7y, -y$	$7y+(-y)=6y$
$-7y, y$	$-7y+y=-6y$

$x^2-6xy-7y^2$ factors as $(x-7y)(x+y)$

Check Use the FOIL method.

$$(x-7y)(x+y)=x^2+xy-7xy-7y^2$$
$$=x^2-6xy-7y^2$$

6. Factor $p^2-5pq-14q^2$. __________

Objective 1 Practice Exercises

For extra help, see Examples 1–6 on pages 404–406 of your text.

Factor completely. If a polynomial cannot be factored, write prime.

1. r^2+r+3 **1.** __________

2. $x^2-11x+28$ **2.** __________

3. $x^2-8x-33$ **3.** __________

Name: Date:
Instructor: Section:

Objective 2 Factor such trinomials after factoring out the greatest common factor.

Video Examples

Review this example for Objective 2:

7. Factor $5x^5 - 45x^4 + 90x^3$.

There is no second-degree term. Look for a common factor.

$$5x^5 - 45x^4 + 90x^3 = 5x^3(x^2 - 9x + 18)$$

Now factor $x^2 - 9x + 18$. The integers –3 and –6 have a product of 18 and a sum of –9.

$5x^5 - 45x^4 + 90x^3$ factors as $5x^3(x-6)(x-3)$

Check Use the FOIL method.

$$5x^3(x-6)(x-3)$$
$$= 5x^3(x^2 - 3x - 6x + 18)$$
$$= 5x^3(x^2 - 9x + 18)$$
$$= 5x^5 - 45x^4 + 90x^3$$

Now Try:

7. Factor $7x^6 - 49x^5 + 70x^4$.

____

Objective 2 Practice Exercises

For extra help, see Example 7 on page 406 of your text.

Factor completely. If a polynomial cannot be factored, write **prime**.

4. $2n^4 - 16n^3 + 30n^2$ **4.** _______________

5. $2a^3b - 10a^2b^2 + 12ab^3$ **5.** _______________

6. $10k^6 + 70k^5 + 100k^4$ **6.** _______________

Name: Date:
Instructor: Section:

Chapter 6 FACTORING AND APPLICATIONS

6.3 More on Factoring Trinomials

Learning Objectives	
1	Factor trinomials by grouping when the coefficient of the second-degree term is not 1.
2	Factor trinomials using the FOIL method.

Key Terms

Use the vocabulary terms listed below to complete each statement in exercises 1–2.

coefficient **trinomial** **FOIL**

outer product **inner product**

1. In the term $6x^2y$, 6 is the ________________________.

2. A polynomial with three terms is a ________________________.

3. The ________________________ of $(2y-5)(y+8)$ is $-5y$.

4. ______________ is a shortcut method for finding the product of two binomials.

5. The ________________________ of $(2y-5)(y+8)$ is $16y$.

Objective 1 Factor trinomials by grouping when the coefficient of the second-degree term is not 1.

Video Examples

Review these examples for Objective 1:

1. Factor $3x^2+11x+10$.

Look for two positive integers whose product is $3\cdot 10=30$ and whose sum is 11.

The integers are 5 and 6, since $5\cdot 6=30$ and $5+6=11$.

$$\begin{aligned}3x^2+11x+10&=3x^2+5x+6x+10\\&=\left(3x^2+5x\right)+(6x+10)\\&=x(3x+5)+2(3x+5)\\&=(3x+5)(x+2)\end{aligned}$$

Check Multiply $(3x+5)(x+2)$ to obtain $3x^2+11x+10$.

Now Try:

1. Factor $5x^2+17x+6$.

____

Name: Date:
Instructor: Section:

2. Factor each trinomial.

a. $8x^2-2x-1$

We must find two integers with a product of $8(-1)=-8$ and a sum of –2. The integers are –4 and 2. We write the middle term as $-4x+2x$.

$$\begin{aligned}8x^2-2x-1&=8x^2-4x+2x-1\\&=(8x^2-4x)+(2x-1)\\&=4x(2x-1)+1(2x-1)\\&=(2x-1)(4x+1)\end{aligned}$$

Check Multiply $(2x-1)(4x+1)$ to obtain $8x^2-2x-1$.

b. $15z^2+z-2$

Look for two integers whose product is $15(-2)=-30$ and whose sum is 1.

The integers are 6 and –5.

$$\begin{aligned}15z^2+z-2&=15z^2-5z+6z-2\\&=(15z^2-5z)+(6z-2)\\&=5z(3z-1)+2(3z-1)\\&=(3z-1)(5z+2)\end{aligned}$$

Check Multiply $(3z-1)(5z+2)$ to obtain $15z^2+z-2$.

c. $12r^2+5rs-2s^2$

Two integers whose product is $12(-2)=-24$ and whose sum is 5 are 8 and –3. Rewrite the trinomial with four terms.

$$\begin{aligned}12r^2+5rs-2s^2&=12r^2+8rs-3rs-2s^2\\&=(12r^2+8rs)+(-3rs-2s^2)\\&=4r(3r+2s)-s(3r+2s)\\&=(3r+2s)(4r-s)\end{aligned}$$

Check Multiply $(3r+2s)(4r-s)$ to obtain $12r^2+5rs-2s^2$.

2. Factor each trinomial.

a. $14x^2-3x-5$

____

b. $3m^2-m-14$

____

c. $10x^2+xy-3y^2$

____

Name: Date:
Instructor: Section:

3. Factor $100x^5 + 140x^4 - 15x^3$.

Factor out the greatest common factor, $5x^3$.
$100x^5 + 140x^4 - 15x^3 = 5x^3(20x^2 + 28x - 3)$

To factor $20x^2 + 28x - 3$, find two integers whose product is $20(-3) = -60$ and whose sum is 28. Factor 60 into prime factors.

$60 = 2 \cdot 2 \cdot 3 \cdot 5$

Combine the prime factors in pairs using one positive factor and one negative factor to get –60. The factors of 30 and –2 have the correct sum, 28.

$$\begin{aligned} &100x^5 + 140x^4 - 15x^3 \\ &= 5x^3(20x^2 + 28x - 3) \\ &= 5x^3(20x^2 + 30x - 2x - 3) \\ &= 5x^3[(20x^2 + 30x) + (-2x - 3)] \\ &= 5x^3[10x(2x + 3) - 1(2x + 3)] \\ &= 5x^3(2x + 3)(10x - 1) \end{aligned}$$

3. Factor $30x^5 + 87x^4 - 63x^3$.

____

Objective 1 Practice Exercises

For extra help, see Examples 1–3 on page 409–411 of your text.

Factor each trinomial by grouping.

1. $8b^2 + 18b + 9$

1. ____________

2. $7a^2b + 18ab + 8b$

2. ____________

3. $10c^2 - 29ct + 21t^2$

3. ____________

Name: Date:
Instructor: Section:

Objective 2 Factor trinomials using the FOIL method.

Video Examples

Review these examples for Objective 2:

5. Factor $6x^2 + 13x + 7$.

The number 6 has several possible pairs of factors, but 7 has only 1 and 7, or –1 and –7. We choose positive factors since all coefficients in the trinomial are positive.

$(\_\_\_ + 7)(\_\_\_ + 1)$

The possible pairs of $6x^2$ are $6x$ and x, or $3x$ and $2x$.

$(3x + 7)(2x + 1)$

gives middle term $3x + 14x = 17x$. Incorrect.

$(2x + 7)(3x + 1)$

gives middle term $2x + 21x = 23x$. Incorrect.

$(6x + 7)(x + 1)$

gives middle term $6x + 7x = 13x$. Correct.

$6x^2 + 13x + 7$ factors as $(6x + 7)(x + 1)$.

Check. Multiply $(6x + 7)(x + 1)$ to obtain $6x^2 + 13x + 7$.

6. Factor $10x^2 - 19x + 7$.

Since 7 has only 1 and 7 or –1 and –7 as factors, it is better to begin by factoring 7. We need two negative factors because the product of two negative factors is positive and their sum is negative, as required.
We try –1 and –7.

$(\_\_\_ - 1)(\_\_\_ - 7)$

The factors of $10x^2$ are $10x$ and x, or $5x$ and $2x$.

$(10x - 1)(x - 7)$

has middle term $-70x - x = -71x$. Incorrect.

$(5x - 1)(2x - 7)$

has middle term $-35x - 2x = -37x$. Incorrect.

$(2x - 1)(5x - 7)$

has middle term $-14x - 5x = -19x$. Correct.

Thus $10x^2 - 19x + 7$ factors as $(2x - 1)(5x - 7)$.

Now Try:

5. Factor $15x^2 + 26x + 7$.

____

6. Factor $20x^2 - 13x + 2$.

____

Name: Date:
Instructor: Section:

7. Factor $6x^2 - x - 15$.

The integer 6 has several possible pairs of factors, as does –15. Since the constant term is negative, one positive factor and one negative factor of –15 are needed. Since the coefficient of the middle term is relatively small, it is wise to avoid large factors. We try 3*x* and 2*x* as factors of $6x^2$ and 5 and –3 as factors of –15.

$(3x+5)(2x-3)$
has middle term $-9x+10x=x$. Incorrect.

$(3x-5)(2x+3)$
has middle term $9x-10x=-x$. Correct.

$6x^2 - x - 15$ factors as $(3x-5)(2x+3)$.

8. Factor $18x^2 - 3xy - 28y^2$.

There are several factors of $18x^2$, including
18*x* and *x*, 9*x* and 2*x*, and 6*x* and 3*x*.
There are many possible pairs of factors of $-28y^2$, including
28*y* and –*y*, –28*y* and *y*, 14*y* and –2*y*,
–14*y* and 2*y*, 7*y* and –4*y*, –7*y* and 4*y*.

Once again, since the coefficient of the middle term is relatively small, avoid the larger factors. Try the factors of 6*x* and 3*x*, and 4*y* and –7*y*.

$(6x+4y)(3x-7y)$ Incorrect

The first binomial has a common factor of 2.

$(6x-7y)(3x+4y)$
has middle term $24xy-21xy=3xy$. Incorrect.

Interchange the signs of the last two terms.

$(6x+7y)(3x-4y)$
has middle term $-24xy+21xy=-3xy$. Correct.

Thus, $18x^2 - 3xy - 28y^2$ factors as $(6x+7y)(3x-4y)$

7. Factor $8x^2 + 2x - 21$.

____

8. Factor $24x^2 - 2xy - 15y^2$.

____

Name: Date:
Instructor: Section:

9. Factor the trinomial.

$-105a^3+65a^2-10a$

The common factor is $-5a$. Then use trial and error.

$$-105a^3+65a^2-10a=-5a\left(21a^2-13a+2\right)$$
$$=-5a(3a-1)(7a-2)$$

Check.

$$-5a(3a-1)(7a-2)=-5a\left(21a^2-13a+2\right)$$
$$=-105a^3+65a^2-10a$$

9. Factor the trinomial.

$-18a^3+66a^2-60a$

____

Objective 2 Practice Exercises

For extra help, see Examples 4–9 on pages 411–414 of your text.

Factor each trinomial completely.

4. $8q^2+10q+3$ — **4.** ____________

5. $3a^2+8ab+4b^2$ — **5.** ____________

6. $4c^2+14cd-8d^2$ — **6.** ____________

Name: Date:
Instructor: Section:

Chapter 6 FACTORING AND APPLICATIONS

6.4 Special Factoring Techniques

Learning Objectives
1 Factor a difference of squares.
2 Factor a perfect square trinomial.
3 Factor a difference of cubes.
4 Factor a sum of cubes.

Key Terms

Use the vocabulary terms listed below to complete each statement in exercises 1–2.

perfect square trinomial **difference**

1. A ______________________ is the result of a subtraction.

2. A ______________________ is a trinomial that can be factored as the square of a binomial.

Objective 1 Factor a difference of squares.

Video Examples

Review these examples for Objective 1:

1. Factor the binomial, if possible.

x^2-64

$x^2-64=x^2-8^2=(x+8)(x-8)$

2. Factor each difference of squares.

a. $36x^2-25$

$36x^2-25=(6x)^2-5^2$
$=(6x+5)(6x-5)$

b. $81y^2-49$

$81y^2-49=(9y)^2-7^2$
$=(9y+7)(9y-7)$

Now Try:

1. Factor the binomial, if possible.

z^2-36

____

2. Factor each difference of squares.

a. $4x^2-81$

____

b. $25t^2-49$

Name: Date:
Instructor: Section:

3. Factor completely.

a. $28y^2-175$

$$28y^2-175=7(4y^2-25)$$
$$=7\left[(2y)^2-5^2\right]$$
$$=7(2y+5)(2y-5)$$

b. p^4-81

$$p^4-81=(p^2)^2-9^2$$
$$=(p^2+9)(p^2-9)$$
$$=(p^2+9)(p+3)(p-3)$$

3. Factor completely.

a. $90x^2-490$

____

b. p^4-256

____

Objective 1 Practice Exercises

For extra help, see Examples 1–3 on pages 417–418 of your text.

Factor each binomial completely. If a binomial cannot be factored, write **prime**.

1. x^2-49 1. __________

2. $81x^4-16$ 2. __________

3. $9x^2+16$ 3. __________

Objective 2 Factor a perfect square trinomial.

Video Examples

Review these examples for Objective 2:

4. Factor $x^2+20x+100$.

The terms x^2 and 100 are perfect squares.

$$x^2+20x+100=(x+10)^2$$

Check the middle term. $2(x)(10)=20x$

The trinomial is a perfect square.

Now Try:

4. Factor $p^2+16p+64$.

____

Name: Date:
Instructor: Section:

5. Factor each trinomial.

a. $36y^2 + 42y + 49$

The first and last terms are perfect squares.

$36y^2 = (6y)^2$ and $49 = 7^2$

Twice the product of the first and last terms of the binomial is $2 \cdot (6y)(7) = 84y$, which is not the middle term of $36y^2 + 42y + 49$.

It is a prime polynomial.

b. $128z^3 + 192z^2 + 72z$

$$128z^3 + 192z^2 + 72z$$
$$= 8z(16z^2 + 24z + 9)$$
$$= 8z[(4z)^2 + 2(4z)(3) + 3^2]$$
$$= 8z(4z + 3)^2$$

c. $49m^2 - 70m + 25$

$$49m^2 - 70m + 25$$
$$= (7m)^2 + 2(7m)(-5) + (-5)^2$$
$$= (7m - 5)^2$$

5. Factor each trinomial.

a. $100y^2 - 70y + 49$

____

b. $20x^3 + 100x^2 + 125x$

____

c. $64m^2 + 48m + 9$

____

Objective 2 Practice Exercises

For extra help, see Examples 4–5 on pages 419–420 of your text.

Factor each trinomial completely. It may be necessary to factor out the greatest common factor first.

4. $z^2 - 26z + 169$

4. ______________

5. $9j^2 + 12j + 4$

5. ______________

6. $-12a^2 + 60ab - 75b^2$

6. ______________

Name: Date:
Instructor: Section:

Objective 3 Factor a difference of cubes.

Video Examples

Review these examples for Objective 3:

6. Factor each difference of cubes.

a. $m^3 - 1000$

Use the pattern for a difference of cubes.

$$m^3 - 1000 = m^3 - 10^3$$
$$= (m-10)(m^2 + 10m + 100)$$

b. $64p^3 - 125$

$$64p^3 - 125 = (4p^3) - 5^3$$
$$= (4p-5)\left[(4p)^2 + (4p)(5) + 5^2\right]$$
$$= (4p-5)(16p^2 + 20p + 25)$$

c. $64y^3 + 1000x^6$

$$64y^3 + 1000x^6$$
$$= (4y)^3 + (10x^2)^3$$
$$= (4y + 10x^2)[(4y)^2 - (4y)(10x^2) + (10x^2)^2]$$
$$= (4y + 10x^2)(16y^2 - 40x^2y + 100x^4)$$

Now Try:

6. Factor each difference of cubes.

a. $t^3 - 216$

____

b. $27k^3 - y^3$

____

c. $27x^3 + 343y^6$

____

Objective 3 Practice Exercises

For extra help, see Example 6 on pages 421–422 of your text.

Factor.

7. $8a^3 - 125b^3$ 7. ____________

8. $216x^3 - 8y^3$ 8. ____________

9. $(m+n)^3 - (m-n)^3$ 9. ____________

Name: Date:
Instructor: Section:

Objective 4 Factor a sum of cubes.

Video Examples

Review these examples for Objective 4:

7. Factor each sum of cubes.

a. $k^3 + 1000$

$$k^3 + 1000 = k^3 + 10^3$$
$$= (k + 10)(k^2 - 10k + 100)$$

b. $2m^3 + 250n^3$

$$2m^3 + 250n^3 = 2(m^3 + 125n^3)$$
$$= 2\left[m^3 + (5n)^3\right]$$
$$= 2(m + 5n)\left[m^2 - m(5n) + (5n)^2\right]$$
$$= 2(m + 5n)(m^2 - 5mn + 25n^2)$$

Now Try:

7. Factor each sum of cubes.

a. $216x^3 + 1$

____

b. $6x^3 + 48y^3$

____

Objective 4 Practice Exercises

For extra help, see Example 7 on page 423 of your text.

Factor.

10. $27r^3 + 8s^3$ **10.** ______________

11. $8a^3 + 64b^3$ **11.** ______________

12. $64x^3 + 343y^3$ **12.** ______________

Name: Date:
Instructor: Section:

Chapter 6 FACTORING AND APPLICATIONS

6.5 Solving Quadratic Equations Using the Zero-Factor Property

Learning Objectives
1 Solve quadratic equations using the zero-factor property.
2 Solve other equations using the zero-factor property.

Key Terms

Use the vocabulary terms listed below to complete each statement in exercises 1−3.

quadratic equation **standard form** **double solution**

1. An equation written in the form $ax^2 + bx + c = 0$ is written in the ____________________ of a quadratic equation.

2. Two factors are identical and both lead to the same solution, called a ____________________.

3. An equation that can written in the form $ax^2 + bx + c = 0,$ with $a \neq 0,$ is a ____________________.

Objective 1 Solve quadratic equations using the zero-factor property.

Video Examples

Review these examples for Objective 1:

1. Solve each equation.

a. $(x+9)(5x-6)=0$

By the zero-factor property, either $x+9=0$ or $5x-6=0$.

$$x+9=0 \quad \text{or} \quad 5x-6=0$$
$$x=-9 \quad \text{or} \quad 5x=6$$
$$x=\frac{6}{5}$$

Check:
Let $x=-9$.

$$(x+9)(5x-6)=0$$
$$(-9+9)[5(-9)-6]\stackrel{?}{=}0$$
$$0(-51)\stackrel{?}{=}0$$
$$0=0 \quad \text{True}$$

Now Try:

1. Solve each equation.

a. $(x+12)(4x-7)=0$

____

Name: Date:
Instructor: Section:

Let $x = \frac{6}{5}$.

$$(x+9)(5x-6) = 0$$

$$\left(\frac{6}{5}+9\right)\left[5\left(\frac{6}{5}\right)-6\right] \stackrel{?}{=} 0$$

$$\left(\frac{51}{5}\right)(6-6) \stackrel{?}{=} 0$$

$$0 = 0 \text{ True}$$

Both values check, so the solution set is $\left\{-9, \frac{6}{5}\right\}$.

b. $x(8x-11) = 0$

Use the zero-factor property.

$$x = 0 \quad \text{or} \quad 8x - 11 = 0$$

$$8x = 11$$

$$x = \frac{11}{8}$$

Check these solutions by substituting each in the original equation. The solution set is $\left\{0, \frac{11}{8}\right\}$.

b. $x(6x-11) = 0$

____

3. Solve $8p^2 + 30 = 46p$.

$$8p^2 + 30 = 46p$$

$8p^2 - 46p + 30 = 0$ Standard form

$2(4p^2 - 23p + 15) = 0$ Factor out 2.

$4p^2 - 23p + 15 = 0$ Divide each side by 2

$(4p-3)(p-5) = 0$ Factor.

$4p - 3 = 0$ or $p - 5 = 0$ Zero-factor

$p = \frac{3}{4}$ or $p = 5$ property

The solution set is $\left\{\frac{3}{4}, 5\right\}$.

3. Solve $15p^2 + 36 = 57p$.

____

4. Solve the equation.

$64m^2 - 49 = 0$

$$64m^2 - 49 = 0$$
$$(8m+7)(8m-7) = 0$$
$$8m+7=0 \quad \text{or} \quad 8m-7=0$$
$$8m=-7 \quad \text{or} \quad 8m=7$$
$$m=-\frac{7}{8} \quad \text{or} \quad m=\frac{7}{8}$$

The solution set is $\left\{-\frac{7}{8}, \frac{7}{8}\right\}$.

4. Solve the equation.

$100x^2 - 9 = 0$ ________________

Objective 1 Practice Exercises

For extra help, see Examples 1–5 on pages 430–434 of your text.

Solve each equation and check your solutions.

1. $2x^2 - 3x - 20 = 0$ **1.** ________________

2. $25x^2 = 20x$ **2.** ________________

3. $c(5c+17) = 12$ **3.** ________________

Name: Date:
Instructor: Section:

Objective 2 Solve other equations using the zero-factor property.

Video Examples

Review these examples for Objective 2:

6. Solve each equation.

a. $12z^3 - 3z = 0$

$$12z^3 - 3z = 0$$
$$3z(4z^2 - 1) = 0$$
$$3z(2z+1)(2z-1) = 0$$

By an extension of the zero-factor property, we have

$3z = 0$ or $2z+1=0$ or $2z-1=0$

$z = 0$ or $z = -\frac{1}{2}$ or $z = \frac{1}{2}$

Check by substituting each value in the original equation. The solution set is $\left\{-\frac{1}{2},\ 0,\ \frac{1}{2}\right\}$.

b. $(3x-1)(x^2 - 13x + 36) = 0$

$$(3x-1)(x^2 - 13x + 36) = 0$$
$$(3x-1)(x-4)(x-9) = 0$$

$3x - 1 = 0$ or $x - 4 = 0$ or $x - 9 = 0$

$x = \frac{1}{3}$ or $x = 4$ or $x = 9$

Check by substituting each value in the original equation. The solution set is $\left\{\frac{1}{3},\ 4,\ 9\right\}$.

7. Solve $x(3x-7) = (x-1)^2 + 11$.

$$x(3x-7) = (x-1)^2 + 11$$
$$3x^2 - 7x = x^2 - 2x + 1 + 11$$
$$3x^2 - 7x = x^2 - 2x + 12$$
$$2x^2 - 5x - 12 = 0$$
$$(2x+3)(x-4) = 0$$

$2x + 3 = 0$ or $x - 4 = 0$

$x = -\frac{3}{2}$ or $x = 4$

Check by substituting each value in the original

Now Try:

6. Solve each equation.

a. $3r^3 = 75r$

____

b. $(5x-2)(x^2 - 11x + 18) = 0$

____

7. Solve $x(2x+5) = (x+2)^2 + 8$.

____

equation. The solution set is $\left\{-\frac{3}{2}, 4\right\}$.

Objective 2 Practice Exercises

For extra help, see Examples 6–7 on pages 434–435 of your text.

Solve each equation and check your solutions.

4. $x^3 + 2x^2 - 8x = 0$ **4.** ________________

5. $z^4 + 8z^3 - 9z^2 = 0$ **5.** ________________

6. $(y^2 - 5y + 6)(y^2 - 36) = 0$ **6.** ________________

Name: Date:
Instructor: Section:

Chapter 6 FACTORING AND APPLICATIONS

6.6 Applications of Quadratic Equations

Learning Objectives

1. Solve problems involving geometric figures.
2. Solve problems involving consecutive integers.
3. Solve problems by applying the Pythagorean theorem.
4. Solve problems using given quadratic models.

Key Terms

Use the vocabulary terms listed below to complete each statement in exercises 1–2.

hypotenuse **legs**

1. In a right triangle, the sides that form the right angle are the ______________.
2. The longest side of a right triangle is the ________________________.

Objective 1 Solve problems involving geometric figures.

Video Examples

Review this example for Objective 1:

1. The length of a rectangle is three times its width. If the width was increased by 4 and the length remained the same, the resulting rectangle would have an area of 231 square inches. Find the dimensions of the original rectangle.

   Step 1 Read the problem carefully. Find the dimensions of the original rectangle.

   Step 2 Assign a variable.
   Let x = width.
   $3x$ = length
   $x + 4$ = new width
   $3x$ = length

   Step 3 Write an equation. The area of the rectangle is given by Area = Length × Width.
   Substitute 231 for area, $3x$ for length, and $x + 4$ for width.
   $$231 = 3x(x+4)$$

Now Try:

1. Mr. Fixxall is building a box which will have a volume of 60 cubic meters. The height of the box will be 4 meters, and the length will be 2 meters more than the width. Find the width and length of the box.

   ____

Step 4 Solve.

$$231 = 3x(x+4)$$
$$231 = 3x^2 + 12x$$
$$3x^2 + 12x - 231 = 0$$
$$3(x^2 + 4x - 77) = 0$$
$$x^2 + 4x - 77 = 0$$
$$(x+11)(x-7) = 0$$
$$x+11=0 \quad \text{or} \quad x-7=0$$
$$x=-11 \quad \text{or} \quad x=7$$

Step 5 State the answer. The solutions are –11 and 7. A rectangle cannot have a side of negative length, so we discard –11. The width is 7 inches. The length is $3(7)=21$ inches.

Step 6 Check. The new width is 7 + 4 = 11. The new area of the rectangle is $11(21)=231$ square inches.

Objective 1 Practice Exercises

For extra help, see Example 1 on page 438 of your text.

Solve each problem. Check your answers to be sure they are reasonable.

1. A book is three times as long as it is wide. Find the length and width of the book in inches if its area is numerically 128 more than its perimeter.

1. width____________

length ____________

2. The area of a triangle is 42 square centimeters. The base is 2 centimeters less than twice the height. Find the base and height of the triangle.

2. base____________

height ____________

3. The volume of a box is 192 cubic feet. If the length of the box is 8 feet and the width is 2 feet more than the height, find the height and width of the box.

3. height ___________

width___________

Objective 2 Solve problems involving consecutive integers.

Video Examples

Review this example for Objective 2:

2. The product of the first and second of three consecutive integers is 2 more than 6 times the third integer. Find the integers.

Step 1 Read carefully. Note that the integers are consecutive integers.

Step 2 Assign a variable.
Let x = the first integer.
Then $x + 1$ = the second integer,
and $x + 2$ = the third integer.

Step 3 Write an equation. The product of the first and second integer is 2 more than 6 times the third integer.

$$x(x+1) = 6(x+2)+2$$

Step 4 Solve.

$$x(x+1) = 6(x+2)+2$$
$$x^2 + x = 6x + 14$$
$$x^2 - 5x - 14 = 0$$
$$(x+2)(x-7) = 0$$
$$x+2=0 \quad \text{or} \quad x-7=0$$
$$x=-2 \quad \text{or} \quad x=7$$

Step 5 State the answer. The value –2 and 7 each lead to a distinct answer.
If $x = -2$, then $x + 1 = -1$, and $x + 2 = 0$.
The integers are –2, –1, 0.
If $x = 7$, then $x + 1 = 8$, and $x + 2 = 9$.
The integers are 7, 8, 9.

Now Try:

2. The product of the second and third of three consecutive integers is 2 more than 8 times the first integer. Find the integers.

____

Step 6 Check. The product of the first and second integers must equal 2 more than 6 times the third. Because

$-2(-1)=6(0)+2$ and $7(8)=6(9)+2$

are both true, both sets of consecutive integers satisfy the statement of the problem.

Objective 2 Practice Exercises

For extra help, see Example 2 on page 439 of your text.

Solve each problem.

4. Find all possible pairs of consecutive odd integers whose sum is equal to their product decreased by 47. **4.** ________________

5. Find two consecutive positive even integers whose product is six more than three times its sum. **5.** ________________

6. Find three consecutive positive odd integers such that four times the sum of all three equals 13 more than the product of the smaller two. **6.** ________________

Name: Date:
Instructor: Section:

Objective 3 Solve problems by applying the Pythagorean theorem.

Video Examples

Review this example for Objective 3:

3. Penny and Carla started biking from the same corner. Penny biked east and Carla biked south. When they were 26 miles apart, Carla had biked 14 miles further than Penny. Find the distance each biked.

Step 1 Read carefully. Find the two distances.

Step 2 Assign a variable.
Let x = Penny's distance.
Then $x + 14$ = Carla's distance.

Step 3 Write an equation. Substitute into the Pythagorean theorem.

$$a^2 + b^2 = c^2$$
$$x^2 + (x+14)^2 = 26^2$$

Step 4 Solve.

$$x^2 + x^2 + 28x + 196 = 676$$
$$2x^2 + 28x - 480 = 0$$
$$2(x^2 + 14x - 240) = 0$$
$$x^2 + 14x - 240 = 0$$
$$(x+24)(x-10) = 0$$
$$x + 24 = 0 \quad \text{or} \quad x - 10 = 0$$
$$x = -24 \quad \text{or} \quad x = 10$$

Step 5 State the answer. Since –24 cannot be a distance, 10 is the distance for Penny, and 10 + 14 = 24 is the distance for Carla.

Step 6 Check. Since $10^2 + 24^2 = 26^2$ is true, the answer is correct.

Now Try:

3. A ladder is leaning against a building. The distance from the bottom of the ladder to the building is 8 feet less than the length of the ladder. How high up the side of the building is the top of the ladder if that distance is 4 feet less than the length of the ladder?

____

Objective 3 Practice Exercises

For extra help, see Example 3 on pages 440–441 of your text.

Solve each problem.

7. A field is in the shape of a right triangle. The shorter leg measures 45 meters. The hypotenuse measures 45 meters less than twice the longer the leg. Find the dimensions of the lot.

7. ____________

Name: Date:
Instructor: Section:

8. A train and a car leave a station at the same time, the train traveling due north and the car traveling west. When they are 100 miles apart, the train has traveled 20 miles farther than the car. Find the distance each has traveled.

8. car ______________

train ______________

9. Two ships left a dock at the same time. When they were 25 miles apart, the ship that sailed due south had gone 10 miles less than twice the distance traveled by the ship that sailed due west. Find the distance traveled by the ship that sailed due south.

9. ______________

Objective 4 Solve problems using given quadratic models.

Video Examples

Review this example for Objective 4:

4. Jeff threw a stone straight upward at 46 feet per second from a dock 6 feet above a lake. The height of the stone above the lake t seconds after it is thrown is given by $h=-16t^2+46t+6$. How long will it take for the stone to reach a height of 39 feet?

Substitute 39 for h.

$$39=-16t^2+46t+6$$

Solve for t.

$$16t^2-46t+33=0$$

$$(8t-11)(2t-3)=0$$

$$8t-11=0 \quad \text{or} \quad 2t-3=0$$

$$t=\frac{11}{8} \quad \text{or} \quad t=\frac{3}{2}$$

Since we have found two acceptable answers,

Now Try:

4. A ball is dropped from the roof of a 19.6 meter high building. Its height h (in meters) t seconds later is given by the equation $h=-4.9t^2+19.6$. After how many second is the height 14.7 meters?

____

the stone will be at height of 39 feet twice (once on its way up and once on its way down) —at $\frac{11}{8}$ sec or $\frac{3}{2}$ sec.

Objective 4 Practice Exercises

For extra help, see Examples 4–5 on pages 441–442 of your text.

Solve each problem.

10. If an object is propelled upward from a height of 16 feet with an initial velocity of 48 feet per second, its height h (in feet) t seconds later is given by the equation $h = -16t^2 + 48t + 16$.

(a) After how many seconds is the height 52 feet?

(b) After how many seconds is the height 48 feet?

10. a. ______________

b. ______________

11. A company determines that its daily revenue R (in dollars) for selling x items is modeled by the equation $R = x(150 - x)$. How many items must be sold for its revenue to be $4400?

11. ______________

12. If a ball is batted at an angle of 35°, the distance that the ball travels is given approximately by $D = 0.029v^2 + 0.021v - 1$, where v is the bat speed in miles per hour and D is the distance traveled in feet. Find the distance a batted ball will travel if the ball is batted with a velocity of 90 miles per hour. Round your answer to the nearest whole number.

12. ______________

Name: Date:
Instructor: Section:

Chapter 7 RATIONAL EXPRESSIONS AND APPLICATIONS

7.1 The Fundamental Property of Rational Expressions

Learning Objectives

1 Find the numerical value of a rational expression.
2 Find the values of the variable for which a rational expression is undefined.
3 Write rational expressions in lowest terms.
4 Recognize equivalent forms of rational expressions.

Key Terms

Use the vocabulary terms listed below to complete each statement in exercises 1–2.

rational expression **lowest terms**

1. The quotient of two polynomials with denominator not 0 is called a ______________________.

2. A rational expression is written in ______________________ if the greatest common factor of its numerator and denominator is 1.

Objective 1 Find the numerical value of a rational expression.

Video Examples

Review this example for Objective 1:

1. Find the numerical value of $\frac{4x+8}{3x-6}$ for the value of x.
$x=-2$

$$\frac{4x+8}{3x-6}=\frac{4(-2)+8}{3(-2)-6}=\frac{0}{-12}=0$$

Now Try:

1. Find the numerical value of $\frac{3x-7}{x+5}$ for the value of x.
$x=-7$

____

Objective 1 Practice Exercises

For extra help, see Example 1 on page 458 of your text.

Find the numerical value of each expression when (a) $x=4$ *and (b)* $x=-1$.

1. $\frac{-3x+1}{2x+1}$

1. (a)______________

(b)______________

Name: Date:
Instructor: Section:

2. $\dfrac{2x^2-4}{x^2-2}$

2. (a) ____________

(b) ____________

3. $\dfrac{2x-5}{2-x-x^2}$

3. (a) ____________

(b) ____________

Objective 2 Find the values of the variable for which a rational expression is undefined.

Video Examples

Review these examples for Objective 2:

2. Find any values of the variable for which each rational expression is undefined.

a. $\dfrac{3x+8}{5x+4}$

Step 1 Set the denominator equal to 0.

$$5x+4=0$$

Step 2 Solve.

$$5x=-4$$

$$x=-\frac{4}{5}$$

Step 3 The given expression is undefined for $-\frac{4}{5}$, so $x \neq -\frac{4}{5}$.

b. $\dfrac{3m^2}{m^2-4m-5}$

Set the denominator equal to 0.

$$m^2-4m-5=0$$

$$(m+1)(m-5)=0$$

$$m+1=0 \quad \text{or} \quad m-5=0$$

$$m=-1 \quad \text{or} \quad m=5$$

The given expression is undefined for –1 and 5, so $m \neq -1$, $m \neq 5$.

Now Try:

2. Find any values of the variable for which each rational expression is undefined.

a. $\dfrac{y+6}{7y-1}$

____

b. $\dfrac{15m^2}{m^2-m-20}$

____

Name: Date:
Instructor: Section:

c. $\frac{6r}{r^2+49}$

This denominator will not equal 0 for any value of r. There are no values for which this expression is undefined.

c. $\frac{12t^2}{t^2+100}$

____

Objective 2 Practice Exercises

For extra help, see Example 2 on pages 459–460 of your text.

Find any value(s) of the variable for which each rational expression is undefined. Write answers with $\neq$.

4. $\frac{4x^2}{x+7}$

4. __________

5. $\frac{2x^2}{x^2+4}$

5. __________

6. $\frac{2y-5}{2y^2+4y-16}$

6. __________

Objective 3 Write rational expressions in lowest terms.

Video Examples

Review these examples for Objective 3:

3. Write the expression in lowest terms.

$\frac{15k^3}{3k^4}$

Write k^3 as $k \cdot k \cdot k$ and k^4 as $k \cdot k \cdot k \cdot k$.

$$\frac{15k^3}{3k^4} = \frac{3 \cdot 5 \cdot k \cdot k \cdot k}{3 \cdot k \cdot k \cdot k \cdot k}$$
$$= \frac{5 \cdot (3 \cdot k \cdot k \cdot k)}{k \cdot (3 \cdot k \cdot k \cdot k)}$$
$$= \frac{5}{k}$$

Now Try:

3. Write the expression in lowest terms.

$\frac{12k^5}{4k^8}$

____

Name: Date:
Instructor: Section:

4. Write each rational expression in lowest terms.

a. $\dfrac{6x-18}{5x-15}$

$$\frac{6x-18}{5x-15}=\frac{6(x-3)}{5(x-3)}=\frac{6}{5}$$

b. $\dfrac{m^2+5m-24}{3m^2-5m-12}$

$$\frac{m^2+5m-24}{3m^2-5m-12}=\frac{(m+8)(m-3)}{(3m+4)(m-3)}=\frac{m+8}{3m+4}$$

5. Write $\dfrac{3x-2y}{2y-3x}$ in lowest terms.

Factor −1 from the denominator.

$$\frac{3x-2y}{2y-3x}=\frac{3x-2y}{-1(-2y+3x)}=\frac{3x-2y}{-1(3x-2y)}=-1$$

4. Write each rational expression in lowest terms.

a. $\dfrac{7x-35}{9x-45}$ ____________

b. $\dfrac{m^2-3m-54}{2m^2-15m-27}$ ____________

5. Write $\dfrac{4y-5x}{5x-4y}$ in lowest terms. ____________

Objective 3 Practice Exercises

For extra help, see Examples 3–6 on pages 460–463 of your text.

Write each rational expression in lowest terms. Assume that no values of any variable make any denominator zero.

7. $\dfrac{15ab^3c^9}{-24ab^2c^{10}}$

7. ________________

8. $\dfrac{16-x^2}{2x-8}$

8. ________________

Name: Date:
Instructor: Section:

9. $\dfrac{9x^2-9x-108}{2x-8}$

9. ______________

Objective 4 Recognize equivalent forms of rational expressions.

Video Examples

Review this example for Objective 4:

7. Write four equal forms of the following rational expression.

$-\dfrac{4x+3}{x-8}$

If we apply the negative sign to the numerator, we obtain the first two equivalent forms.

$\dfrac{-(4x+3)}{x-8}$ and $\dfrac{-4x-3}{x-8}$

If we apply the negative sign to the denominator, we obtain the last two equivalent forms.

$\dfrac{4x+3}{-(x-8)}$ and $\dfrac{4x+3}{-x+8}$

Now Try:

7. Write four equal forms of the following rational expression.

$-\dfrac{10x-7}{4x-3}$

____

Objective 4 Practice Exercises

For extra help, see Example 7 on page 464 of your text.

Write four equivalent forms of the following rational expressions. Assume that no values of any variable make any denominator zero.

10. $-\dfrac{4x+5}{3-6x}$

10. ______________

11. $\dfrac{2p-1}{1-4p}$

11. ______________

12. $-\dfrac{2x-3}{x+2}$

12. ______________

Name: Date:
Instructor: Section:

Chapter 7 RATIONAL EXPRESSIONS AND APPLICATIONS

7.2 Multiplying and Dividing Rational Expressions

Learning Objectives

1. Multiply rational expressions.
2. Divide rational expressions.

Key Terms

Use the vocabulary terms listed below to complete each statement in exercises 1–3.

rational expression **reciprocal** **lowest terms**

1. The ______________________ of the expression $\frac{4x-5}{x+2}$ is $\frac{x+2}{4x-5}$.

2. A ______________________________ is the quotient of two polynomials with denominator not 0.

3. A rational expression is written in _________________________ when the numerator and denominator have no common terms.

Objective 1 Multiply rational expressions.

Video Examples

Review these examples for Objective 1:

1. Multiply. Write each answer in lowest terms.

a. $\frac{7}{12}\cdot\frac{3}{14}$

$$\frac{7}{12}\cdot\frac{3}{14}=\frac{7\cdot 3}{12\cdot 14}$$
$$=\frac{7\cdot 3}{2\cdot 2\cdot 3\cdot 2\cdot 7}$$
$$=\frac{1}{8}$$

b. $\frac{9}{x^2}\cdot\frac{x^3}{15}$

$$\frac{9}{x^2}\cdot\frac{x^3}{15}=\frac{9\cdot x^3}{x^2\cdot 15}$$
$$=\frac{3\cdot 3\cdot x\cdot x\cdot x}{3\cdot 5\cdot x\cdot x}$$
$$=\frac{3x}{5}$$

Now Try:

1. Multiply. Write each answer in lowest terms.

a. $\frac{5}{9}\cdot\frac{12}{25}$

____

b. $\frac{8}{x^3}\cdot\frac{x^2}{6}$

____

Name: Date:
Instructor: Section:

2. Multiply. Write the answer in lowest terms.

$$\frac{3x+2y}{5x}\cdot\frac{x^2}{(3x+2y)^2}$$

Multiply numerators, multiply denominators, factor, and identify the common factors.

$$\frac{3x+2y}{5x}\cdot\frac{x^2}{(3x+2y)^2}=\frac{(3x+2y)x^2}{5x(3x+2y)^2}$$

$$=\frac{(3x+2y)\cdot x\cdot x}{5x(3x+2y)(3x+2y)}$$

$$=\frac{x}{5(3x+2y)}$$

2. Multiply. Write the answer in lowest terms.

$$\frac{r-s}{6s}\cdot\frac{s^3}{(r-s)^2}$$

____

3. Multiply. Write the answer in lowest terms.

$$\frac{x^2+10x+9}{x^2-5x}\cdot\frac{x^2+3x-40}{x^2+9x+8}$$

$$\frac{x^2+10x+9}{x^2-5x}\cdot\frac{x^2+3x-40}{x^2+9x+8}$$

$$=\frac{(x^2+10x+9)(x^2+3x-40)}{(x^2-5x)(x^2+9x+8)}$$

$$=\frac{(x+9)(x+1)(x+8)(x-5)}{x(x-5)(x+1)(x+8)}$$

$$=\frac{x+9}{x}$$

The quotients $\frac{x+1}{x+1}$, $\frac{x+8}{x+8}$, and $\frac{x-5}{x-5}$ are all equal to 1, justifying the final product $\frac{x+9}{x}$.

3. Multiply. Write the answer in lowest terms.

$$\frac{7x^2-7}{x^2+x-2}\cdot\frac{5x+10}{x^2+x}$$

____

Objective 1 Practice Exercises

For extra help, see Examples 1–3 on pages 468–469 of your text.

Multiply. Write each answer in lowest terms.

1. $\frac{8m^4n^3}{3}\cdot\frac{5}{4mn^2}$

1. ____________

Name: Date:
Instructor: Section:

2. $\dfrac{m^2-16}{m-3}\cdot\dfrac{9-m^2}{4-m}$

2. ______________

3. $\dfrac{3x+12}{6x-30}\cdot\dfrac{x^2-x-20}{x^2-16}$

3. ______________

Objective 2 Divide rational expressions.

Video Examples

Review these examples for Objective 2:

4. Divide. Write each answer in lowest terms.

a. $\dfrac{7}{9}\div\dfrac{4}{27}$

Multiply the first expression by the reciprocal of the second.

$$\frac{7}{9}\div\frac{4}{27}=\frac{7}{9}\cdot\frac{27}{4}=\frac{7\cdot 27}{9\cdot 4}=\frac{7\cdot 3\cdot 9}{9\cdot 4}=\frac{21}{4}$$

b. $\dfrac{y}{y-5}\div\dfrac{7y}{y+4}$

$$\frac{y}{y-5}\div\frac{7y}{y+4}=\frac{y}{y-5}\cdot\frac{y+4}{7y}=\frac{y(y+4)}{(y-5)(7y)}=\frac{y+4}{7(y-5)}$$

Now Try:

4. Divide. Write each answer in lowest terms.

a. $\dfrac{9}{10}\div\dfrac{3}{25}$

____

b. $\dfrac{y}{y+2}\div\dfrac{6y}{y-2}$

____

Name: Date:
Instructor: Section:

6. Divide. Write the answer in lowest terms.

$$\frac{m^2-16}{(m-5)(m-4)} \div \frac{(m+4)(m-5)}{-6m}$$

Multiply by the reciprocal.

$$\frac{m^2-16}{(m-5)(m-4)} \div \frac{(m+4)(m-5)}{-6m}$$

$$= \frac{m^2-16}{(m-5)(m-4)} \cdot \frac{-6m}{(m+4)(m-5)}$$

$$= \frac{-6m(m^2-16)}{(m-5)(m-4)(m+4)(m-5)}$$

$$= \frac{-6m(m+4)(m-4)}{(m-5)(m-4)(m+4)(m-5)}$$

$$= \frac{-6m}{(m-5)^2}, \text{ or } -\frac{6m}{(m-5)^2}$$

6. Divide. Write the answer in lowest terms.

$$\frac{x^2-49}{(x-7)(x-3)} \div \frac{(x+7)(x-3)}{8x}$$

____

7. Divide. Write the answer in lowest terms.

$$\frac{x^2-25}{x^2-9} \div \frac{3x^2-15x}{3-x}$$

Multiply by the reciprocal.

$$\frac{x^2-25}{x^2-9} \div \frac{3x^2-15x}{3-x}$$

$$= \frac{x^2-25}{x^2-9} \cdot \frac{3-x}{3x^2-15x}$$

$$= \frac{(x^2-25)(3-x)}{(x^2-9)(3x^2-15x)}$$

$$= \frac{(x+5)(x-5)(3-x)}{(x+3)(x-3)(3x)(x-5)}$$

$$= \frac{-1(x+5)}{3x(x+3)}, \text{ or } \frac{-x-5}{3x(x+3)}$$

7. Divide. Write the answer in lowest terms.

$$\frac{m^2-64}{m^2-81} \div \frac{5m^2+40m}{9-m}$$

____

Objective 2 Practice Exercises

For extra help, see Examples 4–7 on pages 470–471 of your text.

Divide. Write each answer in lowest terms.

4. $\frac{b-7}{16} \div \frac{7-b}{8}$

4. ____________

Name: Date:
Instructor: Section:

5. $\dfrac{m^2+2mn+n^2}{m^2+m} \div \dfrac{m^2-n^2}{m^2-1}$

5. ________________

6. $\dfrac{27-3k^2}{3k^2+8k-3} \div \dfrac{k^2-6k+9}{6k^2-19k+3}$

6. ________________

Name: Date:
Instructor: Section:

Chapter 7 RATIONAL EXPRESSIONS AND APPLICATIONS

7.3 Least Common Denominators

Learning Objectives
1 Find the least common denominator for a group of fractions.
2 Write equivalent rational expressions.

Key Terms

Use the vocabulary terms listed below to complete each statement in exercises 1–2.

least common denominator **equivalent expressions**

1. $\frac{24x-8}{9x^2-1}$ and $\frac{8}{3x+1}$ are ______________________________.

2. The simplest expression that is divisible by all denominators is called the ______________________________.

Objective 1 Find the least common denominator for a list of fractions.

Video Examples

Review these examples for Objective 1:

1. Find the LCD for the pair of fractions.

$\frac{1}{35}, \frac{11}{45}$

Step 1 Factor each denominator into prime factors.

$35 = 5 \cdot 7$

$45 = 3 \cdot 3 \cdot 5 = 3^2 \cdot 5$

Step 2 List each denominator the greatest number of times it appears as a factor in any of the denominators.

The factor 3 appears two times, and the factors 5 and 7 each appear once.

Step 3 Multiply to get the LCD.

$$\begin{aligned} \text{LCD} &= 3 \cdot 3 \cdot 5 \cdot 7 \\ &= 3^2 \cdot 5 \cdot 7 \\ &= 315 \end{aligned}$$

Now Try:

1. Find the LCD for the pair of fractions.

$\frac{5}{18}, \frac{13}{24}$

____

Name: Date:
Instructor: Section:

2. Find the LCD for $\frac{9}{28s^3}$ and $\frac{5}{42s^2}$.

Step 1

$28s^3 = 2 \cdot 2 \cdot 7 \cdot s^3$

$42s^2 = 2 \cdot 3 \cdot 7 \cdot s^2$

Step 2

Here s appears three times, 2 appears twice, and 3 and 7 each appear once.

Step 3

$\text{LCD} = 2^2 \cdot 3 \cdot 7 \cdot s^3 = 84s^3$

2. Find the LCD for $\frac{15}{40a^2}$ and $\frac{13}{24a^4}$.

____

3. Find the LCD for the fractions in each list.

a. $\frac{5}{7b}, \frac{8}{b^2 - 5b}$

$7b = 7 \cdot b$

$b^2 - 5b = b(b-5)$

$\text{LCD} = 7 \cdot b(b-5) = 7b(b-5)$

b. $\frac{6}{c^2 - 5c - 6}, \frac{10}{c^2 + 3c - 54}, \frac{4}{c^2 - 12c + 36}$

$c^2 - 5c - 6 = (c-6)(c+1)$

$c^2 + 3c - 54 = (c-6)(c+9)$

$c^2 - 12c + 36 = (c-6)^2$

Use each factor the greatest number of times it appears as a factor.

$\text{LCD} = (c+1)(c+9)(c-6)^2$

c. $\frac{1}{a-8}, \frac{7}{8-a}$

$-(a-8) = -a + 8 = 8 - a$

Therefore either $8-a$ or $a-8$ can be used as the LCD.

3. Find the LCD for the fractions in each list.

a. $\frac{7}{9w}, \frac{13}{w^2 - 2w}$

____

b. $\frac{12}{b^2 - 16}, \frac{6}{b^2 - 3b - 4}, \frac{9}{b^2 - 8b + 16}$

____

c. $\frac{13}{p-14}, \frac{12}{14-p}$

____

Objective 1 Practice Exercises

For extra help, see Examples 1–3 on pages 475–476 of your text.

Find the least common denominator for each list of rational expressions.

1. $\frac{13}{36b^4}, \frac{17}{27b^2}$

1. __________

Name: Date:
Instructor: Section:

2. $\dfrac{-7}{a^2-2a}, \dfrac{3a}{2a^2+a-10}$ **2.** ____________

3. $\dfrac{8}{w^3-9w}, \dfrac{4w}{w^2+w-6}$ **3.** ____________

Objective 2 Write equivalent rational expressions.

Video Examples

Review these examples for Objective 2:

4. Write the rational expression as an equivalent expression with the indicated denominator.

$$\frac{7}{9}=\frac{?}{45}$$

Step 1 Factor both denominators.

$$\frac{7}{9}=\frac{?}{5\cdot 9}$$

Step 2 A factor of 5 is missing.

Step 3 Multiply $\frac{7}{9}$ by $\frac{5}{5}$.

$$\frac{7}{9}=\frac{7}{9}\cdot\frac{5}{5}=\frac{35}{45}$$

5. Write each rational expression as an equivalent expression with the indicated denominator.

a. $\dfrac{9}{5x+2}=\dfrac{?}{15x+6}$

Factor the denominator on the right.

$$\frac{9}{5x+2}=\frac{?}{3(5x+2)}$$

The missing factor is 3, so multiply by $\frac{3}{3}$.

$$\frac{9}{5x+2}\cdot\frac{3}{3}=\frac{27}{15x+6}$$

Now Try:

4. Write the rational expression as an equivalent expression with the indicated denominator.

$$\frac{13}{6}=\frac{?}{30}$$

____

5. Write each rational expression as an equivalent expression with the indicated denominator.

a. $\dfrac{19}{6c-5}=\dfrac{?}{24c-20}$

____

Name: Date:
Instructor: Section:

b. $\dfrac{7}{q^2+6q}=\dfrac{?}{q^3+q^2-30q}$

Factor the denominator in each rational expression.

$$\frac{7}{q(q+6)}=\frac{?}{q(q+6)(q-5)}$$

The missing factor is $(q-5)$, so multiply by $\dfrac{q-5}{q-5}$.

$$\frac{7}{q(q+6)}=\frac{7}{q(q+6)}\cdot\frac{q-5}{q-5}$$
$$=\frac{7(q-5)}{q^3-q^2-30q}$$
$$=\frac{7q-35}{q^3-q^2-30q}$$

b. $\dfrac{3}{z^2-7z}=\dfrac{?}{z^3-5z^2-14z}$

____

Objective 2 Practice Exercises

For extra help, see Examples 4–5 on pages 476–477 of your text.

Rewrite each rational expression with the indicated denominator. Give the numerator of the new fraction.

4. $\dfrac{5a}{8a-3}=\dfrac{?}{6-16a}$

4. ____________

5. $\dfrac{3}{5r-10}=\dfrac{?}{50r^2-100r}$

5. ____________

6. $\dfrac{3}{k^2+3k}=\dfrac{?}{k^3+10k^2+21k}$

6. ____________

Name: Date:
Instructor: Section:

Chapter 7 RATIONAL EXPRESSIONS AND APPLICATIONS

7.4 Adding and Subtracting Rational Expressions

Learning Objectives
1. Add rational expressions having the same denominator.
2. Add rational expressions having different denominators.
3. Subtract rational expressions.

Key Terms

Use the vocabulary terms listed below to complete each statement in exercises 1–2.

least common multiple **greatest common factor**

1. The ______________________________ of $2m^2 - 5m - 3$ and $2m - 6$ is $m - 3$.

2. The ______________________________ of $2m^2 - 5m - 3$ and $2m - 6$ is $2(m-3)(2m+1)$.

Objective 1 Add rational expressions having the same denominator.

Video Examples

Review these examples for Objective 1:

1. Add. Write each answer in lowest terms.

a. $\frac{1}{8} + \frac{3}{8}$

The denominators are the same, so add the numerators and keep the common denominator.

$$\frac{1}{8} + \frac{3}{8} = \frac{1+3}{8} = \frac{4}{8} = \frac{4 \cdot 1}{4 \cdot 2} = \frac{1}{2}$$

b. $\frac{4x}{x+2} + \frac{8}{x+2}$

$$\frac{4x}{x+2} + \frac{8}{x+2} = \frac{4x+8}{x+2} = \frac{4(x+2)}{x+2} = 4$$

Now Try:

1. Add. Write each answer in lowest terms.

a. $\frac{9}{20} + \frac{7}{20}$

____

b. $\frac{2x^2}{x+4} + \frac{8x}{x+4}$

____

Name: Date:
Instructor: Section:

Objective 1 Practice Exercises

For extra help, see Example 1 on page 480 of your text.

Add. Write each answer in lowest terms.

1. $\dfrac{5}{3w^2}+\dfrac{7}{3w^2}$ **1.** ________________

2. $\dfrac{b}{b^2-4}+\dfrac{2}{b^2-4}$ **2.** ________________

3. $\dfrac{2x+3}{x^2+3x-10}+\dfrac{2-x}{x^2+3x-10}$ **3.** ________________

Objective 2 Add rational expressions having different denominators.

Video Examples

Review these examples for Objective 2:

2. Add. Write each answer in lowest terms.

a. $\dfrac{5}{18}+\dfrac{7}{24}$

Step 1 Find the LCD.

$$18=2\cdot3\cdot3=2\cdot3^2$$

$$24=2\cdot2\cdot2\cdot3=2^3\cdot3$$

$$\text{LCD}=2^3\cdot3^2=72$$

Step 2 Now write each expression as an equivalent expression with the LCD as the denominator.

$$\frac{5}{18}+\frac{7}{24}=\frac{5(4)}{18(4)}+\frac{7(3)}{24(3)}$$

$$=\frac{20}{72}+\frac{21}{72}$$

Step 3 Add the numerators.
Step 4 Write in lowest terms, if necessary.

$$=\frac{20+21}{72}$$

$$=\frac{41}{72}$$

Now Try:

2. Add. Write each answer in lowest terms.

a. $\dfrac{9}{35}+\dfrac{8}{45}$

____

b. $\dfrac{5}{6z}+\dfrac{7}{9z}$

Step 1 Find the LCD.

$6z = 2\cdot 3\cdot z$

$9z = 3\cdot 3\cdot z = 3^2\cdot z$

$\text{LCD} = 2\cdot 3^2\cdot z = 18z$

Step 2 Now write each expression as an equivalent expression with the LCD as the denominator.

$$\frac{5}{6z}+\frac{7}{9z}=\frac{5(3)}{6z(3)}+\frac{7(2)}{9z(2)}$$

$$=\frac{15}{18z}+\frac{14}{18z}$$

Step 3 Add the numerators.

Step 4 Write in lowest terms, if necessary.

$$=\frac{15+14}{18z}$$

$$=\frac{29}{18z}$$

b. $\dfrac{4}{9y}+\dfrac{2}{7y}$

____

4. Add. Write the answer in lowest terms.

$$\frac{3x}{x^2-x-20}+\frac{5}{x^2-2x-15}$$

The LCD is $(x+3)(x+4)(x-5)$.

$$=\frac{3x}{(x+4)(x-5)}+\frac{5}{(x+3)(x-5)}$$

$$=\frac{3x(x+3)}{(x+3)(x+4)(x-5)}+\frac{5(x+4)}{(x+3)(x+4)(x-5)}$$

$$=\frac{3x(x+3)+5(x+4)}{(x+3)(x+4)(x-5)}$$

$$=\frac{3x^2+9x+5x+20}{(x+3)(x+4)(x-5)}$$

$$=\frac{3x^2+14x+20}{(x+3)(x+4)(x-5)}$$

4. Add. Write the answer in lowest terms.

$$\frac{7x}{x^2-2x-8}+\frac{4}{x^2-3x-4}$$

____

Objective 2 Practice Exercises

For extra help, see Examples 2–5 on pages 481–483 of your text.

Add. Write each answer in lowest terms.

4. $\dfrac{7}{x-5}+\dfrac{4}{x+5}$

4. ____________

Name: Date:
Instructor: Section:

5. $\dfrac{3z}{z^2-4}+\dfrac{4z-3}{z^2-4z+4}$

5. ____________

6. $\dfrac{4z}{z^2+6z+8}+\dfrac{2z-1}{z^2+5z+6}$

6. ____________

Objective 3 Subtract rational expressions.

Video Examples

Review these examples for Objective 3:

9. Subtract. Write the answer in lowest terms.

$$\frac{7x}{x^2-4x+4}-\frac{2}{x^2-4}$$

$$=\frac{7x}{(x-2)^2}-\frac{2}{(x+2)(x-2)}$$

The LCD is $(x+2)(x-2)^2$.

$$=\frac{7x}{(x-2)^2}-\frac{2}{(x+2)(x-2)}$$

$$=\frac{7x(x+2)}{(x+2)(x-2)^2}-\frac{2(x-2)}{(x+2)(x-2)^2}$$

$$=\frac{7x(x+2)-2(x-2)}{(x+2)(x-2)^2}$$

$$=\frac{7x^2+14x-2x+4}{(x+2)(x-2)^2}$$

$$=\frac{7x^2+12x+4}{(x+2)(x-2)^2}$$

Now Try:

9. Subtract. Write the answer in lowest terms.

$$\frac{8x}{x^2-10x+25}-\frac{3}{x^2-25}$$

____

Name: Date:
Instructor: Section:

8. Subtract. Write the answer in lowest terms.

$$\frac{5x}{x-7}-\frac{x-42}{7-x}$$

The denominators are opposites.

$$\frac{5x}{x-7}-\frac{x-42}{7-x}=\frac{5x}{x-7}-\frac{(x-42)(-1)}{(7-x)(-1)}$$
$$=\frac{5x}{x-7}-\frac{-x+42}{x-7}$$
$$=\frac{5x+x-42}{x-7}$$
$$=\frac{6x-42}{x-7}$$
$$=\frac{6(x-7)}{x-7}$$
$$=6$$

8. Subtract. Write the answer in lowest terms.

$$\frac{4x}{x-9}-\frac{3x-63}{9-x}$$

____

Objective 3 Practice Exercises

For extra help, see Examples 6–10 on pages 484–486 of your text.

Subtract. Write each answer in lowest terms.

7. $\frac{z+2}{z-2}-\frac{z-2}{z+2}$

7. ____________

8. $\frac{-4}{x^2-4}-\frac{3}{4-2x}$

8. ____________

9. $\frac{m}{m^2-4}-\frac{1-m}{m^2+4m+4}$

9. ____________

Name: Date:
Instructor: Section:

Chapter 7 RATIONAL EXPRESSIONS AND APPLICATIONS

7.5 Complex Fractions

Learning Objectives

1 Define and recognize a complex fraction.
2 Simplify a complex fraction by writing it as a division problem (Method 1).
3 Simplify a complex fraction by multiplying numerator and denominator by the least common denominator (Method 2).

Key Terms

Use the vocabulary terms listed below to complete each statement in exercises 1–2.

complex fraction **LCD**

1. A ________________________ is a rational expression with one or more fractions in the numerator, denominator, or both.

2. To simplify a complex fraction, multiply the numerator and denominator by the ________________________ of all the fractions within the complex fraction.

Objective 1 Define and recognize a complex fraction.

Objective 2 Simplify a complex fraction by writing it as a division problem (Method 1).

Video Examples

Review these examples for Objective 2:

1. Simplify the complex fraction.

$$\frac{8+\frac{4}{x}}{\frac{x}{6}+\frac{1}{12}}$$

Step 1 Write the numerator as a single fraction.

$$8+\frac{4}{x}=\frac{8}{1}+\frac{4}{x}=\frac{8x}{x}+\frac{4}{x}=\frac{8x+4}{x}$$

Do the same with each denominator.

$$\frac{x}{6}+\frac{1}{12}=\frac{x(2)}{6(2)}+\frac{1}{12}=\frac{2x}{12}+\frac{1}{12}=\frac{2x+1}{12}$$

Step 2 Write the equivalent complex fraction as a division problem.

$$\frac{\frac{8x+4}{x}}{\frac{2x+1}{12}}=\frac{8x+4}{x}\div\frac{2x+1}{12}$$

Now Try:

1. Simplify the complex fraction.

$$\frac{9+\frac{3}{x}}{\frac{x}{10}+\frac{1}{30}}$$

____

Step 3 Divide by multiplying by the reciprocal.

$$\frac{8x+4}{x}\div\frac{2x+1}{12}=\frac{8x+4}{x}\cdot\frac{12}{2x+1}$$
$$=\frac{4(2x+1)}{x}\cdot\frac{12}{2x+1}$$
$$=\frac{48}{x}$$

2. Simplify the complex fraction.

$$\frac{\frac{rs^2}{t^3}}{\frac{s^3}{r^3t}}$$

Use the definition of division and then the fundamental property.

$$=\frac{rs^2}{t^3}\div\frac{s^3}{r^3t}$$
$$=\frac{rs^2}{t^3}\cdot\frac{r^3t}{s^3}$$
$$=\frac{r^4}{st^2}$$

2. Simplify the complex fraction.

$$\frac{\frac{a^3b^2}{c}}{\frac{a^5b}{c^3}}$$

____

3. Simplify the complex fraction.

$$\frac{\frac{30}{x+4}-6}{\frac{4}{x+4}+1}=\frac{\frac{30}{x+4}-\frac{6(x+4)}{x+4}}{\frac{4}{x+4}+\frac{1(x+4)}{x+4}}$$
$$=\frac{\frac{30-6(x+4)}{x+4}}{\frac{4+1(x+4)}{x+4}}$$
$$=\frac{\frac{30-6x-24}{x+4}}{\frac{4+x+4}{x+4}}$$
$$=\frac{\frac{6-6x}{x+4}}{\frac{x+8}{x+4}}$$
$$=\frac{6-6x}{x+4}\cdot\frac{x+4}{x+8}$$
$$=\frac{6-6x}{x+8}$$

3. Simplify the complex fraction.

$$\frac{\frac{20}{x-5}-9}{\frac{5}{x-5}+2}$$

____

Name: Date:
Instructor: Section:

Objective 2 Practice Exercises

For extra help, see Examples 1–3 on pages 490–491 of your text.

Simplify each complex fraction by writing it as a division problem.

1. $\dfrac{\dfrac{49m^3}{18n^5}}{\dfrac{21m}{27n^2}}$

1. ________________

2. $\dfrac{\dfrac{p}{2}-\dfrac{1}{3}}{\dfrac{p}{3}+\dfrac{1}{6}}$

2. ________________

3. $\dfrac{3+\dfrac{4}{s}}{2s+\dfrac{2}{3}}$

3. ________________

Objective 3 Simplify a complex fraction by multiplying numerator and denominator by the least common denominator (Method 2).

Video Examples

Review these examples for Objective 3:

4. Simplify the complex fraction.

$$\frac{12+\dfrac{4}{x}}{\dfrac{x}{5}+\dfrac{1}{15}}$$

Step 1 Find the LCD for all the denominators.

The LCD for x, 5, and 15 is $15x$.

Step 2 Multiply the numerator and denominator of the complex fraction by the LCD.

Now Try:

4. Simplify the complex fraction.

$$\frac{4+\dfrac{2}{x}}{\dfrac{x}{3}+\dfrac{1}{6}}$$

____

Name:
Instructor:

Date:
Section:

$$\frac{12+\frac{4}{x}}{\frac{x}{5}+\frac{1}{15}}=\frac{15x\left(12+\frac{4}{x}\right)}{15x\left(\frac{x}{5}+\frac{1}{15}\right)}$$

$$=\frac{180x+60}{3x^2+x}$$

$$=\frac{60(3x+1)}{x(3x+1)}$$

$$=\frac{60}{x}$$

5. Simplify the complex fraction.

$$\frac{\frac{9}{7n}-\frac{3}{n^2}}{\frac{8}{3n}+\frac{5}{6n^2}}$$

The LCD is $42n^2$.

$$=\frac{42n^2\left(\frac{9}{7n}-\frac{3}{n^2}\right)}{42n^2\left(\frac{8}{3n}+\frac{5}{6n^2}\right)}$$

$$=\frac{42n^2\left(\frac{9}{7n}\right)-42n^2\left(\frac{3}{n^2}\right)}{42n^2\left(\frac{8}{3n}\right)+42n^2\left(\frac{5}{6n^2}\right)}$$

$$=\frac{54n-126}{112n+35}, \text{ or } \frac{18(3n-7)}{7(16n+5)}$$

5. Simplify the complex fraction.

$$\frac{\frac{2}{9n}-\frac{2}{5n^2}}{\frac{4}{5n}+\frac{2}{3n^2}}$$

____

Objective 3 Practice Exercises

For extra help, see Examples 4–6 on pages 492–494 of your text.

Simplify each complex fraction by multiplying numerator and denominator by the least common denominator.

4. $$\frac{\frac{9}{x^2}-1}{\frac{3}{x}-1}$$

4. ________________

Name: Date:
Instructor: Section:

5. $\dfrac{\dfrac{x-2}{x+2}}{\dfrac{x}{x-2}}$

5. ________________

6. $\dfrac{\dfrac{6}{k+1}-\dfrac{5}{k-3}}{\dfrac{3}{k-3}+\dfrac{2}{k+2}}$

6. ________________

Name: Date:
Instructor: Section:

Chapter 7 RATIONAL EXPRESSIONS AND APPLICATIONS

7.6 Solving Equations with Rational Expressions

Learning Objectives

1 Distinguish between operations with rational expressions and equations with terms that are rational expressions.
2 Solve equations with rational expressions.
3 Solve a formula for a specified variable.

Key Terms

Use the vocabulary terms listed below to complete each statement in exercises 1–2.

proposed solution **extraneous solution**

1. A solution that is not an actual solution of a given equation is called a(n) ________________________.

2. A value of the variable that appears to be a solution after both sides of an equation with rational expressions are multiplied by a variable expression is called a(n) ____________________ .

Objective 1 Distinguish between operations with rational expressions and equations with terms that are rational expressions.

Video Examples

Review these examples for Objective 1:

1. Identify each of the following as an expression or an equation. Then simplify the expression or solve the equation.

a. $\frac{8}{9}x-\frac{5}{6}x=\frac{2}{3}$

Because there is an equality symbol, this is an equation to be solved. The LCD is 18.

$$18\left(\frac{8}{9}x-\frac{5}{6}x\right)=18\left(\frac{2}{3}\right)$$
$$18\left(\frac{8}{9}x\right)-18\left(\frac{5}{6}x\right)=18\left(\frac{2}{3}\right)$$
$$16x-15x=12$$
$$x=12$$

A check shows that the solution set is $\{12\}$.

Now Try:

1. Identify each of the following as an expression or an equation. Then simplify the expression or solve the equation.

a. $\frac{4}{5}x-\frac{3}{10}x=7$

____

b. $\frac{8}{9}x-\frac{5}{6}x$

This is a difference of two terms. It represents an expression since there is no equality symbol.
The LCD is 18.

$$\frac{8}{9}x-\frac{5}{6}x=\frac{2\cdot 8}{2\cdot 9}x-\frac{3\cdot 5}{3\cdot 6}x$$
$$=\frac{16}{18}x-\frac{15}{18}x$$
$$=\frac{1}{18}x$$

b. $\frac{4}{5}x-\frac{3}{10}x$

____

Objective 1 Practice Exercises

For extra help, see Example 1 on page 498 of your text.

Identify each of the following as an expression or an equation. Then simplify the expression or solve the equation.

1. $\frac{3x}{5}-\frac{4x}{3}=\frac{22}{15}$

1. ______________

2. $\frac{4x}{5}-\frac{5x}{10}$

2. ______________

3. $\frac{2x}{5}+\frac{7x}{3}$

3. ______________

Objective 2 Solve equations with rational expressions.

Video Examples

Review this example for Objective 2:

4. Solve, and check the proposed solution.

$$\frac{x}{x-3}-\frac{3}{x-3}=3$$

Note that x cannot equal 3, since 3 causes both denominators to equal 0.
Multiply by the LCD, $x-3$.

Now Try:

4. Solve, and check the proposed solution.

$$\frac{8x}{x+1}+\frac{8}{x+1}=4$$

____

$$(x-3)\left(\frac{x}{x-3}-\frac{3}{x-3}\right)=(x-3)(3)$$

$$(x-3)\left(\frac{x}{x-3}\right)-(x-3)\left(\frac{3}{x-3}\right)=(x-3)(3)$$

$$x-3=3x-9$$

$$-2x-3=-9$$

$$-2x=-6$$

$$x=3$$

Check $\frac{x}{x-3}-\frac{3}{x-3}=3$

$$\frac{3}{3-3}-\frac{3}{3-3}\stackrel{?}{=}3$$

$$\frac{3}{0}-\frac{3}{0}\stackrel{?}{=}3$$

Division by 0 is undefined.

Thus, the proposed solution 3 must be rejected, and the solution set is $\varnothing$.

7. Solve, and check the proposed solution(s).

$$\frac{6}{x^2-1}=1-\frac{3}{x+1}$$

$x \neq 1,-1$ or a denominator is 0.
Factor the denominator.

$$\frac{6}{(x+1)(x-1)}=1-\frac{3}{x+1}$$

The LCD is $(x+1)(x-1)$.

$$(x+1)(x-1)\frac{6}{(x+1)(x-1)}=(x+1)(x-1)\left(1-\frac{3}{x+1}\right)$$

$$(x+1)(x-1)\frac{6}{(x+1)(x-1)}=(x+1)(x-1)-(x+1)(x-1)\frac{3}{x+1}$$

$$6=(x+1)(x-1)-3(x-1)$$

$$6=x^2-1-3x+3$$

$$0=x^2-3x-4$$

$$0=(x+1)(x-4)$$

$$x+1=0 \quad \text{or} \quad x-4=0$$

$$x=-1 \quad \text{or} \quad x=4$$

Since –1 makes the original denominator equal

7. Solve, and check the proposed solution(s).

$$\frac{x}{x+1}+\frac{4}{x}=\frac{4}{x^2+x}$$

____

0, the proposed solution –1 is an extraneous value.

Check $\frac{6}{x^2-1}=1-\frac{3}{x+1}$

$\frac{6}{4^2-1}\stackrel{?}{=}1-\frac{3}{4+1}$

$\frac{2}{5}=\frac{2}{5}$ True

A check shows that {4} is the solution set.

Objective 2 Practice Exercises

For extra help, see Examples 2–8 on pages 499–504 of your text.

Solve each equation and check your solutions.

4. $\frac{4}{n+2}-\frac{2}{n}=\frac{1}{6}$

4. ________________

5. $\frac{x}{3x+16}=\frac{4}{x}$

5. ________________

6. $\frac{-16}{n^2-8n+12}=\frac{3}{n-2}+\frac{n}{n-6}$

6. ________________

Name: Date:
Instructor: Section:

Objective 3 Solve a formula for a specified variable.

Video Examples

Review these examples for Objective 3:

9. Solve each formula for the specified variable.

a. $r=\frac{s+w}{v}$ for w

Isolate w. Multiply by v.

$$r=\frac{s+w}{v}$$
$$rv=s+w$$
$$rv-s=w$$

Check $r=\frac{s+w}{v}$

$$r\stackrel{?}{=}\frac{s+rv-s}{v}$$
$$r\stackrel{?}{=}\frac{rv}{v}$$
$$r=r \quad \text{True}$$

b. $S=\frac{a_1}{1-r}$ for r

Isolate r. Multiply by $1-r$.

$$S=\frac{a_1}{1-r}$$
$$S(1-r)=a_1$$
$$S-rS=a_1$$
$$-rS=a_1-S$$
$$r=\frac{a_1-S}{-S}, \text{ or } \frac{S-a_1}{S}$$

Now Try:

9. Solve each formula for the specified variable.

a. $a=\frac{b-c}{q}$ for b

____

b. $w=\frac{x}{y+z}$ for y

____

Name: Date:
Instructor: Section:

10. Solve the formula $\frac{1}{p}+\frac{1}{q}=\frac{1}{r}$ for p.

Isolate p, the specified variable. Multiply by the LCD, pqr.

$$pqr\left(\frac{1}{p}+\frac{1}{q}\right)=pqr\left(\frac{1}{r}\right)$$
$$pqr\left(\frac{1}{p}\right)+pqr\left(\frac{1}{q}\right)=pqr\left(\frac{1}{r}\right)$$
$$qr+pr=pq$$
$$qr=pq-pr$$
$$qr=p(q-r)$$
$$\frac{qr}{q-r}=p$$

10. Solve the formula $\frac{1}{x}=\frac{1}{y}+\frac{1}{z}$ for x.

____

Objective 3 Practice Exercises

For extra help, see Examples 9–10 on pages 504–505 of your text.

Solve each formula for the specified variable.

7. $\frac{1}{f}=\frac{1}{d_0}+\frac{1}{d_1}$ for f

7. ____________

8. $m=\frac{y_2-y_1}{x_2-x_1}$ for y_1

8. ____________

9. $A=\frac{2pf}{b(q+1)}$ for q

9. ____________

Name: Date:
Instructor: Section:

Chapter 7 RATIONAL EXPRESSIONS AND APPLICATIONS

7.7 Applications of Rational Expressions

Learning Objectives
1 Solve problems about numbers.
2 Solve problems about distance, rate, and time.
3 Solve problems about work.

Key Terms

Use the vocabulary terms listed below to complete each statement in exercises 1–3.

reciprocal **numerator** **denominator**

1. In the fraction $\frac{x+5}{x-2}$, $x + 5$ is the ______________________.

2. In the fraction $\frac{x+5}{x-2}$, $x - 2$ is the ______________________.

3. The fraction $\frac{x+5}{x-2}$ is the __________________ of the fraction $\frac{x-2}{x+5}$.

Objective 1 Solve problems about numbers.

Video Examples

Review this example for Objective 1:

1. If a certain number is added to the numerator and twice that number is subtracted from the denominator of the fraction $\frac{3}{5}$, the result is equal to 5. Find the number.

Step 1 Read the problem carefully. We are trying to find a number.

Step 2 Assign a variable.
Let x = the number.

Step 3 Write an equation. The fraction $\frac{3+x}{5-2x}$ represents adding the number to the numerator and twice that number is subtracted from the denominator of the fraction $\frac{3}{5}$. The result is equal to 5.

$$\frac{3+x}{5-2x} = 5$$

Now Try:

1. If the same number is added to the numerator and denominator of the fraction $\frac{5}{9}$, the value of the resulting fraction is $\frac{2}{3}$. Find the number.

____

Step 4 Solve. Multiply by the LCD, $5-2x$.

$$(5-2x)\frac{3+x}{5-2x}=(5-2x)5$$
$$3+x=25-10x$$
$$11x=22$$
$$x=2$$

Step 5 State the answer. The number is 2.

Step 6 Check the solution in the original problem. If 2 is added to the numerator, and twice 2 is subtracted from the denominator of $\frac{3}{5}$, the result is $\frac{3+2}{5-2(2)}=\frac{5}{1}$, or 5, as required.

Objective 1 Practice Exercises

For extra help, see Example 1 on page 511 of your text.

Solve each problem. Check your answers to be sure they are reasonable.

1. If three times a number is subtracted from twice its reciprocal, the result is –1. Find the number. **1.** ________________

2. If two times a number is added to one-half of its reciprocal, the result is $\frac{13}{6}$. Find the number. **2.** ________________

3. The denominator of a fraction is 1 less than twice the numerator. If the numerator and the denominator are each increased by 3, the resulting fraction simplifies to $\frac{3}{4}$. Find the original fraction. **3.** ________________

Name: Date:
Instructor: Section:

Objective 2 Solve problems about distance, rate, and time.

Video Examples

Review this example for Objective 2:

2. A boat goes 6 miles per hour in still water. It takes as long to go 40 miles upstream as 80 miles downstream. Find the speed of the current.

Step 1 Read the problem carefully. Find the speed of the current.

Step 2 Assign a variable.
Let x = the speed of the current.
The rate of traveling upstream is $6 - x$.
The rate of traveling downstream is $6 + x$.

	d	r	t
Upstream	40	$6-x$	$\frac{40}{6-x}$
Downstream	80	$6+x$	$\frac{80}{6+x}$

The times are equal.
Step 3 Write an equation.

$$\frac{40}{6-x}=\frac{80}{6+x}$$

Step 4 Solve. The LCD is $(6+x)(6-x)$.

$$(6+x)(6-x)\frac{40}{6-x}=(6+x)(6-x)\frac{80}{6+x}$$

$$40(6+x)=80(6-x)$$

$$240+40x=480-80x$$

$$240+120x=480$$

$$120x=240$$

$$x=2$$

Step 5 State the answer. The speed of the current is 2 miles per hour.

Step 6 Check.

Upstream: $\frac{40}{6-2}=10$ hr

Downstream: $\frac{80}{6+2}=10$ hr

The time upstream is the same as the time downstream, as required.

Now Try:

2. The Cuyahoga River has a current of 2 miles per hour. Ali can paddle 10 miles downstream in the time it takes her to paddle 2 miles upstream. How fast can Ali paddle?

____

Name: Date:
Instructor: Section:

Objective 2 Practice Exercises

For extra help, see Example 2 on pages 512–513 of your text.

Solve each problem.

4. A boat travels 15 miles per hour in still water. The boat travels 20 miles downstream in the same time it takes the boat to travel 10 miles upstream. How fast is the current?

4. ________________

5. A ship goes 120 miles downriver in $2\frac{2}{3}$ hours less than it takes to go the same distance upriver. If the speed of the current is 6 miles per hour, find the speed of the ship.

5. ________________

6. On Saturday, Pablo jogged 6 miles. On Monday, jogging at the same speed, it took him 30 minutes longer to cover 10 miles. How fast did Pablo jog?

6. ________________

Name: Date:
Instructor: Section:

Objective 3 Solve problems about work.

Video Examples

Review this example for Objective 3:

3. Skip can paint a house in 8 hours and Phil can paint a house in 12 hours. How long will it take to paint a house if they work together?

Step 1 Read the problem carefully. Find the time working together.

Step 2 Assign a variable.
Let x = the number of hours working together.

The rate for Skip is $\frac{1}{8}$.

The rate for Phil is $\frac{1}{12}$.

Step 3 Write an equation. The sum of the fractional part for each multiplied by the time working together is the whole job.

$$\frac{1}{8}x+\frac{1}{12}x=1$$

Step 4 Solve. The LCD is 24.

$$24\left(\frac{1}{8}x+\frac{1}{12}x\right)=24(1)$$
$$24\left(\frac{1}{8}x\right)+24\left(\frac{1}{12}x\right)=24$$
$$3x+2x=24$$
$$5x=24$$
$$x=\frac{24}{5}=4\frac{4}{5}$$

Step 5 State the answer. Working together, it takes $4\frac{4}{5}$ hours to paint the house.

Step 6 Check. Substitute $\frac{24}{5}$ for x in the equation from Step 3.

$$\frac{1}{8}x+\frac{1}{12}x=1$$
$$\frac{1}{8}\left(\frac{24}{5}\right)+\frac{1}{12}\left(\frac{24}{5}\right)=1$$
$$\frac{3}{5}+\frac{2}{5}=1 \quad \text{True}$$

Now Try:

3. Chuck can weed the garden in $\frac{1}{2}$ hour, but David takes 2 hours. How long does it take them to weed the garden if they work together?

____

Name: Date:
Instructor: Section:

Objective 3 Practice Exercises

For extra help, see Example 3 on pages 513–514 of your text.

Solve each problem.

7. Kelly can clean the house in 6 hours, but it takes Linda 4 hours. How long would it take them to clean the house if they worked together?

7. ________________

8. Michael can type twice as fast as Sharon. Together they can type a certain job in 2 hours. How long would it take Michael to type the entire job by himself?

8. ________________

Name: Date:
Instructor: Section:

Chapter 7 RATIONAL EXPRESSIONS AND APPLICATIONS

7.8 Variation

Learning Objectives
1 Solve direct variation problems.
2 Solve inverse variation problems.

Key Terms

Use the vocabulary terms listed below to complete each statement in exercises 1–3.

direct variation **constant of variation** **inverse variation**

1. In the equation $y = kx$, the number k is called the ______________________.

2. If two positive quantities x and y are in ______________________ and the constant of variation is positive, then as x increases, y also increases.

3. If two positive quantities x and y are in ______________________ and the constant of variation is positive, then as x increases, y decreases.

Objective 1 Solve direct variation problems.

Video Examples

Review these examples for Objective 1:

1. If w varies directly as v, and $w = 24$ when $v = 20$, find w when $v = 25$.

Step 1 Since w varies directly as v, there is a constant k such that $w = kv$.

Step 2 We let $w = 24$, $v = 20$, and solve for k.

$$w = kv$$
$$24 = k \cdot 20$$
$$\frac{6}{5} = k$$

Step 3 Since $w = kv$ and $k = \frac{6}{5}$, we have

$$w = \frac{6}{5}v.$$

Step 4 Now we can find the value of w when $v = 25$.

$$w = \frac{6}{5} \cdot 25 = 30$$

Thus, $w = 30$ when $v = 25$.

Now Try:

1. If a varies directly as b, and $a = 61.5$ when $b = 82$, find a when $b = 224$.

____

Name: Date:
Instructor: Section:

2. The force required to compress a spring varies directly as the change in the length of the spring. If a force of 25 pounds is required to compress a spring 4 inches, how much force is required to compress the spring 8 inches?

Step 1 If F represents the force and l represents the length of the spring, then there is a constant k such that $F = kl$.

Step 2 We let $F = 25$, $l = 4$, and solve for k.

$$F = kl$$
$$25 = k \cdot 4$$
$$\frac{25}{4} = k$$

Step 3 Since $F = kl$ and $k = \frac{25}{4}$, we have $F = \frac{25}{4} l$.

Step 4 Now we can find the value of F when $l = 8$.

$$F = \frac{25}{4} \cdot 8 = 50$$

Thus, $F = 50$ lbs when $l = 8$.

2. For a given height, the area of a triangle varies directly as its base. Find the area of a triangle with a base of 4 centimeters, if the area is 9.6 square centimeters when the base is 3 centimeters.

____

Objective 1 Practice Exercises

For extra help, see Examples 1–2 on pages 519–520 of your text.

Solve each problem involving direct variation.

1. If y varies directly as x, and $x = 14$ when $y = 42$, find y when $x = 4$.

1. __________

2. If c varies directly as d, and $c = 100$ when $d = 5$, find c when $d = 3$.

2. __________

3. For a given rate, the distance that an object travels varies directly with time. Find the distance an object travels in 5 hours if the object travels 165 miles in 3 hours.

3. __________

Name: Date:
Instructor: Section:

Objective 2 Solve inverse variation problems.

Video Examples

Review these examples for Objective 2:

3. If g varies inversely as f, and $g = 6$ when $f = 12$, find g when $f = 18$.

Since g varies inversely as f, there is a constant k such that $g = \frac{k}{f}$. We know that $g = 6$ when $f = 12$, so we can find k.

$$g = \frac{k}{f}$$

$$6 = \frac{k}{12}$$

$$72 = k$$

Since $g = \frac{72}{f}$, we let $f = 18$ and solve for g.

$$g = \frac{72}{f} = \frac{72}{18} = 4$$

Therefore, when $f = 18$, $g = 4$.

Now Try:

3. If y varies inversely as x, and $y = 10$ when $x = 2$, find y when $x = 4$.

____

4. For a specified distance, time varies inversely with speed. If Ramona walks a certain distance on a treadmill in 40 minutes at 4.2 miles per hour, how long will it take her to walk the same distance at 3.5 miles per hour?

Let t = time and s = speed.
Since t varies inversely as s, there is a constant k such that $t = \frac{k}{s}$. Recall that $40 \text{ min} = \frac{40}{60} \text{ hr}$.

$$t = \frac{k}{s}$$

$$\frac{40}{60} = \frac{k}{4.2}$$

$$2.8 = k$$

Now use $t = \frac{k}{s}$ to find the value of t when $s = 3.5$.

$$t = \frac{2.8}{3.5} = \frac{4}{5}$$

It takes $\frac{4}{5}$ hr, or 48 min to walk the same distance.

Now Try:

4. If the temperature is constant, the pressure of a gas in a container varies inversely as the volume of the container. If the pressure is 9 pounds per square foot in a container of 6 cubic feet, what is the pressure in a container of 7.5 cubic feet?

____

Name: Date:
Instructor: Section:

Objective 2 Practice Exercises

For extra help, see Examples 3–4 on page 522 of your text.

Solve each problem involving indirect variation.

4. If y varies inversely as x, and $y = 20$ when $x = 4$, find y when $x = 10$. **4.** ______________

5. If n varies inversely as m, and $n = 10.5$ when $m = 1.2$, find n when $m = 5.6$. **5.** ______________

Solve the problem

6. The length of a violin string varies inversely with the frequency of its vibrations. A 10-inch violin string vibrates at a frequency of 512 cycles per second. Find the frequency of an 8-inch string. **6.** ______________

Name: Date:
Instructor: Section:

Chapter 8 ROOTS AND RADICALS

8.1 Evaluating Roots

Learning Objectives
1 Find square roots.
2 Decide whether a given root is rational, irrational, or not a real number.
3 Find decimal approximations for irrational square roots.
4 Use the Pythagorean theorem.
5 Use the distance formula.
6 Find cube, fourth, and other roots.

Key Terms

Use the vocabulary terms listed below to complete each statement in exercises 1–9.

square root	**principal square root**	**radicand**
radical	**radical expression**	**perfect square**
irrational number	**cube root**	**index (order)**

1. The number or expression inside a radical sign is called the ________________.
2. A number with a rational square root is called a ________________.
3. In a radical of the form $\sqrt[n]{a}$, the number n is the ________________.
4. The number b is a ________________ of a if $b^2 = a$.
5. The expression $\sqrt[n]{a}$ is called a ________________.
6. The positive square root of a number is its ________________.
7. A real number that is not rational is called an ________________.
8. A ________________ is a radical sign and the number or expression in it.
9. The number b is a ________________ of a if $b^3 = a$.

Objective 1 Find square roots.

Video Examples

Review these examples for Objective 1:

1. Find the square roots of 64.

   What number multiplied by itself equals 64?

   $8^2 = 64$ and $(-8)^2 = 64$.

   Thus, 64 has two square roots: 8 and –8.

Now Try:

1. Find the square roots of 81.

   ____

Name: Date:
Instructor: Section:

2. Find each square root.

a. $\sqrt{121}$

$11^2 = 121$, so $\sqrt{121} = 11$.

b. $-\sqrt{\frac{16}{25}}$

$-\sqrt{\frac{16}{25}} = -\frac{4}{5}$

3. Find the square of each radical expression.

a. $\sqrt{17}$

The square of $\sqrt{17}$ is $\left(\sqrt{17}\right)^2 = 17$.

b. $\sqrt{w^2+3}$

$\left(\sqrt{w^2+3}\right)^2 = w^2+3$

2. Find each square root.

a. $\sqrt{169}$ __________

b. $-\sqrt{\frac{9}{49}}$ __________

3. Find the square of each radical expression.

a. $\sqrt{19}$ __________

b. $\sqrt{n^2+5}$ __________

Objective 1 Practice Exercises

For extra help, see Examples 1–3 on pages 536–537 of your text.

Find all square roots of each number.

1. 625 **1.** __________

2. $\frac{121}{196}$ **2.** __________

Find the square root.

3. $\sqrt{\frac{900}{49}}$ **3.** __________

Name: Date:
Instructor: Section:

Objective 2 Decide whether a given root is rational, irrational, or not a real number.

Video Examples

Review these examples for Objective 2:

4. Tell whether each square root is rational, irrational, or not a real number.

a. $\sqrt{81}$

81 is a perfect square, 9^2, so $\sqrt{81}=9$ is a rational number.

b. $\sqrt{5}$

Because 5 is not a perfect square, $\sqrt{5}$ is irrational.

c. $\sqrt{-16}$

There is no real number whose square is –16. Therefore, $\sqrt{-16}$ is not a real number.

Now Try:

4. Tell whether each square root is rational, irrational, or not a real number.

a. $\sqrt{100}$

____

b. $\sqrt{11}$

____

c. $\sqrt{-9}$

____

Objective 2 Practice Exercises

For extra help, see Example 4 on page 538 of your text.

Tell whether each square root is rational, irrational, *or* not a real number.

4. $\sqrt{72}$ — **4.** ____________

5. $\sqrt{-36}$ — **5.** ____________

6. $\sqrt{6400}$ — **6.** ____________

Objective 3 Find decimal approximations for irrational square roots.

Video Examples

Review this example for Objective 3:

5. Find a decimal approximation for the square root. Round answers to the nearest thousandth.

$-\sqrt{596}$

$-\sqrt{596}\approx -24.413$

Now Try:

5. Find a decimal approximation for the square root. Round answers to the nearest thousandth.

$-\sqrt{678}$

____

Name: Date:
Instructor: Section:

Objective 3 Practice Exercises

For extra help, see Example 5 on page 539 of your text.

Use a calculator to find a decimal approximation for each square root. Round answers to the nearest thousandth.

7. $\sqrt{32}$ **7.** ____________

8. $-\sqrt{131}$ **8.** ____________

9. $\sqrt{210}$ **9.** ____________

Objective 4 Use the Pythagorean theorem.

Video Examples

Review these examples for Objective 4:

6. Find the length of the unknown side of the right triangle with sides a, b, and c, where c is the hypotenuse.

$a = 5,\ b = 12$

Use the Pythagorean theorem.

$$a^2 + b^2 = c^2$$
$$5^2 + 12^2 = c^2$$
$$25 + 144 = c^2$$
$$169 = c^2$$

Since the length of a side of a triangle must be a positive number, find the positive square root of 169 to get c.

$$c = \sqrt{169} = 13$$

7. A ladder 25 feet long leans against a wall. The foot of the ladder is 7 feet from the base of the wall. How high up the wall does the top of the ladder rest?

Step 1 Read the problem again.

Step 2 Assign a variable. Let a represent the height of the top of the ladder when measured straight down to the ground.

Now Try:

6. Find the length of the unknown side of the right triangle with sides a, b, and c, where c is the hypotenuse.
$a = 7,\ b = 24$

____

7. Susan started to drive due south at the same time John started to drive due west. John drove 21 miles in the same time that Susan drove 28 miles. How far apart were they at that time? Round to the nearest mile.

____

Name: Date:
Instructor: Section:

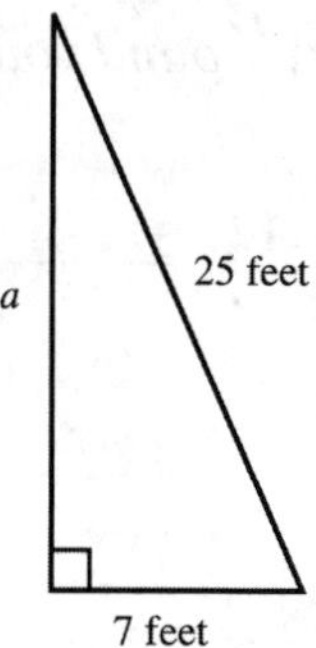

Step 3 Write an equation using the Pythagorean theorem.

$$a^2 + b^2 = c^2$$

$$a^2 + 7^2 = 25^2$$

Step 4 Solve.

$$a^2 + 49 = 625$$

$$a^2 = 576$$ Choose the positive

$$a = 24$$

square root of 576 since *a* represents a length.

Step 5 State the answer. The top of the ladder rests 24 ft up the wall.

Step 6 Check. From the figure, we have the following.

$$24^2 + 7^2 \stackrel{?}{=} 25^2$$

$$576 + 49 = 625$$

The check confirms that the top of the ladder rests 24 ft up the wall.

Objective 4 Practice Exercises

For extra help, see Examples 6–7 on pages 539–540 of your text.

Find the length of the unknown side of each right triangle with sides a, b, and c, where c is the hypotenuse. If necessary, round your answer to the nearest thousandth.

10. $a = 5,\ b = 9$ **10.** ____________

11. $c = 15,\ a = 12$ **11.** ____________

Name: Date:
Instructor: Section:

Use the Pythagorean formula to solve the problem. If necessary, round your answer to the nearest thousandth.

12. A plane flies due east for 35 miles and then due south until it is 37 miles from its starting point. How far south did the plane fly? **12.** ____________

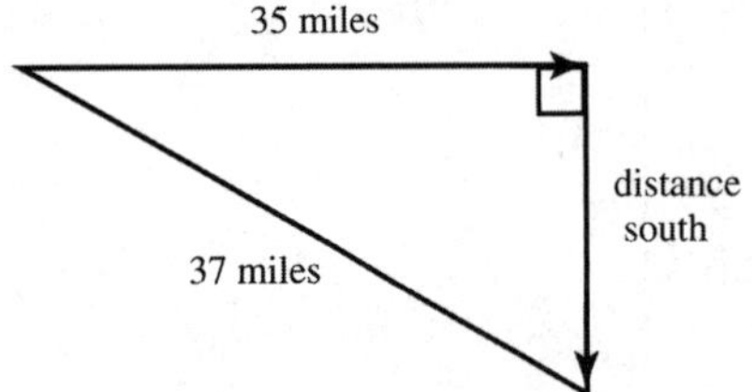

Objective 5 Use the distance formula.

Video Examples

Review this example for Objective 5:

8. Find the distance between (–4, 6) and (1, 8).

$$d=\sqrt{(x_2-x_1)^2+(y_2-y_1)^2}$$
$$=\sqrt{(1-(-4))^2+(8-6)^2}$$
$$=\sqrt{5^2+2^2}$$
$$=\sqrt{29}$$

Now Try:

8. Find the distance between (–2, 2) and (5, 4).

____

Objective 5 Practice Exercises

For extra help, see Example 8 on page 541 of your text.

Find the distance between the given points.

13. (2,–3) and (–1, 2) **13.** ____________

14. (5, 1) and (–6, 2) **14.** ____________

15. (4,–1) and (5,–3) **15.** ____________

Name: Date:
Instructor: Section:

Objective 6 Find cube, fourth, and other roots.

Video Examples

Review these examples for Objective 6:

9. Find each cube root.

a. $\sqrt[3]{729}$

$\sqrt[3]{729} = 9$, because $9^3 = 729$.

b. $\sqrt[3]{-64}$

$\sqrt[3]{-64} = -4$, because $(-4)^3 = -64$.

10. Find each root.

a. $\sqrt[4]{81}$

$\sqrt[4]{81} = 3$, because 3 is positive and $3^4 = 81$.

b. $\sqrt[5]{-1024}$

$\sqrt[5]{-1024} = -4$, because $(-4)^5 = -1024$.

Now Try:

9. Find each cube root.

a. $\sqrt[3]{343}$

____

b. $\sqrt[3]{-125}$

____

10. Find each root.

a. $\sqrt[4]{1296}$

____

b. $\sqrt[5]{-3125}$

____

Objective 6 Practice Exercises

For extra help, see Examples 9–10 on pages 541–542 of your text.

Find each root.

16. $\sqrt[3]{-64}$ **16.** ________

17. $\sqrt[4]{256}$ **17.** ________

18. $\sqrt[7]{-1}$ **18.** ________

Name: Date:
Instructor: Section:

Chapter 8 ROOTS AND RADICALS

8.2 Multiplying, Dividing, and Simplifying Radicals

Learning Objectives
1 Multiply square root radicals.
2 Simplify radicals by using the product rule.
3 Simplify radicals by using the quotient rule.
4 Simplify radicals involving variables.
5 Simplify other roots.

Key Terms

Use the vocabulary terms listed below to complete each statement in exercises 1–3.

perfect cube **radical** **radicand**

1. A root of a number is called a ______________________.

2. A number with a rational cube root is called a ______________________________.

3. The ______________________________ is the number or expression inside a radical sign.

Objective 1 Multiply square root radicals.

Video Examples

Review these examples for Objective 1:

1. Use the product rule for radicals to find each product.

a. $\sqrt{10}\cdot\sqrt{7}$

$\sqrt{10}\cdot\sqrt{7}=\sqrt{10\cdot 7}=\sqrt{70}$

b. $\sqrt{3}\cdot\sqrt{13}$

$\sqrt{3}\cdot\sqrt{13}=\sqrt{39}$

c. $\sqrt{17}\cdot\sqrt{b}\quad (b\geq 0)$

$\sqrt{17}\cdot\sqrt{b}=\sqrt{17b}$

Now Try:

1. Use the product rule for radicals to find each product.

a. $\sqrt{13}\cdot\sqrt{7}$

____

b. $\sqrt{6}\cdot\sqrt{7}$

____

c. $\sqrt{15}\cdot\sqrt{c}\quad (c\geq 0)$

____

Name: Date:
Instructor: Section:

Objective 1 Practice Exercises

For extra help, see Example 1 on page 547 of your text.

Use the product rule for radicals to find each product.

1. $\sqrt{13}\cdot\sqrt{5}$ **1.** ______________

2. $\sqrt{11}\cdot\sqrt{2}$ **2.** ______________

3. $\sqrt{3x}\cdot\sqrt{7},\ x>0$ **3.** ______________

Objective 2 Simplify radicals by using the product rule.

Video Examples

Review these examples for Objective 2:

2. Simplify each radical.

a. $\sqrt{40}$

$$\sqrt{40}=\sqrt{4\cdot 10}$$
$$=\sqrt{4}\cdot\sqrt{10}$$
$$=2\sqrt{10}$$

b. $\sqrt{75}$

$$\sqrt{75}=\sqrt{25\cdot 2}$$
$$=\sqrt{25}\cdot\sqrt{2}$$
$$=5\sqrt{2}$$

c. $\sqrt{48}$

$$\sqrt{48}=\sqrt{16\cdot 3}=\sqrt{16}\cdot\sqrt{3}=4\sqrt{3}$$

3. Find each product and simplify.

a. $\sqrt{16}\cdot\sqrt{20}$

$$\sqrt{16}\cdot\sqrt{20}=4\sqrt{20}$$
$$=4\sqrt{4\cdot 5}$$
$$=4\sqrt{4}\cdot\sqrt{5}$$
$$=4\cdot 2\cdot\sqrt{5}$$
$$=8\sqrt{5}$$

Now Try:

2. Simplify each radical.

a. $\sqrt{12}$

____

b. $\sqrt{98}$

____

c. $\sqrt{80}$

____

3. Find each product and simplify.

a. $\sqrt{25}\cdot\sqrt{18}$

____

b. $\sqrt{27}\cdot\sqrt{50}$

$$\begin{aligned}\sqrt{27}\cdot\sqrt{50} &= \sqrt{27\cdot 50}\\ &= \sqrt{9\cdot 3\cdot 25\cdot 2}\\ &= \sqrt{9}\cdot\sqrt{25}\cdot\sqrt{3\cdot 2}\\ &= 3\cdot 5\cdot\sqrt{6}\\ &= 15\sqrt{6}\end{aligned}$$

b. $\sqrt{6}\cdot\sqrt{12}$

____

Objective 2 Practice Exercises

For extra help, see Examples 2–3 on pages 547–549 of your text.

Simplify the radical.

4. $\sqrt{405}$

4. ______________

Find each product and simplify.

5. $\sqrt{11}\cdot\sqrt{33}$

5. ______________

6. $\sqrt{18}\cdot\sqrt{24}$

6. ______________

Objective 3 Simplify radicals by using the quotient rule.

Video Examples

Review these examples for Objective 3:

4. Use the quotient rule to simplify each radical.

a. $\sqrt{\dfrac{121}{4}}$

$$\sqrt{\frac{121}{4}} = \frac{\sqrt{121}}{\sqrt{4}} = \frac{11}{2}$$

Now Try:

4. Use the quotient rule to simplify each radical.

a. $\sqrt{\dfrac{169}{9}}$

____

Name: Date:
Instructor: Section:

b. $\dfrac{\sqrt{75}}{\sqrt{3}}$

$$\frac{\sqrt{75}}{\sqrt{3}}=\sqrt{\frac{75}{3}}=\sqrt{25}=5$$

c. $\sqrt{\dfrac{5}{16}}$

$$\sqrt{\frac{5}{16}}=\frac{\sqrt{5}}{\sqrt{16}}=\frac{\sqrt{5}}{4}$$

5. Simplify.
$\dfrac{32\sqrt{30}}{8\sqrt{5}}$

$$\frac{32\sqrt{30}}{8\sqrt{5}}=\frac{32}{8}\cdot\frac{\sqrt{30}}{\sqrt{5}}$$
$$=\frac{32}{8}\cdot\sqrt{\frac{30}{5}}$$
$$=4\sqrt{6}$$

b. $\dfrac{\sqrt{405}}{\sqrt{5}}$ __________

c. $\sqrt{\dfrac{7}{64}}$ __________

5. Simplify.
$\dfrac{9\sqrt{30}}{3\sqrt{3}}$ __________

Objective 3 Practice Exercises

For extra help, see Examples 4–6 on pages 549–550 of your text.

Use the quotient rule and product rule, as necessary to simplify each expression.

7. $\sqrt{\dfrac{25}{81}}$ **7.** __________

8. $\dfrac{\sqrt{24}}{2\sqrt{6}}$ **8.** __________

9. $\sqrt{\dfrac{2}{125}}\cdot\sqrt{\dfrac{2}{5}}$ **9.** __________

Name: Date:
Instructor: Section:

Objective 4 Simplify radicals involving variables.

Video Examples

Review these examples for Objective 4:

7. Simplify each radical. Assume that all variables represent nonnegative real numbers.

a. $\sqrt{16m^{10}}$

$\sqrt{16m^{10}} = \sqrt{16} \cdot \sqrt{m^{10}} = 4m^5$

b. $\sqrt{r^7}$

$\sqrt{r^7} = \sqrt{r^6 \cdot r}$

$= \sqrt{r^6}\sqrt{r}$

$= r^3\sqrt{r}$

c. $\sqrt{\dfrac{7}{y^2}}, \quad y \neq 0$

$\sqrt{\dfrac{7}{y^2}} = \dfrac{\sqrt{7}}{\sqrt{y^2}} = \dfrac{\sqrt{7}}{y}$

Now Try:

7. Simplify each radical. Assume that all variables represent nonnegative real numbers.

a. $\sqrt{49x^8}$ ____________

b. $\sqrt{r^{19}}$ ____________

c. $\sqrt{\dfrac{11}{y^4}}$ ____________

Objective 4 Practice Exercises

For extra help, see Example 7 on page 551 of your text.

Simplify each radical. Assume that all variables represent positive real numbers.

10. $\sqrt{p^2q^6}$ **10.** ____________

11. $\sqrt{32x^4y^5}$ **11.** ____________

12. $\sqrt{\dfrac{81}{25x^6}}$ **12.** ____________

Name: Date:
Instructor: Section:

Objective 5 Simplify other roots.

Video Examples

Review these examples for Objective 5:

8. Simplify each radical.

a. $\sqrt[3]{40}$

$\sqrt[3]{40}=\sqrt[3]{8\cdot 5}=\sqrt[3]{8}\cdot\sqrt[3]{5}=2\sqrt[3]{5}$

b. $\sqrt[4]{1250}$

$\sqrt[4]{1250}=\sqrt[4]{625\cdot 2}=\sqrt[4]{625}\cdot\sqrt[4]{2}=5\sqrt[4]{2}$

c. $\sqrt[3]{\frac{64}{27}}$

$\sqrt[3]{\frac{64}{27}}=\frac{\sqrt[3]{64}}{\sqrt[3]{27}}=\frac{4}{3}$

9. Simplify each radical.

a. $\sqrt[4]{m^8}$

$\sqrt[4]{m^8}=m^2$

b. $\sqrt[3]{64x^9}$

$\sqrt[3]{64x^9}=\sqrt[3]{64}\cdot\sqrt[3]{x^9}=4x^3$

c. $\sqrt[3]{40a^5}$

$$\begin{aligned}\sqrt[3]{40a^5}&=\sqrt[3]{8a^3\cdot 5a^2}\\&=\sqrt[3]{8a^3}\cdot\sqrt[3]{5a^2}\\&=2a\sqrt[3]{5a^2}\end{aligned}$$

d. $\sqrt[3]{\frac{y^6}{1000}}$

$\sqrt[3]{\frac{y^6}{1000}}=\frac{\sqrt[3]{y^6}}{\sqrt[3]{1000}}=\frac{y^2}{10}$

Now Try:

8. Simplify each radical.

a. $\sqrt[3]{54}$ __________

b. $\sqrt[4]{162}$ __________

c. $\sqrt[3]{\frac{8}{343}}$ __________

9. Simplify each radical.

a. $\sqrt[4]{m^{12}}$ __________

b. $\sqrt[3]{125x^6}$ __________

c. $\sqrt[4]{80a^5}$ __________

d. $\sqrt[3]{\frac{x^{21}}{216}}$ __________

Name: Date:
Instructor: Section:

Objective 5 Practice Exercises

For extra help, see Examples 8–9 on page 552 of your text.

Simplify each expression.

13. $\sqrt[5]{-64}$ **13.** ________________

14. $\sqrt[3]{\frac{1728}{1000}}$ **14.** ________________

15. $\sqrt[4]{\frac{625}{256}}$ **15.** ________________

Name: Date:
Instructor: Section:

Chapter 8 ROOTS AND RADICALS

8.3 Adding and Subtracting Radicals

Learning Objectives
1 Add and subtract radicals.
2 Simplify radical sums and differences.
3 Simplify more complicated radical expressions.

Key Terms

Use the vocabulary terms listed below to complete each statement in exercises 1−3.

like radicals **index** **unlike radicals**

1. In the expression, $\sqrt[4]{x^2}$, the "4" is called the ______________________.

2. The expressions $2\sqrt{2}$ and $6\sqrt[3]{2}$ are ______________________.

3. The expressions $2\sqrt{2}$ and $7\sqrt{2}$ are ______________________.

Objective 1 Add and subtract radicals.

Video Examples

Review these examples for Objective 1:

1. Add or subtract, as indicated.

a. $7\sqrt{11}+6\sqrt{11}$

$7\sqrt{11}+6\sqrt{11}=(7+6)\sqrt{11}=13\sqrt{11}$

b. $9\sqrt{13}-12\sqrt{13}$

$9\sqrt{13}-12\sqrt{13}=(9-12)\sqrt{13}=-3\sqrt{13}$

c. $\sqrt{5}+\sqrt{10}$

$\sqrt{5}+\sqrt{10}$ cannot be combined. They are unlike radicals.

Now Try:

1. Add or subtract, as indicated.

a. $8\sqrt{21}+2\sqrt{21}$

____

b. $7\sqrt{17}-13\sqrt{17}$

____

c. $\sqrt{10}+\sqrt{15}$

____

Objective 1 Practice Exercises

For extra help, see Example 1 on pages 555–556 of your text.

Add or subtract wherever possible.

1. $5\sqrt{2}+\sqrt{3}$ **1.** ______________

Name: Date:
Instructor: Section:

2. $6\sqrt{3}-2\sqrt{3}+4\sqrt{3}$ **2.** ________________

3. $3\sqrt{5}-9\sqrt{5}+\sqrt{5}$ **3.** ________________

Objective 2 Simplify radical sums and differences.

Video Examples

Review these examples for Objective 2:

2. Add or subtract, as indicated.

a. $4\sqrt{5}+\sqrt{20}$

$$\begin{aligned}4\sqrt{5}+\sqrt{20} &= 4\sqrt{5}+\sqrt{4\cdot 5}\\ &= 4\sqrt{5}+\sqrt{4}\cdot\sqrt{5}\\ &= 4\sqrt{5}+2\sqrt{5}\\ &= 6\sqrt{5}\end{aligned}$$

b. $2\sqrt{28}+8\sqrt{63}$

$$\begin{aligned}2\sqrt{28}+8\sqrt{63} &= 2\left(\sqrt{4}\cdot\sqrt{7}\right)+8\left(\sqrt{9}\cdot\sqrt{7}\right)\\ &= 2\left(2\sqrt{7}\right)+8\left(3\sqrt{7}\right)\\ &= 4\sqrt{7}+24\sqrt{7}\\ &= 28\sqrt{7}\end{aligned}$$

c. $6\sqrt[3]{54}+2\sqrt[3]{3}$

$$\begin{aligned}6\sqrt[3]{54}+2\sqrt[3]{3} &= 6\left(\sqrt[3]{27}\cdot\sqrt[3]{3}\right)+2\sqrt[3]{3}\\ &= 6\left(3\sqrt[3]{3}\right)+2\sqrt[3]{3}\\ &= 18\sqrt[3]{3}+2\sqrt[3]{3}\\ &= 20\sqrt[3]{3}\end{aligned}$$

Now Try:

2. Add or subtract, as indicated.

a. $3\sqrt{6}+\sqrt{150}$

____

b. $3\sqrt{24}+7\sqrt{54}$

____

c. $4\sqrt[3]{128}+7\sqrt[3]{2}$

____

Objective 2 Practice Exercises

For extra help, see Example 2 on page 556 of your text.

Simplify and add or subtract wherever possible.

4. $4\sqrt{128}+2\sqrt{32}$ **4.** ________________

Name: Date:
Instructor: Section:

5. $7\sqrt{162}-9\sqrt{32}$ 5. ______________

6. $5\sqrt{32}-8\sqrt{18}+2\sqrt{20}$ 6. ______________

Objective 3 Simplify more complicated radical expressions.

Video Examples

Review these examples for Objective 3:

3. Simplify each radical expression. Assume that all variables represent nonnegative real numbers.

a. $\sqrt{3}\cdot\sqrt{6}+5\sqrt{2}$

$$\begin{aligned}\sqrt{3}\cdot\sqrt{6}+5\sqrt{2}&=\sqrt{3\cdot 6}+5\sqrt{2}\\&=\sqrt{18}+5\sqrt{2}\\&=\sqrt{9}\cdot\sqrt{2}+5\sqrt{2}\\&=3\sqrt{2}+5\sqrt{2}\\&=8\sqrt{2}\end{aligned}$$

b. $\sqrt{75k}+\sqrt{108k}$

$$\begin{aligned}\sqrt{75k}+\sqrt{108k}&=\sqrt{25\cdot 3k}+\sqrt{36\cdot 3k}\\&=\sqrt{25}\cdot\sqrt{3k}+\sqrt{36}\cdot\sqrt{3k}\\&=5\sqrt{3k}+6\sqrt{3k}\\&=11\sqrt{3k}\end{aligned}$$

c. $7x\sqrt{20}+4\sqrt{5x^2}$

$$\begin{aligned}7x\sqrt{20}+4\sqrt{5x^2}&=7x\sqrt{4\cdot 5}+4\sqrt{5\cdot x^2}\\&=7x\sqrt{4}\cdot\sqrt{5}+4\sqrt{5}\cdot\sqrt{x^2}\\&=7x\cdot 2\sqrt{5}+4x\sqrt{5}\\&=14x\sqrt{5}+4x\sqrt{5}\\&=18x\sqrt{5}\end{aligned}$$

Now Try:

3. Simplify each radical expression. Assume that all variables represent nonnegative real numbers.

a. $\sqrt{3}\cdot\sqrt{15}+7\sqrt{5}$ ______________

b. $\sqrt{28k}+\sqrt{63k}$ ______________

c. $6x\sqrt{75}+2\sqrt{3x^2}$ ______________

Name: Date:
Instructor: Section:

d. $8\sqrt[3]{81x^3} - \sqrt[3]{375x^3}$

$$8\sqrt[3]{81x^3} - \sqrt[3]{375x^3}$$
$$= 8\sqrt[3]{(27x^3)(3)} - \sqrt[3]{(125x^3)(3)}$$
$$= 8(3x)\sqrt[3]{3} - 5x\sqrt[3]{3}$$
$$= 24x\sqrt[3]{3} - 5x\sqrt[3]{3}$$
$$= 19x\sqrt[3]{3}$$

d. $7\sqrt[3]{16m^3} - \sqrt[3]{54m^3}$

____

Objective 3 Practice Exercises

For extra help, see Example 3 on page 557 of your text.

Perform the indicated operations. Assume that all variables represent nonnegative real numbers.

7. $\sqrt{5} \cdot \sqrt{7} + 3\sqrt{35}$ **7.** ____________

8. $3\sqrt{125x} - \sqrt{80x} + 2\sqrt{45x}$ **8.** ____________

9. $11\sqrt{5w} \cdot \sqrt{30w} - 8w\sqrt{24}$ **9.** ____________

Name: Date:
Instructor: Section:

Chapter 8 ROOTS AND RADICALS

8.4 Rationalizing the Denominator

Learning Objectives
1. Rationalize denominators with square roots.
2. Write radicals in simplified form.
3. Rationalize denominators with cube roots.

Key Terms

Use the vocabulary terms listed below to complete each statement in exercises 1–3.

rationalizing the denominator **product rule** **quotient rule**

1. The ________________________ states that $\sqrt{a} \cdot \sqrt{b} = \sqrt{ab}$.

2. The process of ________________________ is changing the denominator of a fraction from a radical to an expression not involving a radical.

3. The ________________________ states that $\sqrt{\frac{a}{b}} = \frac{\sqrt{a}}{\sqrt{b}}$.

Objective 1 Rationalize denominators with square roots.

Video Examples

Review these examples for Objective 1:

1. Rationalize each denominator.

a. $\frac{10}{\sqrt{5}}$

$\frac{10}{\sqrt{5}} = \frac{10 \cdot \sqrt{5}}{\sqrt{5} \cdot \sqrt{5}}$ Multiply by $\frac{\sqrt{5}}{\sqrt{5}} = 1$.

$= \frac{10\sqrt{5}}{5}$

$= 2\sqrt{5}$ Write in lowest terms.

b. $\frac{18}{\sqrt{27}}$

$\frac{18}{\sqrt{27}} = \frac{18}{3\sqrt{3}}$

$= \frac{18 \cdot \sqrt{3}}{3\sqrt{3} \cdot \sqrt{3}}$ Multiply by $\frac{\sqrt{3}}{\sqrt{3}} = 1$.

$= \frac{18\sqrt{3}}{3 \cdot 3}$

$= 2\sqrt{3}$

Now Try:

1. Rationalize each denominator.

a. $\frac{14}{\sqrt{7}}$

____

b. $\frac{5}{\sqrt{75}}$

____

Name: Date:
Instructor: Section:

Objective 1 Practice Exercises

For extra help, see Example 1 on page 560 of your text.

Rationalize each denominator.

1. $\dfrac{15}{\sqrt{10}}$ **1.** ________________

2. $\dfrac{6}{\sqrt{28}}$ **2.** ________________

3. $\dfrac{3\sqrt{5}}{\sqrt{125}}$ **3.** ________________

Objective 2 Write radicals in simplified form.

Video Examples

Review these examples for Objective 2:

2. Simplify.

$$\sqrt{\frac{75}{7}}$$

$$\begin{aligned}\sqrt{\frac{75}{7}} &= \frac{\sqrt{75}\cdot\sqrt{7}}{\sqrt{7}\cdot\sqrt{7}}\\ &= \frac{\sqrt{25\cdot 3}\cdot\sqrt{7}}{7}\\ &= \frac{\sqrt{25}\cdot\sqrt{3}\cdot\sqrt{7}}{7}\\ &= \frac{5\sqrt{3}\cdot\sqrt{7}}{7}\\ &= \frac{5\sqrt{21}}{7}\end{aligned}$$

Now Try:

2. Simplify.

$$\sqrt{\frac{7}{12}}$$

____

Name: Date:
Instructor: Section:

3. Simplify.

$$\sqrt{\frac{7}{15}}\cdot\sqrt{\frac{1}{10}}$$

$$\sqrt{\frac{7}{15}}\cdot\sqrt{\frac{1}{10}}=\sqrt{\frac{7}{15}\cdot\frac{1}{10}}$$
$$=\sqrt{\frac{7}{150}}$$
$$=\frac{\sqrt{7}}{\sqrt{150}}$$
$$=\frac{\sqrt{7}}{\sqrt{25}\cdot\sqrt{6}}$$
$$=\frac{\sqrt{7}}{5\sqrt{6}}$$
$$=\frac{\sqrt{7}\cdot\sqrt{6}}{5\sqrt{6}\cdot\sqrt{6}}$$
$$=\frac{\sqrt{42}}{5\cdot 6}$$
$$=\frac{\sqrt{42}}{30}$$

3. Simplify.

$$\sqrt{\frac{1}{3}}\cdot\sqrt{\frac{2}{15}}$$

____

4. Simplify. Assume that all variables represent positive real numbers.

$$\frac{\sqrt{9m}}{\sqrt{n}}$$

$$\frac{\sqrt{9m}}{\sqrt{n}}=\frac{\sqrt{9m}\cdot\sqrt{n}}{\sqrt{n}\cdot\sqrt{n}}$$
$$=\frac{\sqrt{9mn}}{n}$$
$$=\frac{\sqrt{9}\cdot\sqrt{mn}}{n}$$
$$=\frac{3\sqrt{mn}}{n}$$

4. Simplify. Assume that all variables represent positive real numbers.

$$\frac{\sqrt{25a}}{\sqrt{b}}$$

____

Name: Date:
Instructor: Section:

Objective 2 Practice Exercises

For extra help, see Examples 2–4 on pages 561–562 of your text.

Perform the indicated operations and write all answers in simplest form. Rationalize all denominators. Assume that all variables represent positive real numbers.

4. $\sqrt{\frac{3}{2}} \cdot \sqrt{\frac{5}{6}}$ **4.** ______________

5. $\frac{\sqrt{k^2 m^4}}{\sqrt{k^5}}$ **5.** ______________

6. $\sqrt{\frac{5a^2b^3}{6}}$ **6.** ______________

Objective 3 Rationalize denominators with cube roots.

Video Examples

Review these examples for Objective 3:

5. Rationalize each denominator.

a. $\sqrt[3]{\frac{2}{5}}$

$$\sqrt[3]{\frac{2}{5}} = \frac{\sqrt[3]{2} \cdot \sqrt[3]{5 \cdot 5}}{\sqrt[3]{5} \cdot \sqrt[3]{5 \cdot 5}} = \frac{\sqrt[3]{2 \cdot 5 \cdot 5}}{\sqrt[3]{5 \cdot 5 \cdot 5}} = \frac{\sqrt[3]{50}}{5}$$

b. $\frac{\sqrt[3]{7}}{\sqrt[3]{25}}$

$$\frac{\sqrt[3]{7}}{\sqrt[3]{25}} = \frac{\sqrt[3]{7} \cdot \sqrt[3]{5}}{\sqrt[3]{25} \cdot \sqrt[3]{5}} = \frac{\sqrt[3]{35}}{\sqrt[3]{125}} = \frac{\sqrt[3]{35}}{5}$$

Now Try:

5. Rationalize each denominator.

a. $\sqrt[3]{\frac{5}{3}}$

____

b. $\frac{\sqrt[3]{11}}{\sqrt[3]{9}}$

____

Name: Date:
Instructor: Section:

c. $\dfrac{\sqrt[3]{13}}{\sqrt[3]{5x^2}}$

$$\frac{\sqrt[3]{13}}{\sqrt[3]{5x^2}}=\frac{\sqrt[3]{13}\cdot\sqrt[3]{5\cdot5\cdot x}}{\sqrt[3]{5x^2}\cdot\sqrt[3]{5\cdot5\cdot x}}=\frac{\sqrt[3]{325x}}{5x}$$

c. $\dfrac{\sqrt[3]{5}}{\sqrt[3]{4x}}$ ____________

Objective 3 Practice Exercises

For extra help, see Example 5 on pages 562–563 of your text.

Rationalize each denominator. Assume that all variables in the denominator represent nonzero real numbers.

7. $\dfrac{\sqrt[3]{6}}{\sqrt[3]{9}}$ **7.** ____________

8. $\sqrt[3]{\dfrac{8t}{125u}}$ **8.** ____________

9. $\sqrt[3]{\dfrac{5}{49x}}$ **9.** ____________

Name: Date:
Instructor: Section:

Chapter 8 ROOTS AND RADICALS

8.5 More Simplifying and Operations with Radicals

Learning Objectives
1. Simplify products of radical expressions.
2. Use conjugates to rationalize denominators of radical expressions.
3. Write radical expressions with quotients in lowest terms.

Key Terms

Use the vocabulary terms listed below to complete each statement in exercises 1–2.

conjugate **rationalize the denominator**

1. To ______________________________ of $\frac{3}{\sqrt{5}}$, multiply both the numerator and the denominator by $\sqrt{5}$.

2. The ____________________________ of $a + b$ is $a - b$.

Objective 1 Simplify products of radical expressions.

Video Examples

Review these examples for Objective 1:

1. Find each product, and simplify.

a. $\sqrt{7}\left(\sqrt{27}-\sqrt{48}\right)$

$$\begin{aligned}\sqrt{7}\left(\sqrt{27}-\sqrt{48}\right)&=\sqrt{7}\left(3\sqrt{3}-4\sqrt{3}\right)\\&=\sqrt{7}\left(-\sqrt{3}\right)\\&=-\sqrt{7\cdot 3}\\&=-\sqrt{21}\end{aligned}$$

b. $\left(\sqrt{7}+5\sqrt{2}\right)\left(\sqrt{7}-8\sqrt{2}\right)$

Use the FOIL method to multiply.

$$\begin{aligned}&\left(\sqrt{7}+5\sqrt{2}\right)\left(\sqrt{7}-8\sqrt{2}\right)\\&=\sqrt{7}\left(\sqrt{7}\right)+\sqrt{7}\left(-8\sqrt{2}\right)+5\sqrt{2}\left(\sqrt{7}\right)+5\sqrt{2}\left(-8\sqrt{2}\right)\\&=7-8\sqrt{14}+5\sqrt{14}-40\cdot 2\\&=7-3\sqrt{14}-80\\&=-73-3\sqrt{14}\end{aligned}$$

Now Try:

1. Find each product, and simplify.

a. $\sqrt{2}\left(\sqrt{45}-\sqrt{20}\right)$

____

b. $\left(\sqrt{6}+3\sqrt{5}\right)\left(\sqrt{6}-5\sqrt{5}\right)$

____

Name: Date:
Instructor: Section:

c. $(\sqrt{6}+\sqrt{15})(\sqrt{6}-\sqrt{5})$

$$\begin{aligned}&(\sqrt{6}+\sqrt{15})(\sqrt{6}-\sqrt{5})\\&=\sqrt{6}(\sqrt{6})+\sqrt{6}(-\sqrt{5})+\sqrt{15}(\sqrt{6})+\sqrt{15}(-\sqrt{5})\\&=6-\sqrt{30}+\sqrt{90}-\sqrt{75}\\&=6-\sqrt{30}+\sqrt{9}\cdot\sqrt{10}-\sqrt{25}\cdot\sqrt{3}\\&=6-\sqrt{30}+3\sqrt{10}-5\sqrt{3}\end{aligned}$$

c. $(\sqrt{7}+\sqrt{27})(\sqrt{7}-\sqrt{3})$

____

2. Find each product.

a. $(\sqrt{15}-6)^2$

$$\begin{aligned}(\sqrt{15}-6)^2&=(\sqrt{15})^2-2(\sqrt{15})(6)+6^2\\&=15-12\sqrt{15}+36\\&=51-12\sqrt{15}\end{aligned}$$

b. $(7-\sqrt{a})^2$ Assume that $a\neq 0$.

$$\begin{aligned}(7-\sqrt{a})^2&=7^2-2(7)\sqrt{a}+(\sqrt{a})^2\\&=49-14\sqrt{a}+a\end{aligned}$$

2. Find each product.

a. $(\sqrt{11}-5)^2$

____

b. $(12-\sqrt{m})^2$
Assume that $m\neq 0$.

____

3. Find each product.

a. $(7+\sqrt{5})(7-\sqrt{5})$

Recall $(a+b)(a-b)=a^2-b^2$.

$$\begin{aligned}(7+\sqrt{5})(7-\sqrt{5})&=7^2-(\sqrt{5})^2\\&=49-5\\&=44\end{aligned}$$

b. $(\sqrt{a}-\sqrt{10})(\sqrt{a}+\sqrt{10})$ Assume that $a\neq 0$.

$$\begin{aligned}(\sqrt{a}-\sqrt{10})(\sqrt{a}+\sqrt{10})&=(\sqrt{a})^2-(\sqrt{10})^2\\&=a-10\end{aligned}$$

3. Find each product.

a. $(11+\sqrt{4})(11-\sqrt{4})$

____

b. $(\sqrt{b}-\sqrt{20})(\sqrt{b}+\sqrt{20})$
Assume that $b\neq 0$.

____

Objective 1 Practice Exercises

For extra help, see Examples 1–3 on pages 566–568 of your text.

Find each product, and simplify.

1. $\sqrt{7}(2\sqrt{8}-9\sqrt{7})$

1. __________

Name: Date:
Instructor: Section:

2. $(4\sqrt{5}+\sqrt{3})(\sqrt{2}-\sqrt{7})$ 2. ____________

3. $(2\sqrt{3}-5\sqrt{2})(2\sqrt{3}+5\sqrt{2})$ 3. ____________

Objective 2 Use conjugates to rationalize denominators of radical expressions.

Video Examples

Review these examples for Objective 2:

4. Simplify by rationalizing each denominator.

a. $\dfrac{7}{4+\sqrt{7}}$

$$\frac{7}{4+\sqrt{7}}=\frac{7(4-\sqrt{7})}{(4+\sqrt{7})(4-\sqrt{7})}$$
$$=\frac{7(4-\sqrt{7})}{4^2-(\sqrt{7})^2}$$
$$=\frac{7(4-\sqrt{7})}{16-7}, \text{ or } \frac{7(4-\sqrt{7})}{9}$$

b. $\dfrac{8+\sqrt{3}}{\sqrt{3}-4}$

$$\frac{8+\sqrt{3}}{\sqrt{3}-4}=\frac{(8+\sqrt{3})(\sqrt{3}+4)}{(\sqrt{3}-4)(\sqrt{3}+4)}$$
$$=\frac{8\sqrt{3}+32+3+4\sqrt{3}}{3-16}$$
$$=\frac{12\sqrt{3}+35}{-13}$$
$$=\frac{-12\sqrt{3}-35}{13}$$

Now Try:

4. Simplify by rationalizing each denominator.

a. $\dfrac{10}{7+\sqrt{10}}$ ____________

b. $\dfrac{5+\sqrt{6}}{\sqrt{6}-3}$ ____________

Name: Date:
Instructor: Section:

c. $\dfrac{5}{2-\sqrt{z}}$ Assume that $z \neq 4$ and $z \geq 0$.

$$\frac{5}{2-\sqrt{z}} = \frac{5(2+\sqrt{z})}{(2-\sqrt{z})(2+\sqrt{z})}$$
$$= \frac{5(2+\sqrt{z})}{4-z}$$

c. $\dfrac{7}{5-\sqrt{b}}$
Assume that $b \neq 25$ and $b \geq 0$.

____

Objective 2 Practice Exercises

For extra help, see Example 4 on page 569 of your text.

Rationalize each denominator. Write quotients in lowest terms.

4. $\dfrac{\sqrt{2}}{\sqrt{5}-2}$

4. ______________

5. $\dfrac{\sqrt{6}+2}{\sqrt{2}-4}$

5. ______________

6. $\dfrac{\sqrt{5}-2}{\sqrt{3}+2}$

6. ______________

Name: Date:
Instructor: Section:

Objective 3 Write radical expressions with quotients in lowest terms.

Video Examples

Review this example for Objective 3:

5. Write $\frac{5\sqrt{2}+10}{35}$ in lowest terms.

$$\frac{5\sqrt{2}+10}{35}=\frac{5(\sqrt{2}+2)}{5(7)}$$

$$=1\cdot\frac{\sqrt{2}+2}{7} \quad \text{Divide out the common factor; } \tfrac{5}{5}=1$$

$$=\frac{\sqrt{2}+2}{7}$$

Now Try:

5. Write $\frac{4\sqrt{5}+20}{36}$ in lowest terms.

____

Objective 3 Practice Exercises

For extra help, see Example 5 on page 570 of your text.

Write each quotient in lowest terms.

7. $\frac{3+\sqrt{27}}{9}$ **7.** ______________

8. $\frac{12+6\sqrt{6}}{8}$ **8.** ______________

9. $\frac{135\sqrt{3}+25}{5}$ **9.** ______________

Name: Date:
Instructor: Section:

Chapter 8 ROOTS AND RADICALS

8.6 Solving Equations with Radicals

Learning Objectives
1 Solve radical equations having square root radicals.
2 Identify equations with no solutions.
3 Solve equations by squaring a binomial.
4 Solve radical equations having cube root radicals.

Key Terms

Use the vocabulary terms listed below to complete each statement in exercises 1–2.

radical equation **extraneous solution**

1. An ____________________ is a potential solution to an equation that does not satisfy the equation.

2. An equation with a variable in the radicand is a ____________________.

Objective 1 Solve radical equations having square root radicals.

Video Examples

Review these examples for Objective 1:

1. Solve $\sqrt{p+2}=5$.

$$\left(\sqrt{p+2}\right)^2=5^2$$
$$p+2=25$$
$$p=23$$

Check $\sqrt{p+2}=5$
$$\sqrt{23+2}\stackrel{?}{=}5$$
$$\sqrt{25}\stackrel{?}{=}5$$
$$5=5 \quad \text{True}$$

The solution set is $\{23\}$.

2. Solve $4\sqrt{x}=\sqrt{x+30}$.

$$\left(4\sqrt{x}\right)^2=\left(\sqrt{x+30}\right)^2$$
$$4^2\left(\sqrt{x}\right)^2=\left(\sqrt{x+30}\right)^2$$
$$16x=x+30$$
$$15x=30$$
$$x=2$$

Now Try:

1. Solve $\sqrt{p+5}=4$.

____

2. Solve $5\sqrt{x}=\sqrt{x+72}$.

____

Name: Date:
Instructor: Section:

Check $4\sqrt{x} = \sqrt{x+30}$

$4\sqrt{2} \stackrel{?}{=} \sqrt{2+30}$

$4\sqrt{2} \stackrel{?}{=} \sqrt{32}$

$4\sqrt{2} = 4\sqrt{2}$ True

The solution set is $\{2\}$.

Objective 1 Practice Exercises

For extra help, see Examples 1–2 on pages 574–575 of your text.

Solve each equation.

1. $\sqrt{3x+1} = 3$ **1.** ________________

2. $\sqrt{2+4k} = 3\sqrt{k}$ **2.** ________________

3. $\sqrt{4x+3} = \sqrt{3x+5}$ **3.** ________________

Objective 2 Identify equations with no solutions.

Video Examples

Review these examples for Objective 2:

3. Solve $\sqrt{x} = -25$.

$\left(\sqrt{x}\right)^2 = (-25)^2$

$x = 625$

Check $\sqrt{x} = -25$

$\sqrt{625} \stackrel{?}{=} -25$

$25 = -25$ False

Because the statement $25 = -25$ is false, the number 625 is not a solution. It is an extraneous solution and must be rejected. There is no solution. The solution set is $\varnothing$.

Now Try:

3. Solve $\sqrt{x} = -10$.

____

Name: Date:
Instructor: Section:

4. Solve $x = \sqrt{x^2 + 7x + 21}$.

Step 1 The radical is already isolated on the right side of the equation.

Step 2 Square each side.

$$x^2 = \left(\sqrt{x^2 + 7x + 21}\right)^2$$

$$x^2 = x^2 + 7x + 21$$

Step 3 $0 = 7x + 21$

Step 4 This step is not needed.

Step 5 $-21 = 7x$

$-3 = x$

Step 6 Check

$$x = \sqrt{x^2 + 7x + 21}$$

$$-3 \stackrel{?}{=} \sqrt{(-3)^2 + 7(-3) + 21}$$

$$-3 \stackrel{?}{=} \sqrt{9 - 21 + 21}$$

$-3 = 3$ False

Since substituting –3 for x leads to a false result, the equation has no solution, and the solution set is $\varnothing$.

4. Solve $x = \sqrt{x^2 + 8x + 40}$.

____

Objective 2 Practice Exercises

For extra help, see Examples 3–4 on pages 575–576 of your text.

Solve each equation.

4. $\sqrt{x+2} + 7 = 0$

4. ______________

5. $\sqrt{2m+3} = 3\sqrt{m+5}$

5. ______________

6. $r = \sqrt{r^2 - 6r + 12}$

6. ______________

Name: Date:
Instructor: Section:

Objective 3 Solve equations by squaring a binomial.

Video Examples

Review these examples for Objective 3:

5. Solve $\sqrt{2x-4}=2-x$.

$$\left(\sqrt{2x-4}\right)^2=(2-x)^2$$
$$2x-4=4-4x+x^2$$
$$0=x^2-6x+8$$
$$0=(x-2)(x-4)$$
$$x-2=0 \quad \text{or} \quad x-4=0$$
$$x=2 \quad \text{or} \quad x=4$$

Check Let $x=2$. Let $x=4$.

$$\sqrt{2x-4}=2-x \qquad \sqrt{2x-4}=2-x$$
$$\sqrt{2(2)-4}\stackrel{?}{=}2-2 \qquad \sqrt{2(4)-4}\stackrel{?}{=}2-4$$
$$\sqrt{4-4}\stackrel{?}{=}0 \qquad \sqrt{8-4}\stackrel{?}{=}-2$$
$$0=0 \quad \text{True} \qquad \sqrt{4}\stackrel{?}{=}-2$$
$$2=-2 \quad \text{False}$$

Only 2 is a valid solution. (4 is extraneous.)
The solution set is $\{2\}$.

7. Solve $\sqrt{40+x}=4+\sqrt{x}$.

$$\left(\sqrt{40+x}\right)^2=\left(4+\sqrt{x}\right)^2$$
$$40+x=16+8\sqrt{x}+x$$
$$24=8\sqrt{x}$$
$$3=\sqrt{x}$$
$$9=x$$

Check $\sqrt{40+x}=4+\sqrt{x}$
$$\sqrt{40+9}\stackrel{?}{=}4+\sqrt{9}$$
$$\sqrt{49}\stackrel{?}{=}4+3$$
$$7=7 \quad \text{True}$$

The solution set is $\{9\}$.

Now Try:

5. Solve $\sqrt{3x-5}=x-3$.

____

7. Solve $\sqrt{p+4}-\sqrt{p-1}=1$.

____

Name: Date:
Instructor: Section:

Objective 3 Practice Exercises

For extra help, see Examples 5–7 on pages 577–578 of your text.

Solve each equation.

7. $\sqrt{b-4}=b-6$

7. ______________

8. $3\sqrt{p+6}=p+6$

8. ______________

9. $q-1=\sqrt{q^2-4q+7}$

9. ______________

Objective 4 Solve radical equations having cube root radicals.

Video Examples

Review these examples for Objective 4:

8. Solve the equation.

$$\sqrt[3]{5r-6}=\sqrt[3]{3r+4}$$

$$\left(\sqrt[3]{5r-6}\right)^3=\left(\sqrt[3]{3r+4}\right)^3$$

$$5r-6=3r+4$$

$$2r=10$$

$$r=5$$

Check $\sqrt[3]{5r-6}=\sqrt[3]{3r+4}$

$$\sqrt[3]{5(5)-6}\stackrel{?}{=}\sqrt[3]{3(5)+4}$$

$$\sqrt[3]{19}=\sqrt[3]{19} \quad \text{True}$$

The solution set is $\{5\}$.

Now Try:

8. Solve the equation.

$$\sqrt[3]{8x+5}=\sqrt[3]{7x+7}$$

____

Name: Date:
Instructor: Section:

Objective 4 Practice Exercises

For extra help, see Example 8 on page 579 of your text.

Solve each equation.

10. $\sqrt[3]{2a-63}+5=0$ **10.** ________________

11. $\sqrt[3]{5a+1}-\sqrt[3]{2a-11}=0$ **11.** ________________

12. $\sqrt[3]{8x+5}=\sqrt[3]{7x+7}$ **12.** ________________

Name: Date:
Instructor: Section:

Chapter 9 QUADRATIC EQUATIONS

9.1 Solving Quadratic Equations by the Square Root Property

Learning Objectives

1. Review the zero-factor property.
2. Solve equations of the form $x^2 = k$, where $k > 0$.
3. Solve equations of the form $(ax+b)^2 = k$, where $k > 0$.
4. Use formulas involving second-degree variables.

Key Terms

Use the vocabulary terms listed below to complete each statement in exercises 1–2.

quadratic equation **zero-factor property**

1. An equation that can be written in the form $ax^2 + bx + c = 0$ is a ________________________.

2. The ________________________ states that if a product equals 0, then at least one of the factors of the product also equals zero.

Objective 1 Review the zero-factor property.

Video Examples

Review these examples for Objective 1:

1. Solve each equation by the zero-factor property.

a. $x^2 + 5x + 4 = 0$

$$x^2 + 5x + 4 = 0$$
$$(x+4)(x+1) = 0$$
$$x+4 = 0 \quad \text{or} \quad x+1 = 0$$
$$x = -4 \quad \text{or} \quad x = -1$$

The solution set is $\{-4, -1\}$.

b. $x^2 = 64$

$$x^2 = 64$$
$$x^2 - 64 = 0$$
$$(x+8)(x-8) = 0$$
$$x+8 = 0 \quad \text{or} \quad x-8 = 0$$
$$x = -8 \quad \text{or} \quad x = 8$$

The solution set is $\{-8, 8\}$.

Now Try:

1. Solve each equation by the zero-factor property.

a. $x^2 + 8x + 7 = 0$

____

b. $x^2 = 100$

____

Name: Date:
Instructor: Section:

Objective 1 Practice Exercises

For extra help, see Example 1 on page 592 of your text.

Solve each equation by using the zero-factor property.

1. $x^2 + 6x + 8 = 0$

1. ____________

2. $x^2 = 121$

2. ____________

3. $x^2 + 2x - 35 = 0$

3. ____________

Objective 2 Solve equations of the form $x^2 = k$, where $k > 0$.

Video Examples

Review these examples for Objective 2:

2. Solve each equation. Write radicals in simplified form.

a. $x^2 = 36$

By the square root property, if $x^2 = 36$, then $x = \sqrt{36} = 6$ or $x = -\sqrt{36} = -6$

The solution set is {–6, 6}, or $\{\pm 6\}$.

b. $x^2 = 13$

The solutions are $x = \sqrt{13}$ or $x = -\sqrt{13}$

The solution set is $\{-\sqrt{13}, \sqrt{13}\}$, or $\{\pm\sqrt{13}\}$.

c. $p^2 = -64$

Because –64 is a negative number and because the square of a real number cannot be negative, there is no real number solution of this equation. The solution set is $\varnothing$.

Now Try:

2. Solve each equation. Write radicals in simplified form.

a. $x^2 = 81$

____

b. $x^2 = 23$

____

c. $n^2 = -25$

____

d. $5p^2 - 100 = 0$

$5p^2 - 100 = 0$

$5p^2 = 100$

$p^2 = 20$

$p = \sqrt{20}$ or $p = -\sqrt{20}$

$p = 2\sqrt{5}$ or $p = -2\sqrt{5}$

Check

$5p^2 - 100 = 0$	$5p^2 - 100 = 0$
$5(2\sqrt{5})^2 - 100 \stackrel{?}{=} 0$	$5(-2\sqrt{5})^2 - 100 \stackrel{?}{=} 0$
$5(20) - 100 \stackrel{?}{=} 0$	$5(20) - 100 \stackrel{?}{=} 0$
$0 = 0$ True	True $0 = 0$

The solution set is $\{2\sqrt{5}, -2\sqrt{5}\}$, or $\{\pm 2\sqrt{5}\}$.

d. $3x^2 - 54 = 0$

Objective 2 Practice Exercises

For extra help, see Example 2 on pages 593–594 of your text.

Solve each equation by using the square root property. Express all radicals in simplest form.

4. $r^2 = 900$ **4.** ________________

5. $s^2 - 98 = 0$ **5** ________________

6. $p^2 = -144$ **6.** ________________

Name: Date:
Instructor: Section:

Objective 3 Solve equations of the form $(ax+b)^2 = k$, where $k > 0$.

Video Examples

Review these examples for Objective 3:

3. Solve the equation.

$$(x-4)^2 = 64$$

$$x-4=\sqrt{64} \quad \text{or} \quad x-4=-\sqrt{64}$$
$$x-4=8 \quad \text{or} \quad x-4=-8$$
$$x=12 \quad \text{or} \quad x=-4$$

Check

$(x-4)^2 = 64$	$(x-4)^2 = 64$
$(12-4)^2 \stackrel{?}{=} 64$	$(-4-4)^2 \stackrel{?}{=} 64$
$8^2 \stackrel{?}{=} 64$	$(-8)^2 \stackrel{?}{=} 64$
$64 = 64$ True	$64 = 64$ True

The solution set is {–4, 12}.

4. Solve $(5r-3)^2 = 12$.

$$5r-3=\sqrt{12} \quad \text{or} \quad 5r-3=-\sqrt{12}$$
$$5r-3=2\sqrt{3} \quad \text{or} \quad 5r-3=-2\sqrt{3}$$
$$5r=3+2\sqrt{3} \quad \text{or} \quad 5r=3-2\sqrt{3}$$
$$r=\frac{3+2\sqrt{3}}{5} \quad \text{or} \quad r=\frac{3-2\sqrt{3}}{5}$$

Check

$$(5r-3)^2 = 12$$
$$\left[5\cdot\frac{3+2\sqrt{3}}{5}-3\right]^2 \stackrel{?}{=} 12$$
$$\left(3+2\sqrt{3}-3\right)^2 \stackrel{?}{=} 12$$
$$\left(2\sqrt{3}\right)^2 \stackrel{?}{=} 12$$
$$12 = 12 \quad \text{True}$$

The check of the other solution is similar. The solution set is

The solution set is $\left\{\frac{3+2\sqrt{3}}{5}, \frac{3-2\sqrt{3}}{5}\right\}$.

Now Try:

3. Solve the equation.

$$(x-5)^2 = 121$$

____

4. Solve $(7r-3)^2 = 32$.

____

Name: Date:
Instructor: Section:

5. Solve $(x+2)^2 = -16$.

Because the square root of -16 is not a real number, there is no real number solution for this equation. The solution set is $\varnothing$.

5. Solve $(x-7)^2 = -49$.

____

Objective 3 Practice Exercises

For extra help, see Examples 3–5 on pages 594–595 of your text.

Solve each equation by using the square root property. Express all radicals in simplest form.

7. $(y+2)^2 = 16$

7. __________

8. $(7p-4)^2 = 289$

8. __________

9. $(10m-5)^2 - 9 = 0$

9. __________

Objective 4 Use formulas involving second-degree variables.

Video Examples

Review this example for Objective 4:

6. We can approximate the weight of a bass, in pounds, given its length L and its girth (distance around) g, where both are measured in inches, using the following formula.

$$w = \frac{L^2 g}{1200}$$

Approximate the length of a bass weighing 2.40 lb and having girth 9 in.

Start with the formula and substitute 2.40 for w and 9 for g.

Now Try:

6. We can approximate the weight of a bass, in pounds, given its length L and its girth (distance around) g, where both are measured in inches, using the following formula.

$$w = \frac{L^2 g}{1200}$$

Approximate the length of a bass weighing 2.50 lb and having girth 10.5 in.

____

$$w = \frac{L^2 g}{1200}$$

$$2.40 = \frac{L^2 \cdot 9}{1200}$$

$$2880 = 9L^2$$

$$L^2 = 320$$

$$L = \sqrt{320} \text{ or } L = -\sqrt{320}$$

The calculator shows that $\sqrt{320} \approx 17.89$, so the length of the bass is almost 18 in. (We discard the solution $-\sqrt{320} \approx -17.89$, since L represents length.)

Objective 4 Practice Exercises

For extra help, see Example 6 on pages 595–596 of your text.

Solve each problem.

10. The formula $A = P(1+r)^2$ gives the amount A that P dollars invested at an annual rate of interest r will grow to in two years. Mary invests $1500, and after two years, she has $1653.75. What is the interest rate?

10. ________________

11. The volume of a cylinder is given by the formula $V = \pi r^2 h$, where V = the volume, r = the radius of the base of the cylinder, and h = the height of the cylinder. If the volume of a can is 20π in.^{3}, and its height is 5 inches, find the radius of the can.

11. ________________

12. One leg of a right triangle has length 5 cm, and the hypotenuse has length 10 cm. Find the length of the other leg. (Hint: Use the Pythagorean theorem.)

12. ________________

Name: Date:
Instructor: Section:

Chapter 9 QUADRATIC EQUATIONS

9.2 Solving Quadratic Equations by Completing the Square

Learning Objectives

1. Solve quadratic equations by completing the square when the coefficient of the second-degree term is 1.
2. Solve quadratic equations by completing the square when the coefficient of the second-degree term is not 1.
3. Simplify the terms of an equation before solving.
4. Solve applied problems that require quadratic equations.

Key Terms

Use the vocabulary terms listed below to complete each statement in exercises 1–3.

completing the square **perfect square trinomial** **square root property**

1. A ______________________________ can be written in the form $x^2+2kx+k^2$ or $x^2-2kx+k^2$

2. The ______________________________ says that, if k is positive and $a^2=k$, then $a=\pm\sqrt{k}$.

3. Use the process called ________________________ in order to rewrite an equation so it can be solved using the square root property.

Video Examples

Objective 1 Solve quadratic equations by completing the square when the coefficient of the second-degree term is 1.

Review these examples for Objective 1:

1. Complete each trinomial so that it is a perfect square. Then factor the trinomial.

a. x^2+24x+ ______

The perfect square trinomial will have the form $x^2+2kx+k^2$. Thus, the middle term $24x$, must equal $2kx$.

$$24x=2kx$$
$$12=k$$

Therefore, $k=12$ and $k^2=12^2=144$. The perfect square trinomial is $x^2+24x+144$, which factors as $(x+12)^2$.

Now Try:

1. Complete each trinomial so that it is a perfect square. Then factor the trinomial.

a. x^2+10x+ ______

____

b. $x^2 - 36x +$ ______

The perfect square trinomial will have the form $x^2 - 2kx + k^2$. Thus, the middle term $-36x$, must equal $-2kx$.

$$-36x = -2kx$$
$$18 = k$$

Therefore, $k = 18$ and $k^2 = 18^2 = 324$. The required perfect square trinomial is $x^2 - 36x + 324$, which factors as $(x-18)^2$.

2. Solve $x^2 + 8x + 3 = 0$.

$$x^2 + 8x + 3 = 0$$
$$x^2 + 8x = -3$$

The expression on the left must be written as a perfect square trinomial in the form $x^2 + 2kx + k^2$.

$$x^2 + 8x + \_\_\_\_\_$$

Here, $2kx = 8x$, so $k = 4$ and $k^2 = 16$. The required perfect square trinomial is $x^2 + 8x + 16$ which factors as $(x+4)^2$.

Therefore, if we add 16 to each side of $x^2 + 8x = -3$, the equation will have a perfect square trinomial on the left side, as needed.

$$x^2 + 8x + 16 = -3 + 16$$
$$(x+4)^2 = 13$$

Use the square root property.

$$x + 4 = \sqrt{13} \quad \text{or} \quad x + 4 = -\sqrt{13}$$
$$x = -4 + \sqrt{13} \quad \text{or} \quad x = -4 - \sqrt{13}$$

Check by substituting $-4 + \sqrt{13}$ and then $-4 - \sqrt{13}$ for *x* in the original equation. The solution set is $\{-4 + \sqrt{13}, -4 - \sqrt{13}\}$.

b. $x^2 - 22x +$ ______

____

2. Solve $x^2 + 10x - 7 = 0$.

____

Name: Date:
Instructor: Section:

Objective 1 Practice Exercises

For extra help, see Examples 1–4 on pages 598–600 of your text.

Solve each equation by completing the square.

1. $r^2 + 8r = -4$ **1.** ________________

2. $x^2 - 4x = 2$ **2.** ________________

3. $x^2 + 2x = 63$ **3.** ________________

Objective 2 Solve quadratic equations by completing the square when the coefficient of the second-degree term is not 1.

Video Examples

Review this example for Objective 2:

7. Solve $5p^2 - 20p + 21 = 0$.

$$5p^2 - 20p + 21 = 0$$
$$p^2 - 4p + \frac{21}{5} = 0$$
$$p^2 - 4p = -\frac{21}{5}$$

The coefficient of p is –4. Take half of –4, square the result, and add it to each side.

$$p^2 - 4p + 4 = -\frac{21}{5} + 4$$
$$(p-2)^2 = -\frac{1}{5}$$

We cannot use the square root property to solve this equation, because the square root of $-\frac{1}{5}$ is not a real number. This equation has no real number solution. The solution set is $\varnothing$.

Now Try:

7. Solve $6p^2 - 36p + 55 = 0$.

____

Name: Date:
Instructor: Section:

Objective 2 Practice Exercises

For extra help, see Examples 5–7 on pages 601–603 of your text.

Solve each equation by completing the square.

4. $6x^2 - x = 15$

4. ________________

5. $3x^2 - 2x + 4 = 0$

5. ________________

6. $3t^2 + t - 2 = 0$

6. ________________

Objective 3 Simplify the terms of an equation before solving.

Video Examples

Review this example for Objective 3:

8. Solve $(x-6)(x+2) = 7$.

$$(x-6)(x+2) = 7$$

$$x^2 - 4x - 12 = 7 \quad \text{FOIL}$$

$$x^2 - 4x = 19 \quad \text{Add 12.}$$

$$x^2 - 4x + 4 = 19 + 4 \quad \text{Add } \left[\tfrac{1}{2}(-4)\right]^2 = 4$$

$$(x-2)^2 = 23$$

$$x - 2 = \sqrt{23} \quad \text{or} \quad x - 2 = -\sqrt{23}$$

$$x = 2 + \sqrt{23} \quad \text{or} \quad x = 2 - \sqrt{23}$$

The solution set is $\{2+\sqrt{23},\ 2-\sqrt{23}\}$.

Now Try:

8. Solve $(x+8)(x-2) = 5$.

____

Name: Date:
Instructor: Section:

Objective 3 Practice Exercises

For extra help, see Example 8 on page 603 of your text.

Simplify each of the following equations and then solve by completing the square.

7. $6y^2 + 3y = 4y^2 + y - 5$ **7.** ________

8. $(b-1)(b+7) = 9$ **8.** ________

9. $(s+3)(s+1) = 1$ **9.** ________

Objective 4 Solve applied problems that require quadratic equations.

Video Examples

Review this example for Objective 4:

9. If James throws an object upward from ground level with an initial velocity of 80 feet per second, its height s (in feet) after t seconds is given by the formula $s = -16t^2 + 80t$. After how many seconds will the object reach a height of 64 feet?

Since s represents the height, we substitute 64 for s in the formula, and then solve for t using completing the square.

$$64 = -16t^2 + 80t$$

$$-4 = t^2 - 5t \qquad \text{Divide by } -16.$$

$$t^2 - 5t = -4$$

$$t^2 - 5t + \frac{25}{4} = -4 + \frac{25}{4} \qquad \text{Add } \left[\frac{1}{2}(-5)\right]^2 = \frac{25}{4}.$$

$$\left(t - \frac{5}{2}\right)^2 = \frac{9}{4}$$

Now Try:

9. A certain projectile is located at a distance of $d = 3t^2 - 6t + 1$ feet from its starting point after t seconds. How many seconds will it take the projectile to travel 10 feet?

____

$$t-\frac{5}{2}=\frac{3}{2} \quad \text{or} \quad t-\frac{5}{2}=-\frac{3}{2}$$

$$t=4 \quad \text{or} \quad t=1$$

The ball reaches a height of 64 twice, once on the way up and again on the way down. It takes 1 second to reach 64 ft on the way up and then after 4 sec, the object reaches 64 ft on the way down.

Objective 4 Practice Exercises

For extra help, see Example 9 on page 604 of your text.

Solve.

10. A rule for estimating the number of board feet of lumber that can be cut from a log depends on the diameter and length of the log. To find the diameter d (in inches) needed to get x board feet of lumber from an 8-foot log, use the formula, $\left(\frac{d-4}{4}\right)^2=x$. Find the diameter needed to get 20 board feet of lumber.

10. ____________

11. The commodities market is very unstable; money can be made or lost quickly on investments in soybeans, wheat, pork bellies, and so on. Suppose that an investor kept track of his total profit, P (in thousands of dollars), at time t (in months), after he began investing, and found that his profit was given by the formula $P=4t^2-24t+32$. Find the times at which he broke even on his investment.

11. ____________

12. George and Albert have found that the profit (in dollars) from their cigar shop is given by the formula $P=-10x^2+100x+300$, where x is the number of units of cigars sold daily. How many units should be sold for a profit of \$460?

12. ____________

Name: Date:
Instructor: Section:

Chapter 9 QUADRATIC EQUATIONS

9.3 Solving Quadratic Equations by the Quadratic Formula

Learning Objectives
1. Identify the values of *a*, *b*, and *c* in a quadratic equation.
2. Use the quadratic formula to solve quadratic equations.
3. Solve quadratic equations with a double solution.
4. Solve quadratic equations with fractions as coefficients.

Key Terms

Use the vocabulary terms listed below to complete each statement in exercises 1–3.

discriminant **quadratic formula** **double solution**

1. The formula $x = \dfrac{-b \pm \sqrt{b^2 - 4ac}}{2a}$ is called the ____________________.

2. A quadratic equation that has one distinct rational number solution is called a ____________________.

3. In the formula $x = \dfrac{-b \pm \sqrt{b^2 - 4ac}}{2a}$, $b^2 - 4ac$ is called the ____________________.

Objective 1 Identify the values of *a*, *b*, and *c* in a quadratic equation.

Video Examples

Review these examples for Objective 1:

1. Identify the values of the variables *a*, *b*, and *c* in each quadratic equation $ax^2 + bx + c = 0$.

a. $7x^2 - 6x + 4 = 0$

Here $a = 7$, $b = -6$, and $c = 4$.

b. $-8x^2 + 3 = 5x$

Write the equation in standard form.

$-8x^2 - 5x + 3 = 0$

Here, $a = -8$, $b = -5$, and $c = 3$.

c. $9x^2 - 32 = 0$

The *x*-term is missing.

$9x^2 + 0x - 32 = 0$

Then, $a = 9$, $b = 0$, and $c = -32$.

Now Try:

1. Identify the values of the variables *a*, *b*, and *c* in each quadratic equation $ax^2 + bx + c = 0$.

a. $12x^2 - 10x + 7 = 0$

____

b. $-4x^2 + 8 = 110x$

____

c. $15x^2 - 8 = 0$

____

Name: Date:
Instructor: Section:

d. $(3x-5)(x+6)=-29$

$$3x^2+13x-30=-29$$
$$3x^2+13x-1=0$$

Here, $a=3$, $b=13$, and $c=-1$.

d. $(5x-4)(6x+7)=11$

____

Objective 1 Practice Exercises

For extra help, see Example 1 on page 607 of your text.

Write each equation in standard form, if necessary, and then identify the values of a, b, and c. Do not actually solve the equation.

1. $10x^2=-4x$

1. ______________

2. $4p=-4p^2+7$

2. ______________

3. $(z+1)(z+2)=-7$

3. ______________

Objective 2 Use the quadratic formula to solve quadratic equations.

Video Examples

Review these examples for Objective 2:

2. Solve $2x^2+3x-20=0$.

Substitute $a=2$, $b=3$, and $c=-20$.

$$x=\frac{-b\pm\sqrt{b^2-4ac}}{2a}$$
$$x=\frac{-3\pm\sqrt{3^2-4(2)(-20)}}{2(2)}$$
$$x=\frac{-3\pm\sqrt{9+160}}{4}$$
$$x=\frac{-3\pm\sqrt{169}}{4}$$
$$x=\frac{-3\pm13}{4}$$

Find the two solutions by first using the plus symbol, and then using the minus symbol.

$$x=\frac{-3+13}{4}=\frac{10}{4}=\frac{5}{2}\text{ or}$$
$$x=\frac{-3-13}{4}=\frac{-16}{4}=-4$$

Now Try:

2. Solve $3x^2-x-14=0$.

____

Name: Date:
Instructor: Section:

Check each solution in the original equation.

The solution set is $\left\{-4, \frac{5}{2}\right\}$.

3. Solve $x^2 = 4x - 1$.

Write the given equation in standard form as $x^2 - 4x + 1 = 0$.

$$x = \frac{-b \pm \sqrt{b^2 - 4ac}}{2a}$$

$$x = \frac{-(-4) \pm \sqrt{(-4)^2 - 4(1)(1)}}{2(1)}$$

$$x = \frac{4 \pm \sqrt{12}}{2}$$

$$x = \frac{4 \pm 2\sqrt{3}}{2}$$

$$x = \frac{2(2 \pm \sqrt{3})}{2}$$

$$x = 2 \pm \sqrt{3}$$

The solution set is $\{2 \pm \sqrt{3}\}$.

3. Solve $x^2 = 6x - 4$.

____

Objective 2 Practice Exercises

For extra help, see Examples 2–3 on pages 608–609 of your text.

Use the quadratic formula to solve each equation. Write all radicals in simplified form, and write all answers in lowest terms.

4. $y^2 = 13 - 12y$

4. _______________

5. $-7r^2 = 5r + 3$

5. _______________

6. $5k^2 + 4k - 2 = 0$

6. _______________

Name: Date:
Instructor: Section:

Objective 3 Solve quadratic equations with a double solution.

Video Examples

Review this example for Objective 3:

4. Solve $9x^2 + 49 = 42x$.

Write the equation in standard form.

$9x^2 - 42x + 49 = 0$

Here, $a = 9$, $b = -42$, and $c = 49$.

$$x = \frac{-(-42) \pm \sqrt{(-42)^2 - 4(9)(49)}}{2(9)}$$

$$x = \frac{42 \pm \sqrt{1764 - 1764}}{18}$$

$$x = \frac{42 \pm 0}{18}$$

$$x = \frac{7}{3}$$

In this case, $b^2 - 4ac = 0$, and the trinomial $9x^2 - 42x + 49$ is a perfect square. There is one distinct solution in the solution set $\left\{\frac{7}{3}\right\}$.

Now Try:

4. Solve $4x^2 - 44x + 121 = 0$.

____

Objective 3 Practice Exercises

For extra help, see Example 4 on page 609 of your text.

Use the quadratic formula to solve each equation. Write all radicals in simplified form, and write all answers in lowest terms.

7. $m^2 + 4 = 4m$

7. ____________

8. $4q^2 + 12q + 9 = 0$

8. ____________

9. $49a^2 - 126a = -81$

9. ____________

Name: Date:
Instructor: Section:

Objective 4 Solve quadratic equations with fractions as coefficients.

Video Examples

Review this example for Objective 4:

5. Solve $\frac{1}{5}t^2 = \frac{1}{5}t + \frac{3}{10}$.

Clear fractions. Multiply by the LCD, 10.

$$\frac{1}{5}t^2 = \frac{1}{5}t + \frac{3}{10}$$
$$10\left(\frac{1}{5}t^2\right) = 10\left(\frac{1}{5}t + \frac{3}{10}\right)$$
$$10\left(\frac{1}{5}t^2\right) = 10\left(\frac{1}{5}t\right) + 10\left(\frac{3}{10}\right)$$
$$2t^2 = 2t + 3$$
$$2t^2 - 2t - 3 = 0$$

Identify $a = 2$, $b = -2$, and $c = -3$.

$$t = \frac{-(-2) \pm \sqrt{(-2)^2 - 4(2)(-3)}}{2(2)}$$
$$t = \frac{2 \pm \sqrt{28}}{4}$$
$$t = \frac{2 \pm 2\sqrt{7}}{4}$$
$$t = \frac{2(1 \pm \sqrt{7})}{2(2)}$$
$$t = \frac{1 \pm \sqrt{7}}{2}$$

A check confirms the solution set is $\left\{\frac{1 \pm \sqrt{7}}{2}\right\}$.

Now Try:

5. Solve $\frac{5}{14}x^2 = \frac{2}{7}x + \frac{1}{7}$.

____

Objective 4 Practice Exercises

For extra help, see Example 5 on page 610 of your text.

Use the quadratic formula to solve each equation. Write all radicals in simplified form, and write all answers in lowest terms.

10. $\frac{1}{6}y^2 + \frac{1}{2}y = \frac{2}{3}$

10. ________________

Name: Date:
Instructor: Section:

11. $\frac{1}{4}t^2 - \frac{1}{3}t + \frac{5}{12} = 0$ **11.** ________________

12. $-\frac{1}{4}x^2 + 4 = \frac{1}{2}x$ **12.** ________________

Name: Date:
Instructor: Section:

Chapter 9 QUADRATIC EQUATIONS

9.4 Graphing Quadratic Equations

Learning Objectives

1 Graph quadratic equations of the form $y = ax^2 + bx + c$ $(a \neq 0)$.

Key Terms

Use the vocabulary terms listed below to complete each statement in exercises 1–4.

parabola **vertex** **axis** **line of symmetry**

1. If a graph is folded on its______________________________, the two sides coincide.

2. The ______________________________ of a parabola that opens upward or downward is the lowest or highest point on the graph.

3. The ______________________________ of a parabola that opens upward or downward is a vertical line through the vertex.

4. The graph of the quadratic equation $y = ax^2 + bx + c$ is called a ____________________.

Objective 1 Graph quadratic equations of the form $a^2x + bx + c = 0$ $(a \neq 0)$.

Video Examples

Review these examples for Objective 1:

1. Graph $y = x^2 + x - 2$.

Find the x-intercepts of the parabola. Let $y = 0$.

$$0 = x^2 + x - 2$$
$$0 = (x+2)(x-1)$$
$$x+2=0 \quad \text{or} \quad x-1=0$$
$$x=-2 \quad \text{or} \quad x=1$$

There are two intercepts, (–2, 0) and (1, 0). Since the x-value of the vertex is halfway between the x-values of the x-intercepts, it is half their sum.

$$x = \frac{1}{2}(-2+1) = -\frac{1}{2}$$

We find the corresponding y-value.

$$y = \left(-\frac{1}{2}\right)^2 + \left(-\frac{1}{2}\right) - 2 = -\frac{9}{4}$$

Now Try:

1. Graph $y = x^2 + 4x - 5$.

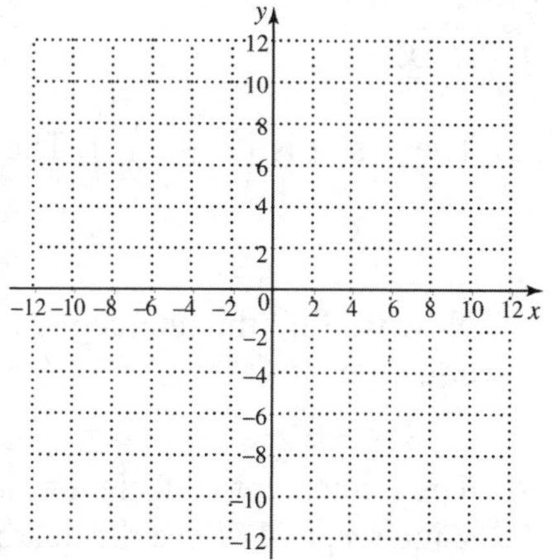

Name: Date:
Instructor: Section:

The vertex is $\left(-\frac{1}{2},-\frac{9}{4}\right)$. The axis of symmetry is the vertical line $x=-\frac{1}{2}$.

To find the y-intercept, we substitute 0 for x in the equation.

$$y=0^2+0-2=-2$$

We plot the three intercepts and the vertex, and find additional ordered pairs as needed.

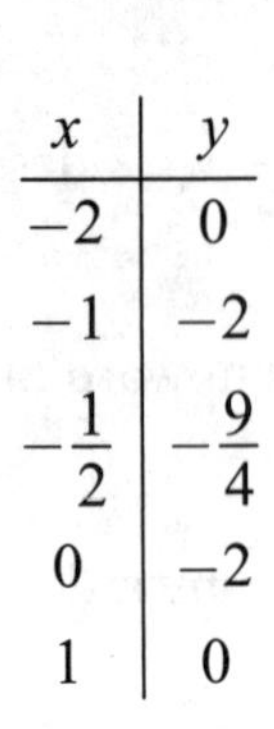

x	y
-2	0
-1	-2
$-\frac{1}{2}$	$-\frac{9}{4}$
0	-2
1	0

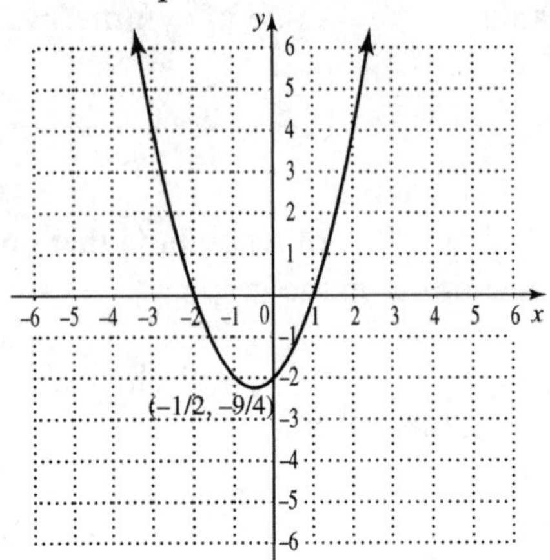

2. Graph $y=-x^2-3x+1$.

Here $a=-1$ and $b=-3$, so we substitute to find the x-value of the vertex.

$$x=-\frac{b}{2a}=-\frac{-3}{2(-1)}=-\frac{3}{2}$$

We find the y-value from the original equation.

$$y=-\left(-\frac{3}{2}\right)^2-3\left(-\frac{3}{2}\right)+1$$

$$y=\frac{13}{4}$$

The vertex is $\left(-\frac{3}{2},\frac{13}{4}\right)$. The axis is the line $x=-\frac{3}{2}$.

Now we find the intercepts. Let $x=0$ in the equation.

$$y=-0^2-3(0)+1=1$$

The y-intercept is (0, 1). Let $y=0$ to get the x-intercepts. We use the quadratic formula to solve for x.

2. Graph $y=x^2+8x+14$.

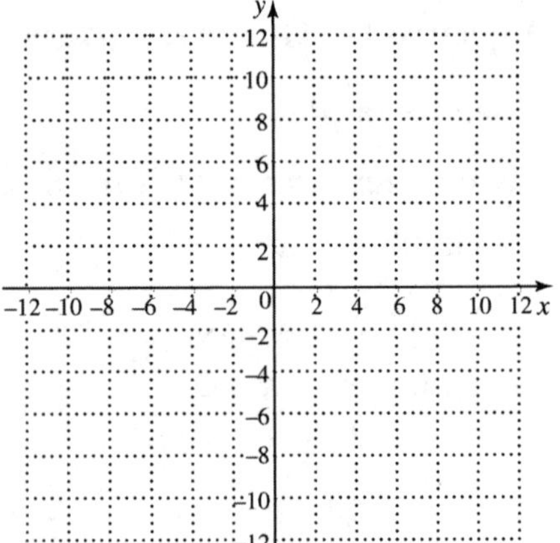

$$x=\frac{3\pm\sqrt{(-3)^2-4(-1)(1)}}{2(-1)}$$

$$x=\frac{3\pm\sqrt{13}}{-2}$$

Using a calculator, we find that the x-intercepts are (–3.3, 0) and (0.3, 0).

We plot the intercepts, vertex, and other ordered pairs, and join these points with a smooth curve.

x	y
-4	-3
-3.3	0
$-\frac{3}{2}$	$\frac{13}{4}$
0	1
0.3	0
1	-3

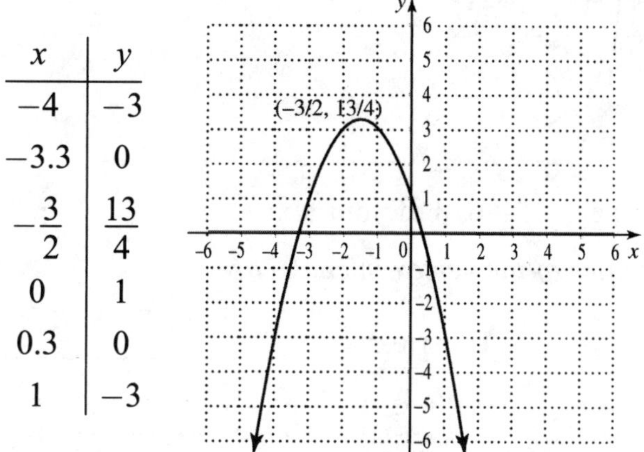

3. Graph $y=9-x^2$. Give the vertex and intercepts.

Here $a = –1$ and $b = 0$, find the vertex.

$$x=-\frac{b}{2a}=-\frac{0}{2(-1)}=0$$

We find the y-value from the original equation.

$$y=9-x^2=9-0^2=9$$

The vertex is (0, 9), which is also the y-intercept. Let $y = 0$ to get the x-intercepts. We use the quadratic formula to solve for x.

$$x=\frac{0\pm\sqrt{0^2-4(-1)(9)}}{2(-1)}=\frac{\pm\sqrt{36}}{-2}$$

$$x=\pm3$$

The x-intercepts are (–3, 0) and (3, 0).

Select several values for x and find the corresponding values for y. Plot the ordered pairs and join them with a smooth curve.

3. Graph $y=-x^2-1$. Give the vertex and intercepts.

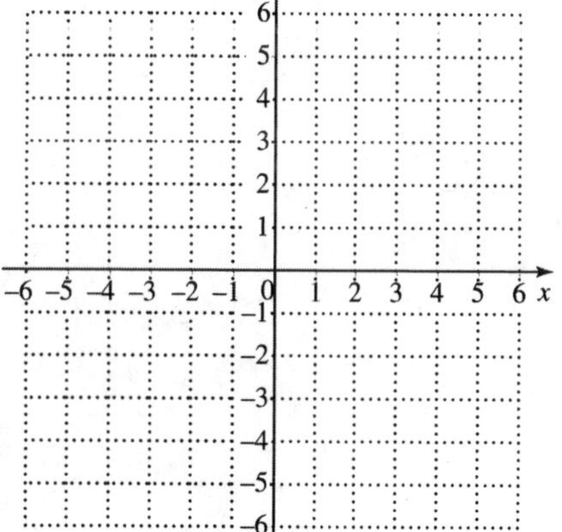

Name: Date:
Instructor: Section:

x	y
-4	-7
-3	0
0	9
3	0
4	-7

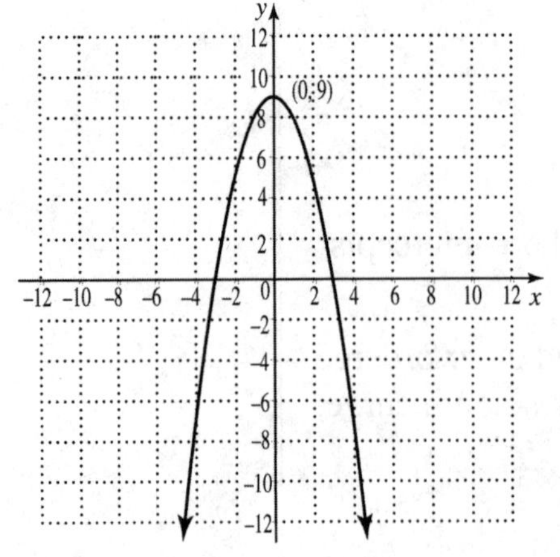

Objective 1 Practice Exercises

For extra help, see Examples 1–3 on pages 614–618 of your text.

Graph each equation. Give the coordinates of the vertex in each case.

1. $y = -x^2 - 2x - 1$

1. vertex: ______________________

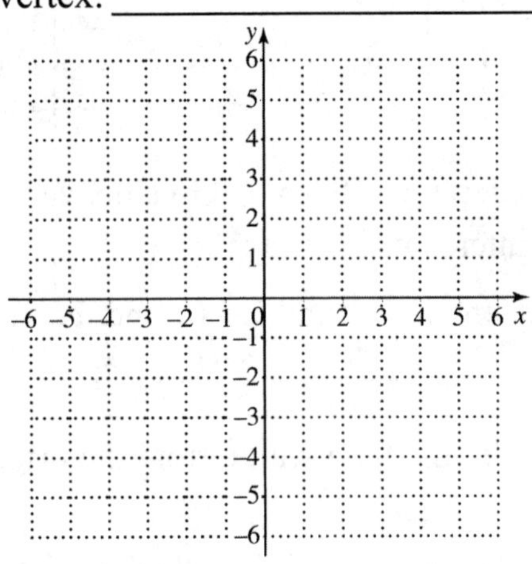

2. $y = x^2 + 2x - 2$

2. vertex: ______________________

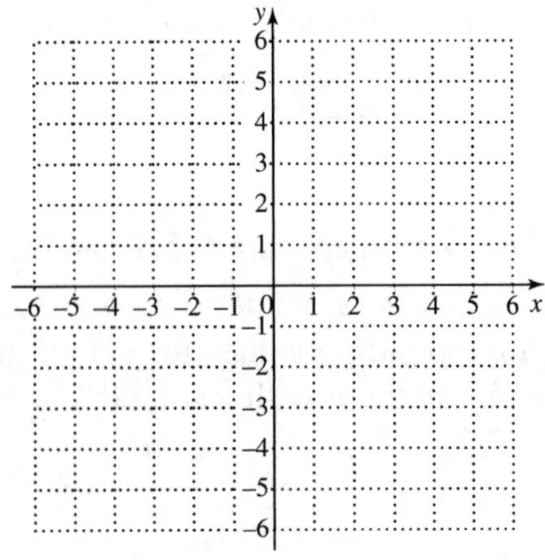

Name: Date:
Instructor: Section:

3. $y = -x^2 - 2x + 8$

3. vertex: ____________________

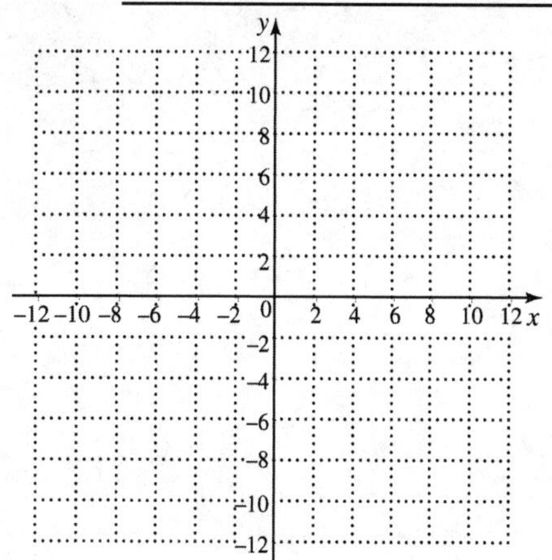

Chapter R PREALGEBRA REVIEW

R.1 Fractions

Key Terms

1. equivalent fractions
2. improper fraction
3. numerator
4. proper fraction
5. denominator
6. composite number
7. prime factorization
8. prime number
9. lowest terms

Objective 1

Now Try

1. $2 \cdot 3 \cdot 5 \cdot 7$

Practice Exercises

1. $2 \cdot 7 \cdot 7$
3. $2 \cdot 3 \cdot 7 \cdot 13$

Objective 2

Now Try

2. $\frac{3}{5}$

Practice Exercises

5. $\frac{5}{6}$

Objective 3

Now Try

3. $14\frac{4}{5}$
4. $\frac{86}{7}$

Practice Exercises

7. $21\frac{2}{5}$
9. $\frac{244}{11}$

Objective 4

Now Try

5. $\frac{1}{10}$
6. $\frac{16}{21}$

Practice Exercises

11. $\frac{7}{5}$ or $1\frac{2}{5}$

Objective 5

Now Try

7. $\frac{3}{4}$
8. $\frac{23}{24}$
9. $\frac{2}{9}$

Practice Exercises

13. $\frac{256}{225}$ or $1\frac{31}{225}$

15. $\frac{119}{24}$ or $4\frac{23}{24}$

Objective 6

Now Try

10. $8\frac{3}{4}$ cups

Practice Exercises

17. $11\frac{3}{4}$ yards

Objective 7

Now Try

11a. Home; $\frac{1}{50}$

11b. 104 workers

Practice Exercises

19. Lunch Room

21. 754 workers

R.2 Decimals and Percents

Key Terms

1. decimals
2. place value
3. percent

Objective 1

Now Try

1a. $\frac{72}{100}$

1b. $\frac{53}{1000}$

1c. $\frac{37{,}058}{10{,}000}$

Practice Exercises

1. $\frac{7}{1000}$

3. $\frac{300{,}005}{10{,}000}$

Objective 2

Now Try

2a. 37.871

2b. 27.282

Practice Exercises

5. 755.098

Objective 3

Now Try

3a. 251.116

3b. 0.028

4a. 32.4

4b. 1.43

Practice Exercises

7. 2.3424
9. 0.4292

Objective 4

Now Try

6a. 0.35
6b. $3.\overline{5}$ or 3.555…

Practice Exercises

11. $0.\overline{4}$ or 0.444...

Objective 5

Now Try

9a. 0.91
9b. 0.06
9c. 43%
9d. 520%

Practice Exercises

13. 3.62
15. 8.4%

Objective 6

Now Try

10a. $\frac{3}{10}$
10b. $1\frac{1}{4}$
11a. 15%
11b. $5\frac{5}{9}\%$ exact, or 5.6% rounded

Practice Exercises

17. $\frac{139}{250}$

Objective 7

Now Try

12. discount: $10.64, sale price: $8.36

Practice Exercises

19. $14; sale price: $56
21. $42; sale price: $14

Chapter 1 THE REAL NUMBER SYSTEM

1.1 Exponents, Order of Operations, and Inequality

Key Terms

1. exponential expression
2. base
3. exponent

Objective 1

Now Try

1. 49

Practice Exercises

1. 27
3. 0.16

Objective 2

Now Try

2. 23

Practice Exercises

5. 45

Objective 3

Now Try

3. 215

Practice Exercises

7. –8
9. 48

Objective 4

Now Try

4. true

Practice Exercises

11. false

Objective 5

Now Try

5. $19 \le 11 + 8$

Practice Exercises

13. $7 = 13 - 6$
15. $20 \ge 2 \cdot 7$

Objective 6

Now Try

6. $11 < 15$

Practice Exercises

17. $8 \le 12$

1.2 Variables, Expressions, and Equations

Key Terms

1. equation
2. elements
3. variable
4. algebraic expression
5. set
6. Solution
7. constant

Objective 1

Now Try

1. 252, 567
2. 65

Practice Exercises

1. 8
3. $\frac{28}{13}$

Objective 2

Now Try

3. $20x$

Practice Exercises

5. $8x-11$

Objective 3

Now Try

4. no

Practice Exercises

7. no
9. yes

Objective 4

Now Try

5. $x+7=11$; 4

Practice Exercises

11. $5+x=14$; 9

Objective 5

Now Try

6. expression

Practice Exercises

13. expression
15. equation

1.3 Real Numbers and the Number Line

Key Terms

1. whole numbers
2. additive inverse
3. integers
4. natural numbers
5. absolute value
6. number line
7. irrational number
8. coordinate
9. negative number
10. positive number
11. real numbers
12. set-builder notation
13. rational number
14. signed numbers

Objective 1

Now Try

1. –282
2. –5 –4 –3 –2 –1 0 1 2 3 4 5

3a. 0, 7

3b. –10, 0, 7

3c. $-10, -\frac{5}{8}, 0, 0.\overline{4}, 5\frac{1}{2}, 7, 9.9$

3d. $\sqrt{5}$

Practice Exercises

1. –75 pounds
3. –5 –4 –3 –2 –1 0 1 2 3 4 5

Objective 2

Now Try

4. true

Practice Exercises

5. false

Objective 3

Practice Exercises

7. 25
9. –4.5

Objective 4

Now Try

5a. 10

5b. –10

5c. 3

Practice Exercises

11. 1.22

Objective 5

Now Try

6. Milk and Electricity

Practice Exercises

13. Gasoline
15. Eggs in 2012 to 2013

1.4 Adding and Subtracting Real Numbers

Key Terms

1. minuend
2. subtrahend
3. difference
4. sum
5. addends

Objective 1

Now Try

1. –5
2. –12

Practice Exercises

1. –18
3. $-5\frac{5}{8}$

Objective 2

Now Try

3. 3
4. $-\frac{7}{10}$

Practice Exercises

5. $-\frac{1}{6}$

Objective 3

Now Try

6a. –3
6b. 2

Practice Exercises

7. –25
9. 0

Objective 4

Now Try

7. 1

Practice Exercises

11. –6

Objective 5

Now Try

8. $-10+11+2$; 3
9. $-17-9$; –26
10. 5464 ft

Practice Exercises

13. $-4-4$; -8
15. –51.2°C

Objective 6

Now Try

11. –$0.006

Practice Exercises

17. –$0.061

1.5 Multiplying and Dividing Real Numbers

Key Terms

1. quotient   2. reciprocals   3. product

Objective 1

Now Try

1. –56

Practice Exercises

1. –28   3. –13.12

Objective 2

Now Try

2a. 30

Practice Exercises

5. $\frac{4}{5}$

Objective 3

Now Try

2b. –12, –6, –4, –3, –2, –1, 1, 2, 3, 4, 6, and 12

Practice Exercises

7. –8, –4, –2, –1, 1, 2, 4, 8

9. –42, –21, –14, –7, –6, –3, –2, –1, 1, 2, 3, 6, 7, 14, 21, 42

Objective 4

Now Try

3a. 2   3b. 3

Practice Exercises

11. 0

Objective 5

Now Try

4a. 28   4b. $-\frac{5}{4}$

Practice Exercises

13. 64   15. $\frac{16}{21}$

Objective 6

Now Try

5. –432

Practice Exercises

17. 25

Objective 7

Now Try

6. $\frac{5}{6}[-8+(-4)];\ -10$

Practice Exercises

19. $(-7)(3)+(-7);\ -28$

21. $-12+\frac{49}{-7};\ -19$

Objective 8

Now Try

8. $\frac{36}{x}=-4$

Practice Exercises

23. $-8x=72$

1.6 Properties of Real Numbers

Key Terms

1. identity element for addition
2. identity element for multiplication

Objective 1

Now Try

1a. –12

1b. 2

Practice Exercises

1. 4

3. $(4+z)$

Objective 2

Now Try

2a. 4

2b. $[(-3)\cdot 4]$

3a. associative

3b. commutative

3c. both

4. $39x+20$

Practice Exercises

5. $[(-4+3y)]$

Objective 3

Now Try

5a. 0

5b. 1

6a. $\frac{7}{9}$

6b. $\frac{2}{3}$

Practice Exercises

7. 4

9. $\frac{6}{7}$

Objective 4

Now Try

7a. $\frac{5}{8}$ 7b. -8 7c. -10

7d. 11 8. 6

Practice Exercises

11. 0; identity

Objective 5

Now Try

9a. $17x-102$ 9b. $-8x+20$ 9c. $3(11+7)$

9d. $12(y+6+x)$ 10a. $-3x-4$ 10b. $4x+5y-z$

10c. $3(x+y+1)$

Practice Exercises

13. $2an-4bn+6cn$ 15. $2k-7$

1.7 Simplifying Expressions

Key Terms

1. numerical coefficient 2. term 3. like terms

Objective 1

Now Try

1a. $35x-21y$ 1b. $11-7x$

Practice Exercises

1. $8x+27$ 3. $10x+3$

Objective 2

Practice Exercises

5. $\frac{7}{9}$

Objective 3

Practice Exercises

7. like 9. unlike

Objective 4

Now Try

2a. $23r$ 2b. $19x$ 2c. $8x^2$

3a. $37k-24$ 3b. $-\frac{5}{4}x+6$

Practice Exercises

11. $2x-14$

Objective 5

Now Try

4. $11 + 10x + 8x + 4x$; $11+22x$

Practice Exercises

13. $6x+12+4x=10x+12$

15. $4(2x-6x)+6(x+9)=-10x+54$

Chapter 2 LINEAR EQUATIONS AND INEQUALITIES IN ONE VARIABLE

2.1 The Addition Property of Equality

Key Terms

1. equivalent equations
2. linear equation
3. solution set

Objective 1

Practice Exercises

1. no
3. yes

Objective 2

Now Try

1. {21}
3. {–19}
5. {13}
6. {–3}

Practice Exercises

5. $\frac{1}{2}$

Objective 3

Now Try

7. {29}
8. {8}

Practice Exercises

7. 7
9. 7.2

2.2 The Multiplication Property of Equality

Key Terms

1. multiplication property of equality
2. addition property of equality

Objective 1

Now Try

1. {14} 3. {5.7} 4. {24}
5. {36} 6. {–3}

Practice Exercises

1. {–17} 3. {6.4}

Objective 2

Now Try

7. {4}

Practice Exercises

5. {–5}

2.3 More on Solving Linear Equations

Key Terms

1. contradiction 2. conditional equation 3. identity

Objective 1

Now Try

1. {–6} 2. {8} 4. $\left\{\frac{15}{2}\right\}$

Practice Exercises

1. $\left\{\frac{5}{2}\right\}$ 3. $\left\{-\frac{1}{5}\right\}$

Objective 2

Now Try

6. {all real numbers} 7. ∅

Practice Exercises

5. infinitely many

Objective 3

Now Try

9. {2} 10. {2}

Practice Exercises

7. {2} 9. {10}

Objective 4

Now Try

11. $67 - t$

Practice Exercises

11. $\frac{17}{p}$

2.4 Applications of Linear Equations

Key Terms

1. supplementary angles
2. complementary angles
3. right angle
4. straight angle
5. consecutive integers

Objective 1

Practice Exercises

1. Read the problem; assign a variable to represent the unknown; write an equation; solve the equation; state the answer; check the answer.

Objective 2

Now Try

2. 16

Practice Exercises

3. $-2(4 - x) = 24$; 16

Objective 3

Now Try

3. 52
6. 8 feet

Practice Exercises

5. $x + (x + 5910) = 34{,}730$; Mt. Rainier: 14,410 ft; Mt. McKinley: 20,320 feet
7. $x + (5 + 3x) + 4x = 29$; Mark: 3 laps; Pablo: 14 laps; Faustino: 12 laps

Objective 4

Now Try

8. –2, 0, 2, 4

Practice Exercises

9. 27, 28

Objective 5

Now Try

10. 66°

Practice Exercises

11. 133°
13. 27°

2.5 Formulas and Additional Applications from Geometry

Key Terms

1. vertical angles
2. formula
3. perimeter
4. area

Objective 1

Now Try

1. $W = 5.5$

Practice Exercises

1. $a = 36$
3. $h = 12$

Objective 2

Now Try

2. 12 ft
3. 15 ft, 20 ft, 30 ft

Practice Exercises

5. 1.5 years

Objective 3

Now Try

5. 54°, 126°

Practice Exercises

7. 35°, 35°
9. 129°, 51°

Objective 4

Now Try

6. $r = \frac{d}{t}$
7. $a = P - b - c$
8. $h = \frac{2A}{b + B}$

Practice Exercises

11. $n = \frac{S}{180} + 2$ or $n = \frac{S + 360}{180}$

2.6 Ratio, Proportion, and Percent

Key Terms

1. proportion
2. ratio
3. terms
4. cross products

Objective 1

Now Try

1a. $\frac{11}{17}$
1b. $\frac{4}{15}$
2. 24-ounce jar, \$0.054 per oz

Practice Exercises

1. $\frac{8}{3}$
3. 45-count box

Objective 2

Now Try

3. True

5. $\left\{-\frac{3}{4}\right\}$

Practice Exercises

5. $\left\{\frac{10}{3}\right\}$

Objective 3

Now Try

6. $259.20

Practice Exercises

7. 15 inches

9. $135

Objective 4

Now Try

7a. 140

7b. 950

7c. 85%

8. 87 male students

Practice Exercises

11. 2%

2.7 Further Applications of Linear Equations

Objective 1

Now Try

1. $117

Practice Exercises

1. 1100 students

3. 15,000 students

Objective 2

Now Try

2. 8 oz

Practice Exercises

5. 60 lb

Objective 3

Now Try

4. 5%: $1600; 7%: $3500

Practice Exercises

7. 7%: $800; 9%: $300

Objective 4

Now Try

5. 32 quarters; 18 nickels

Practice Exercises

9. 25 large jars; 55 small jars

11. 1500 general admission; 750 reserved

Objective 5

Now Try

6. 270 miles

7. 5 hours

Practice Exercises

13. 450 miles

2.8 Solving Linear Inequalities

Key Terms

1. three-part inequality
2. interval
3. linear inequality
4. inequalities
5. interval notation

Objective 1

Now Try

1.

-5 -4 -3 -2 -1 0 1 2 3 4 5

Practice Exercises

1. $(3, \infty)$;

-5 -4 -3 -2 -1 0 1 2 3 4 5

3. $(-\infty, -4)$;

-5 -4 -3 -2 -1 0 1 2 3 4 5

Objective 2

Now Try

2. $[-3, \infty)$

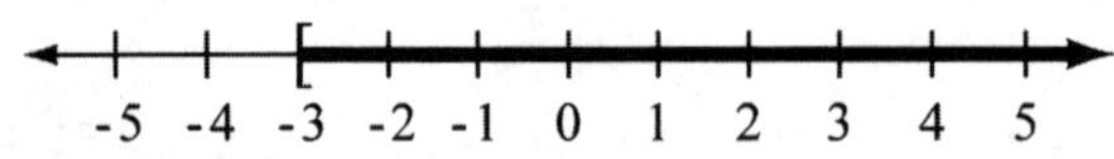

Practice Exercises

5. $(2, \infty)$;

-5 -4 -3 -2 -1 0 1 2 3 4 5

Objective 3

Now Try

3a. $(-\infty, -5]$

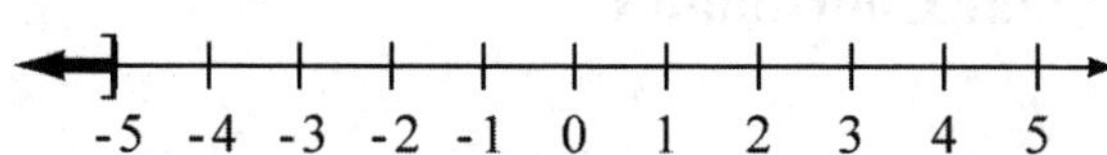

3b. $(-\infty, -4)$

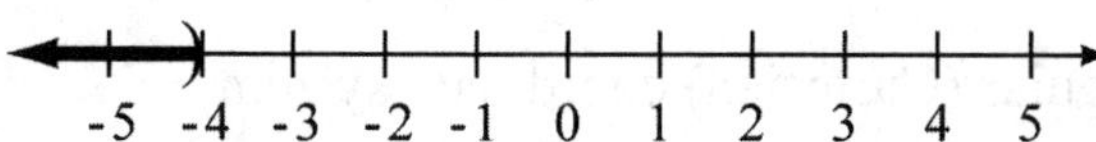

Practice Exercises

7. $(-2, \infty)$;

-5 -4 -3 -2 -1 0 1 2 3 4 5

9. $(-\infty, 4]$

-5 -4 -3 -2 -1 0 1 2 3 4 5

Objective 4

Now Try

4. $[2, \infty)$

-5 -4 -3 -2 -1 0 1 2 3 4 5

Practice Exercises

11. $(-\infty, 5]$;

-5 -4 -3 -2 -1 0 1 2 3 4 5

Objective 5

Now Try

7. 10 feet

Practice Exercises

13. 89

15. all numbers greater than 5

Objective 6

Now Try

9. $[2, 4)$

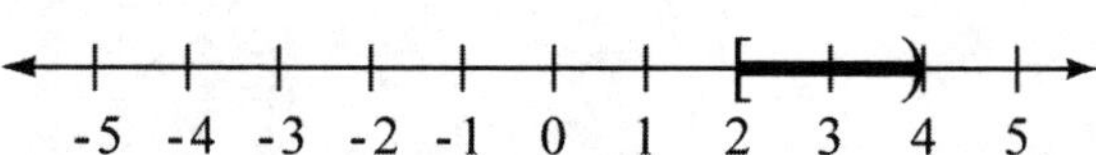

Practice Exercises

17. $[-5, -3)$;

-5 -4 -3 -2 -1 0 1 2 3 4 5

Chapter 3 LINEAR EQUATIONS AND INEQUALITIES IN TWO VARIABLES; FUNCTIONS

3.1 Linear Equations and Rectangular Coordinates

Key Terms

1. line graph
2. linear equation in two variables
3. coordinates
4. x-axis
5. y-axis
6. ordered pair
7. rectangular (Cartesian) coordinate system
8. quadrants
9. origin
10. Plot
11. scatter diagram
12. table of values
13. plane

Objective 1

Now Try

1a. 2000-2001, 2002-2003, 2004-2005

1b. 300 degrees

Practice Exercises

1. 2001-2002, 2003-2004

Objective 2

Practice Exercises

3. (4, 7)

5. (0.2, 0.3)

Objective 3

Now Try

2a. yes

2b. no

Practice Exercises

7. not a solution

Objective 4

Now Try

3. (5, 13)

Practice Exercises

9. (a) (2, −1); (b) (0, −5); (c) (4, 3); (d) (−1, −7); (e) (7, 9)

Objective 5

Now Try

4.

x	y
1	−4
5	12
2	0
3	4

(1, −4), (5, 12), (2, 0), (3, 4)

Practice Exercises

11. $(1,-3)$, $(1, 0)$, $(1, 5)$

13. $(0, 4)$, $(3, 0)$, $\left(\frac{15}{4},-1\right)$

Objective 6

Now Try

5.

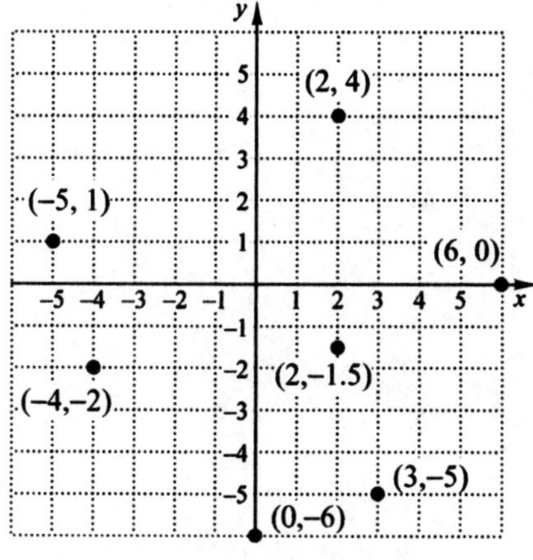

Practice Exercises

15.

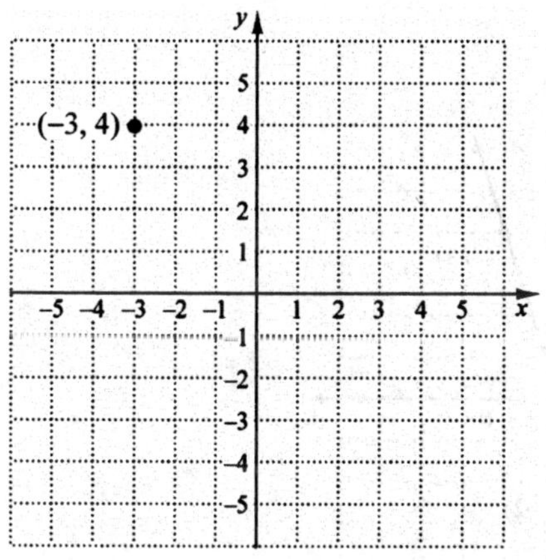

3.2 Graphing Linear Equations in Two Variables

Key Terms

1. *y*-intercept
2. *x*-intercept
3. graphing
4. graph

Objective 1

Now Try

2.

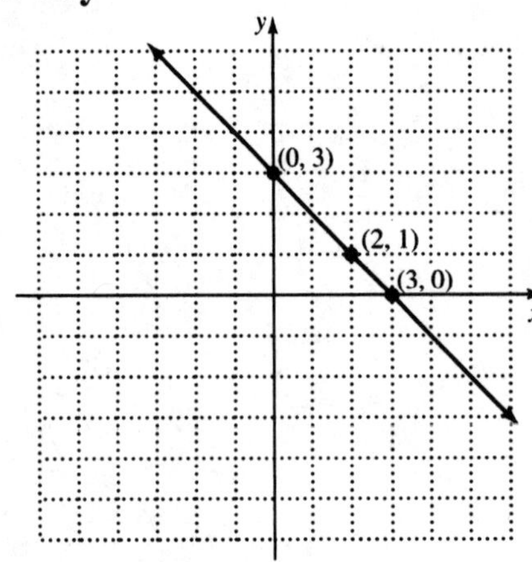

Practice Exercises

1. $(0,-2), \left(\frac{2}{3},0\right), (2,4)$

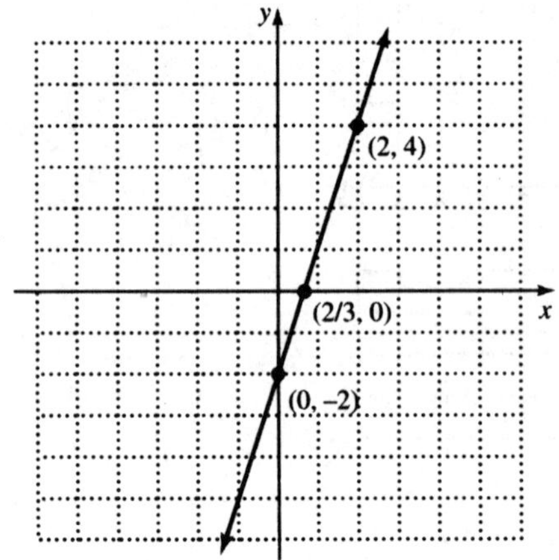

3. $\left(0,-\frac{1}{2}\right), (1,0), (-3,-2)$

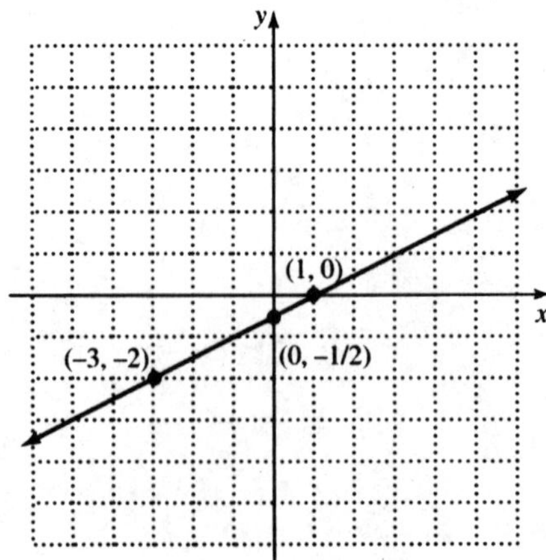

Objective 2

Now Try

3.

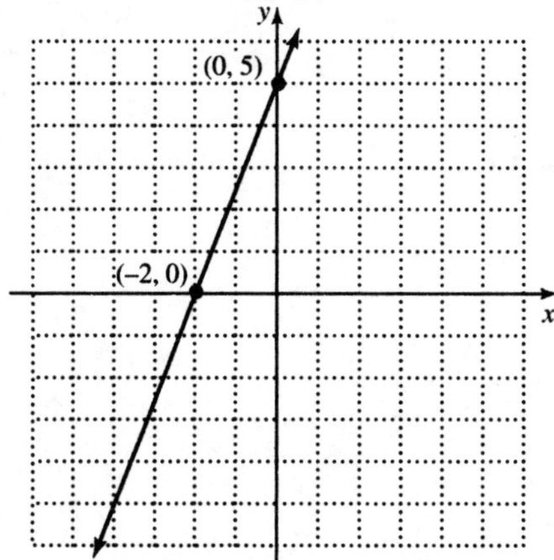

4.

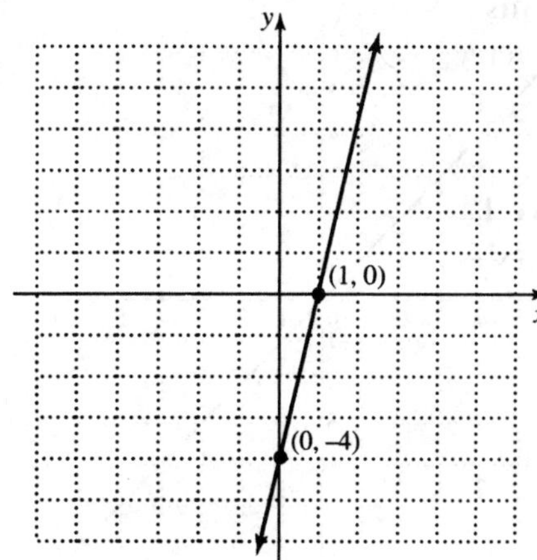

Practice Exercises

5.

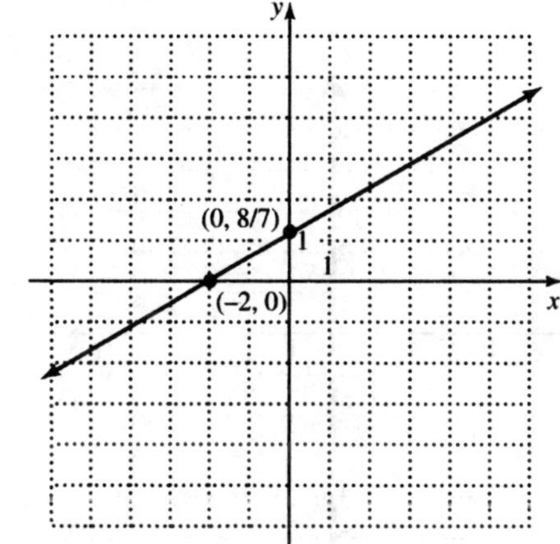

Objective 3

Now Try

5.

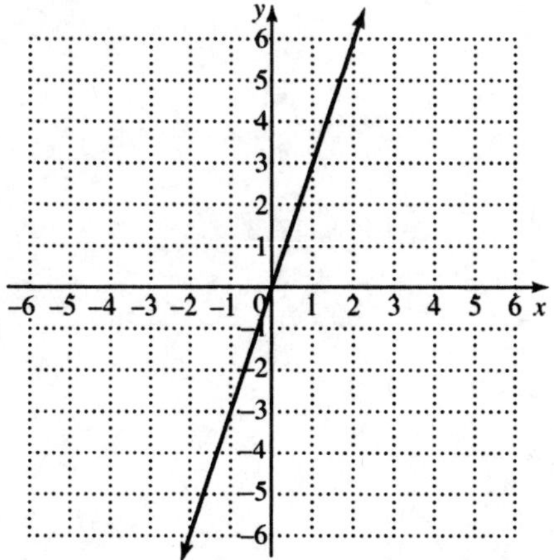

Practice Exercises

7. $x + y = 0$

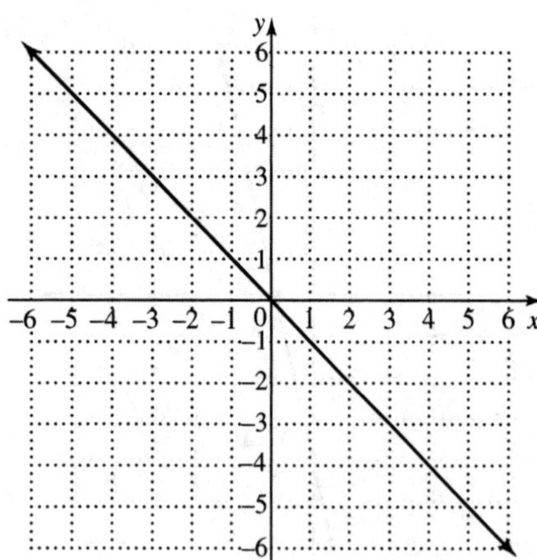

Objective 4

Now Try

6.

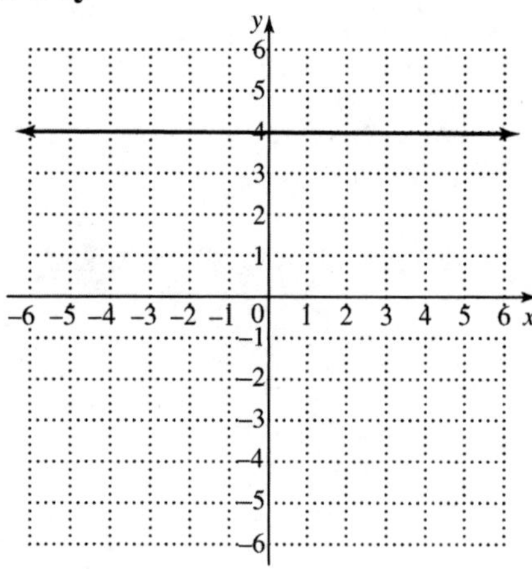

7.

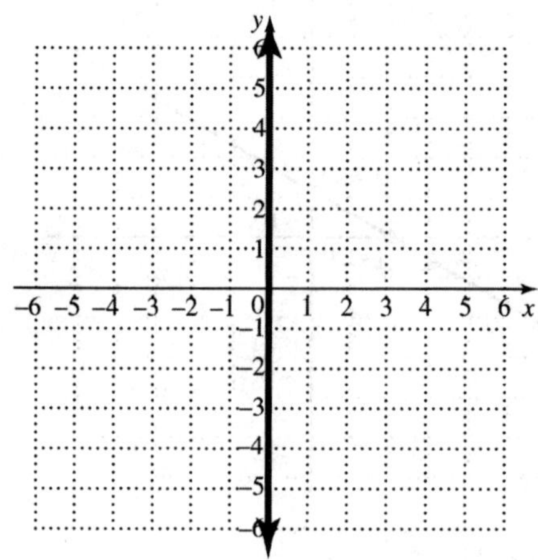

Practice Exercises

9. $x - 1 = 0$

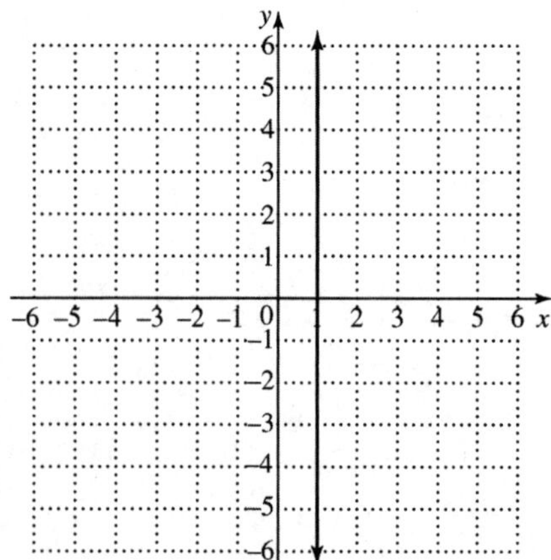

Objective 5

Now Try

8a. 0 calculators, $45
5000 calculators, $42
20,000 calculators, $33
45,000 calculators, $18

8b. (0, $45), (5, $42), (20, $33), (45, $18)

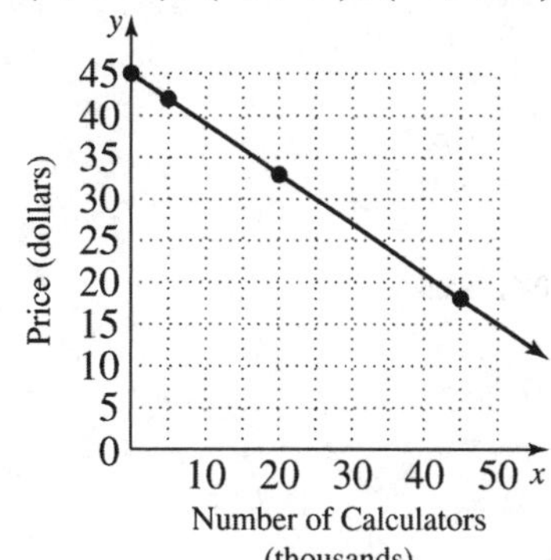

8c. 30,000 calculators, $27

Practice Exercises

11. 2004, 4.9 million;
2005, 5.53 million;
2006, 6.16 million;
2007, 6.79 million

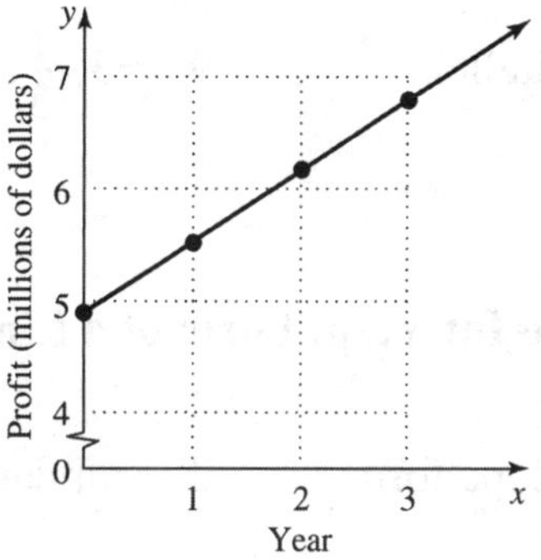

13. 48, 20°C;
54, 22°C;
60, 24°C;
66, 26°C

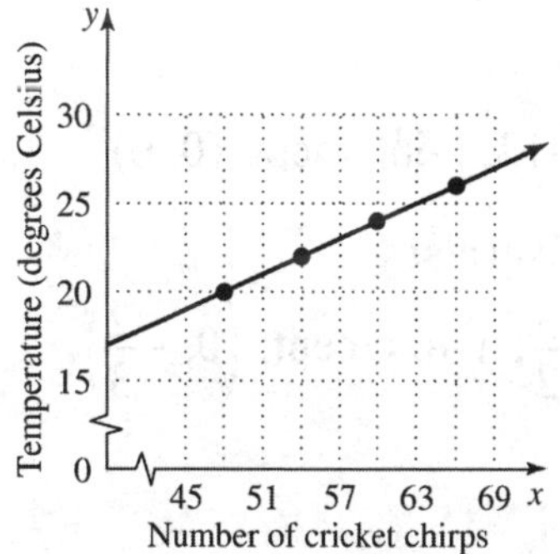

3.3 The Slope of a Line

Key Terms

1. perpendicular lines
2. slope
3. rise
4. parallel lines
5. run

Objective 1

Now Try

1. $\frac{4}{1}$, or 4
2. $-\frac{16}{9}$
3. 0
4. undefined slope

Practice Exercises

1. -2
3. 0

Objective 2

Now Try

5. $\frac{7}{4}$

Practice Exercises

5. $\frac{2}{3}$

Objective 3

Now Try

6a. parallel

6b. perpendicular

Practice Exercises

7. 1; 1; parallel
9. -3; $\frac{1}{3}$; perpendicular

3.4 Slope-Intercept Form of a Linear Equation

Key Terms

1. point-slope form
2. standard form
3. slope-intercept form

Objective 1

Now Try

1a. slope: -12; y-intercept: $(0, 6)$

1b. slope: $-\frac{1}{7}$; y-intercept: $\left(0, -\frac{7}{5}\right)$

Practice Exercises

1. slope: $\frac{3}{2}$; y-intercept: $\left(0, -\frac{2}{3}\right)$

Objective 2

Now Try

2.

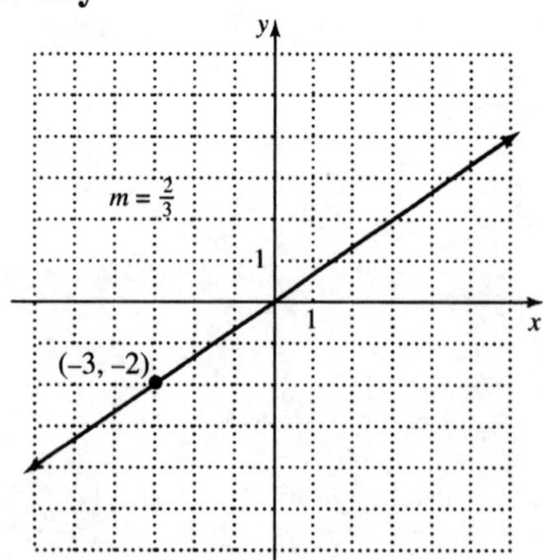

Practice Exercises

3.

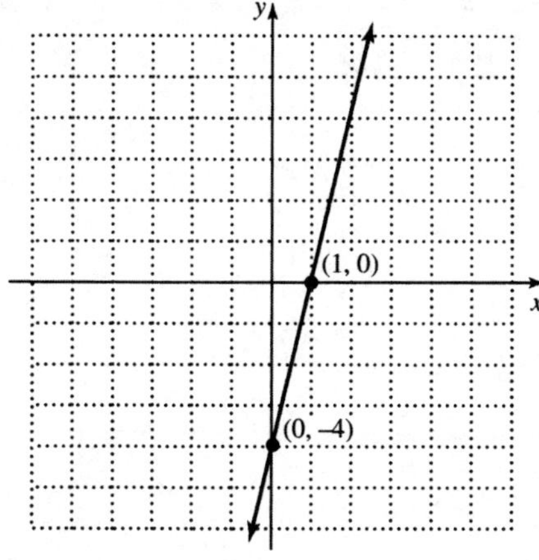

5.

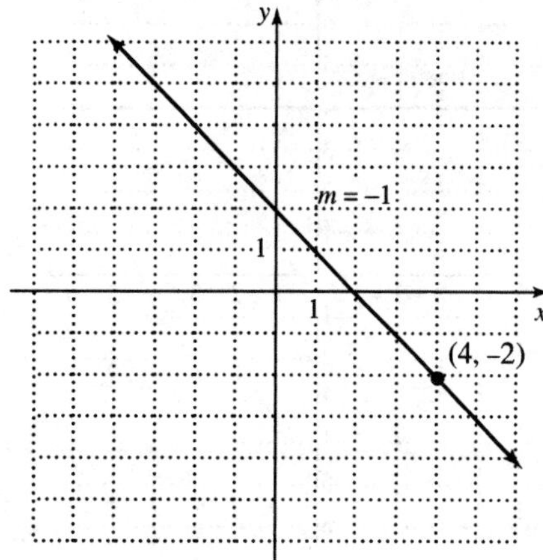

Objective 3

Now Try

4a. $y=\frac{3}{4}x-5$

4b. $y=6x+10$

Practice Exercises

7. $y=-2x+12$

Objective 4

Now Try

5a.

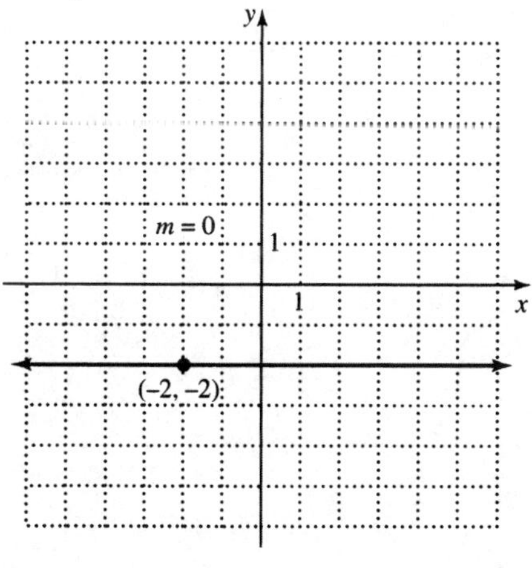

5b.

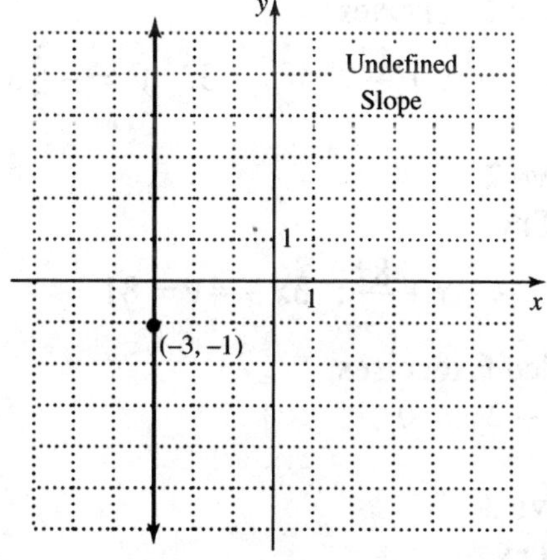

6a. $y=5$

6b. $x=-5$

Practice Exercises

9.

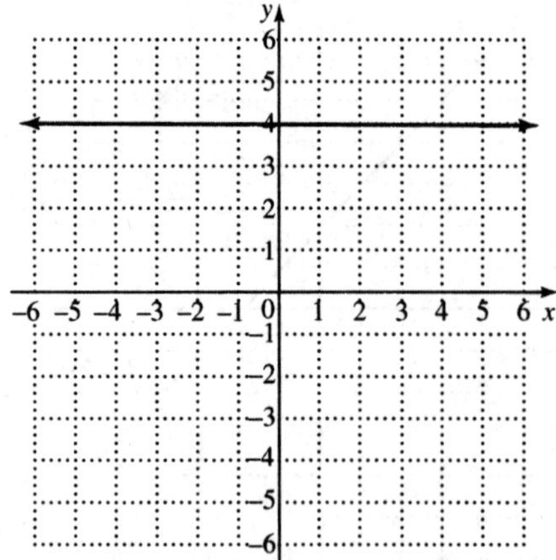

11. $x=-3$

3.5 Point-Slope Form of a Linear Equation and Modeling

Key Terms

1. standard form
2. slope-intercept form
3. point-slope form

Objective 1

Now Try

1. $y=-\frac{2}{3}x+15$

Practice Exercises

1. $y=-\frac{3}{5}x+\frac{11}{5}$

3. $y=-\frac{3}{2}x+5$

Objective 2

Now Try

2. $y=-\frac{3}{4}x+\frac{81}{4};\ 3x+4y=81$

Practice Exercises

5. $3x-2y=0$

Objective 3

Now Try

3.

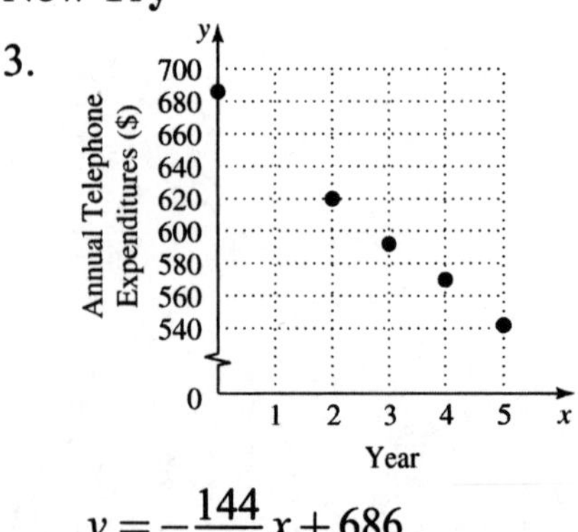

$y=-\frac{144}{5}x+686$

Practice Exercises

7.

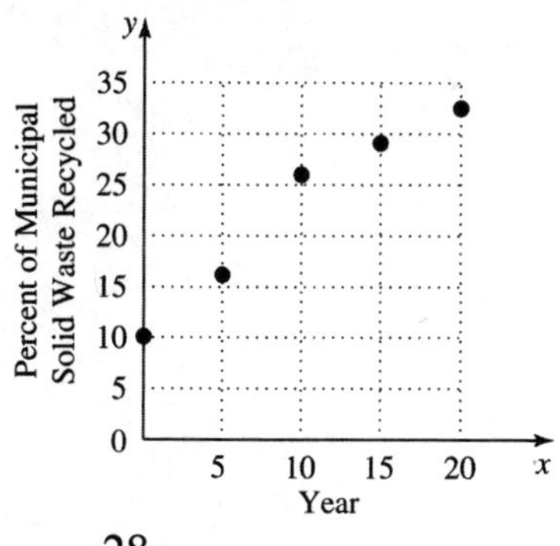

$$y = \frac{28}{25}x + 10.1$$

3.6 Graphing Linear Inequalities in Two Variables

Key Terms

1. boundary line
2. linear inequality in two variables

Objective 1

Now Try

1.

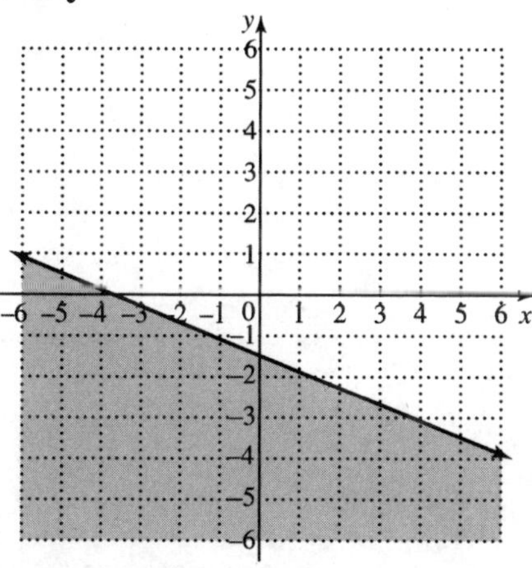

3.

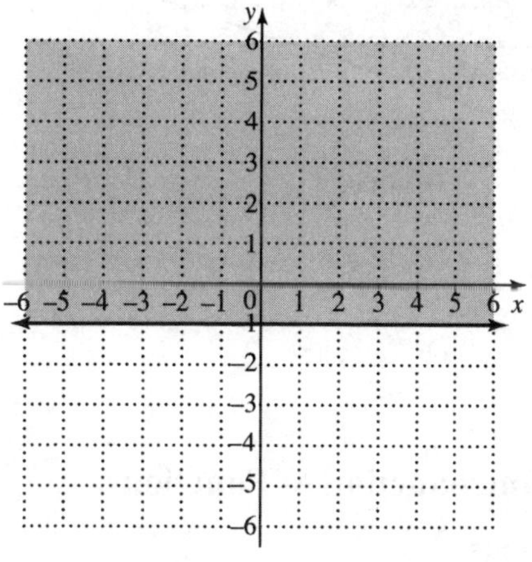

Practice Exercises

1.

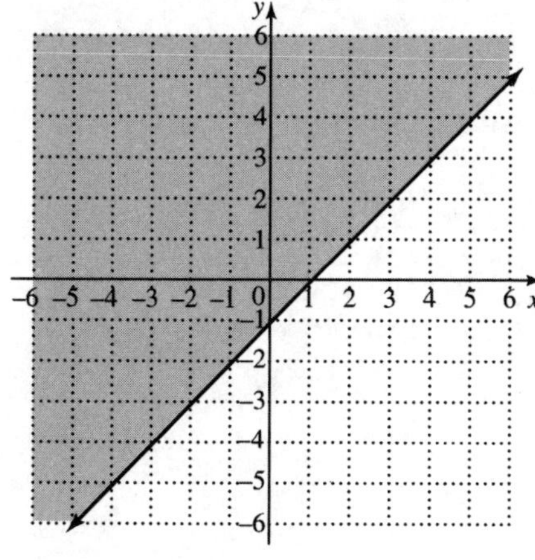

3.

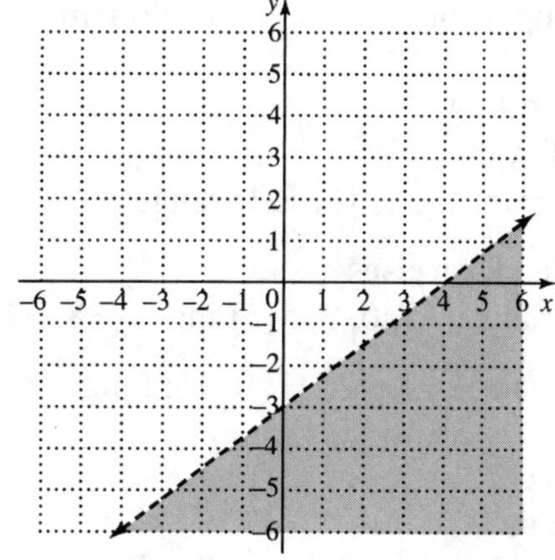

Objective 2

Now Try

4.

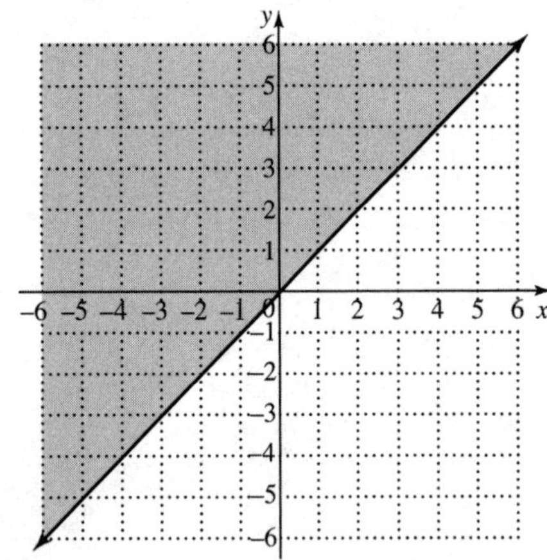

Practice Exercises

5.

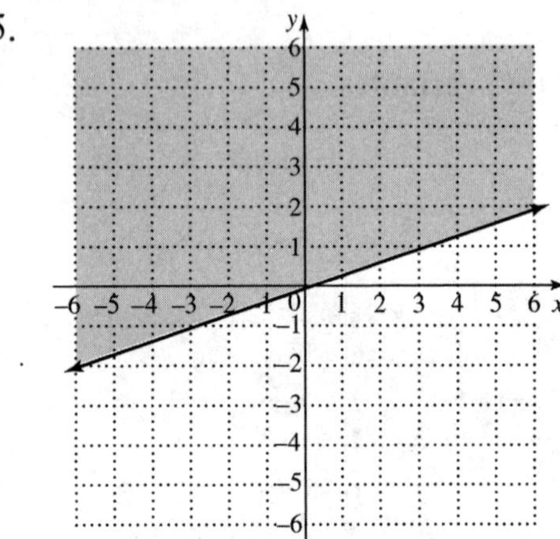

3.7 Introduction to Functions

Key Terms

1. relation
2. range
3. function
4. components
5. domain

Objective 1

Now Try

1. domain: {6, 8, 10, 12}; range: {7, 9, 11, 13}

Practice Exercises

1. domain: {−3, 0, 2, 5}; range: {−8, −4, −1, 2, 7}

3. domain: {−3, −2, −1, 0, 1}; range: {−5, 0, 5}

Objective 2

Now Try

2a. not a function

2b. function

Practice Exercises
5. not a function

Objective 3
Now Try
3a. not a function
3b. function
3c. function

Practice Exercises
7. not a function
9. function

Objective 4
Now Try
4. domain: $(-\infty, \infty)$; range: $(-\infty, \infty)$

Practice Exercises
11. domain: $(-\infty, \infty)$; range: $(-\infty, \infty)$

Objective 5
Now Try
5a. 5
5b. –10
5c. –15

Practice Exercises
13. (a) –13; (b) –7; (c) 5
15. (a) 9; (b) 9; (c) 9

Objective 6
Now Try
6. {(2003, 719 million), (2004, 817 million), (2005, 1018 million), (2006, 1093 million), (2007, 1262 million)}; yes

Practice Exercises
17. Domain: {2008, 2009, 2010, 2011, 2012}
Range: {596 thousand, 625 thousand, 795 thousand, 872 thousand}

Chapter 4 SYSTEMS OF LINEAR EQUATIONS AND INEQUALITIES

4.1 Solving Systems of Linear Equations by Graphing

Key Terms
1. independent equations
2. consistent system
3. solution set of the system
4. solution of the system
5. dependent equations
6. inconsistent system
7. system of linear equations

Objective 1
Now Try
1. no

Practice Exercises

1. no
3. no

Objective 2

Now Try

2.

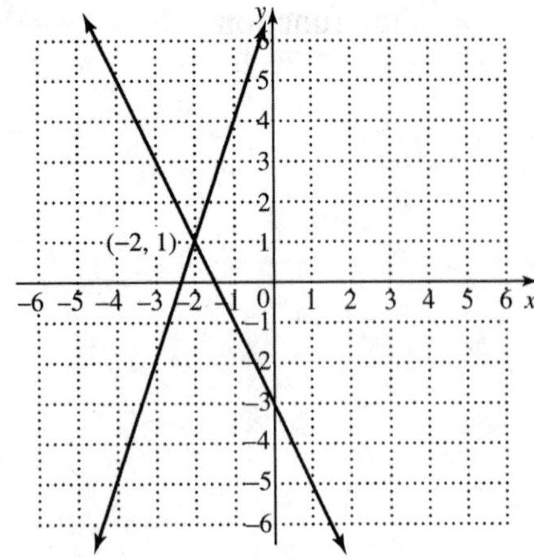

Practice Exercises

5.

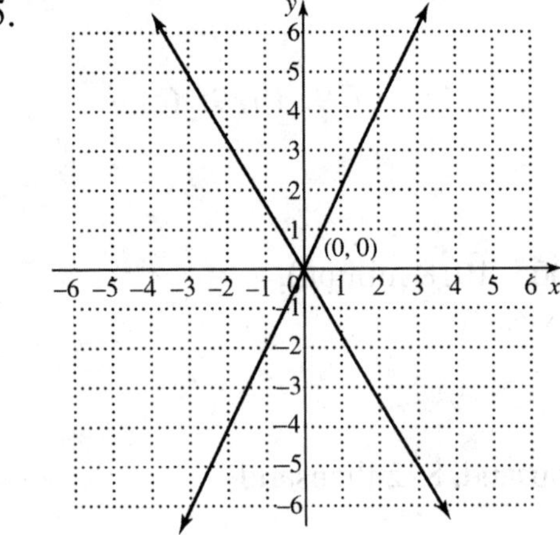

Objective 3

Now Try

3a. ∅

3b. $\{(x, y) | 4x - 2y = 8\}$

Practice Exercises

7. no solution

Objective 4

Now Try

4a. neither

4b. intersecting lines

4c. exactly one solution

Practice Exercises

9. (a) neither (b) intersecting lines (c) one solution

11. (a) dependent (b) one line (c) infinitely many solutions

4.2 Solving Systems of Linear Equations by Substitution

Key Terms

1. ordered pair
2. substitution
3. dependent system
4. inconsistent system

Objective 1

Now Try

1. $\{(1, 6)\}$
2. $\{(9, -4)\}$
3. $\{(8, -2)\}$

Practice Exercises

1. $(2, 4)$
3. $(4, -9)$

Objective 2

Now Try

4. $\varnothing$
5. $\{(x, y)|\ 5x + 4y = 20\}$

Practice Exercises

5. $\{(x, y)\,|\,x - 2y = -6\}$

Objective 3

Now Try

6. $\{(-2, 5)\}$
7. $\{(3, -5)\}$

Practice Exercises

7. $(-9, -11)$
9. $\{(x, y)\,|\,0.3x + 0.4y = 0.5\}$

4.3 Solving Systems of Linear Equations by Elimination

Key Terms

1. elimination method
2. addition property of equality
3. substitution

Objective 1

Now Try

1. $\{(8, 3)\}$

Practice Exercises

1. $(8, 3)$
3. $(5, 0)$

Objective 2

Now Try

3. $\{(-4, 9)\}$

Practice Exercises

5. $\left(\frac{1}{2}, 1\right)$

Objective 3

Now Try

4. $\left\{\left(\frac{9}{17}, \frac{11}{17}\right)\right\}$

Practice Exercises

7. (2, −4)

9. (3, −2)

Objective 4

Now Try

5a. $\{(x, y) \mid 9x - 7y = 5\}$

5b. ∅

Practice Exercises

11. $\{(x, y) \mid -x - 2y = 3\}$

4.4 Applications of Linear Systems

Key Terms

1. $d = rt$

2. system of linear equations

Objective 1

Now Try

1. 56 cm, 26 cm

Practice Exercises

1. 32, 18

3. length: 14 ft; width: 11 ft

Objective 2

Now Try

2. 1500 general admission tickets; 750 reserved seats

Practice Exercises

5. 30 $5 bills; 60 $10 bills

Objective 3

Now Try

3. water: 9 oz; 80% solution: 3 oz

Practice Exercises

7. $6 coffee: 100 lbs; $12 coffee: 50 lb

9. $1.60 candy: 20 lb; $2.50 candy: 10 lb

Objective 4

Now Try

4. Enid: 44 mph; Jerry: 16 mph

Practice Exercises

11. plane A: 400 mph; plane B: 360 mph

4.5 Solving Systems of Linear Inequalities

Key Terms

1. solution set of a system of linear inequalities
2. system of linear inequalities

Objective 1

Now Try

1. 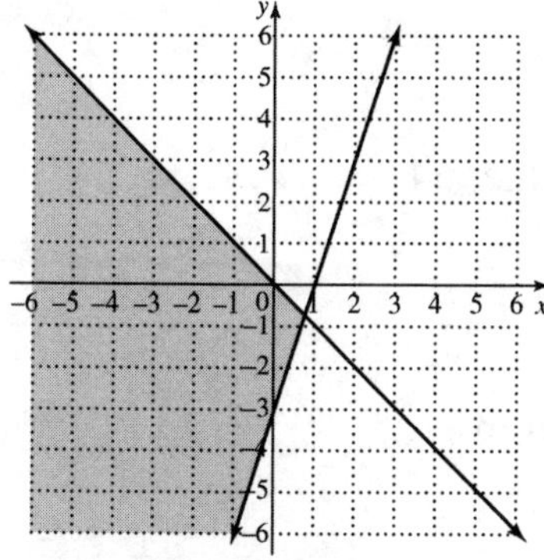

2. 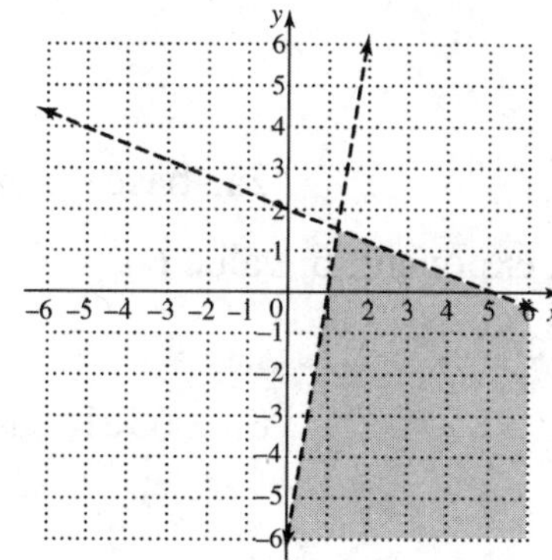

3. 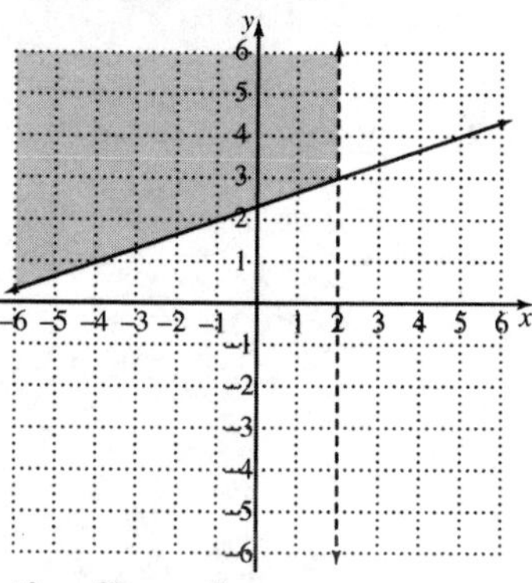

Practice Exercises

1. 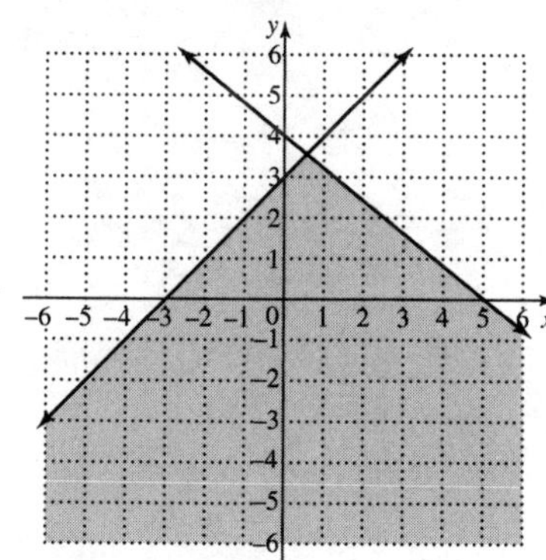

3. 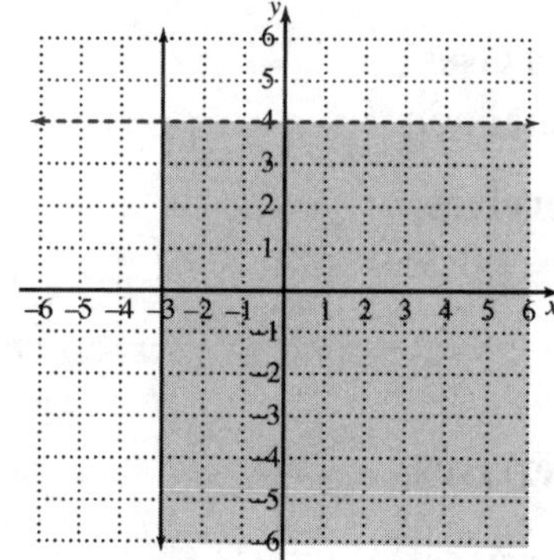

Chapter 5 EXPONENTS AND POLYNOMIALS

5.1 The Product Rule and Power Rules for Exponents

Key Terms

1. power
2. exponential expression
3. base

Objective 1

Now Try

1. 4^5

2a. base: 2; exponent: 6; value: 64

2b. base: –2; exponent: 6; value 64

Practice Exercises

1. $\frac{1}{243}$

3. -6561; base: 3; exponent: 8

Objective 2

Now Try

3a. 9^{13}

3b. m^{27}

3c. $18x^{11}$

3d. 108

Practice Exercises

5. $8c^{15}$

Objective 3

Now Try

4a. 7^8

4b. x^{30}

Practice Exercises

7. 7^{12}

9. $(-3)^{21}$

Objective 4

Now Try

5. $64a^3b^3$

Practice Exercises

11. $-0.008a^{12}b^3$

Objective 5

Now Try

6. $\frac{1}{1024}$

Practice Exercises

13. $-\frac{8x^3}{125}$

15. $-\frac{128a^7}{b^{14}}$

Objective 6

Now Try

7a. $\frac{5^5}{2^3}$, or $\frac{3125}{8}$ 7b. $-x^{27}y^{13}$

Practice Exercises

17. $32a^9b^{14}c^5$

Objective 7

Now Try

8. $28x^5$

Practice Exercises

19. $36x^5$ 21. $28q^{11}$

5.2 Integer Exponents and the Quotient Rule

Key Terms

1. power rule for exponents 2. base; exponent
3. product rule for exponents

Objective 1

Now Try

1a. 1 1b. –1 1c. 1 1d. 0

Practice Exercises

1. –1 3. 0

Objective 2

Now Try

2a. $\frac{1}{27}$ 2b. 25 2c. $\frac{8}{27}$ 2d. $\frac{1}{8}$ 2e. $\frac{1}{p^5}$

3a. $\frac{125}{36}$ 3b. $\frac{y^2}{x^7}$ 3c. $\frac{qr^5}{4p^3}$

Practice Exercises

5. $\frac{1}{m^{18}n^9}$

Objective 3

Now Try

4a. 9 4b. z^{10} 4c. $(a-b)^2$ 4d. $\frac{36b^7}{a^8}$

Practice Exercises

7. $\dfrac{k^4m^5}{2}$ 9. $\dfrac{p^8}{3^5m^3}$ or $\dfrac{p^8}{243m^3}$

Objective 4

Now Try

5a. 6 5b. $3125b^5$ 5c. $\dfrac{243}{32p^{20}}$ 5d. $\dfrac{x^{13}y^2}{343z}$

Practice Exercises

11. $a^{16}b^{22}$

5.3 Scientific Notation

Key Terms

1. scientific notation 2. power rule 3. quotient rule

Objective 1

Now Try

1b. 4.771×10^{10} 1c. 4.63×10^{-2}

Practice Exercises

1. 2.3651×10^4 3. -2.208×10^{-4}

Objective 2

Now Try

2a. 27,960,000 2b. 0.000164

Practice Exercises

5. 0.0064

Objective 3

Now Try

3a. 2.7×10^8, or 270,000,000 3b. 3×10^{-8}, or 0.00000003

4. 9×10^{23} grains of sand

Practice Exercises

7. 2.53×10^2 9. 4.86×10^{19} atoms

5.4 Adding, Subtracting, and Graphing Polynomials

Key Terms

1. degree of a term
2. descending powers
3. term
4. trinomial
5. polynomial
6. monomial
7. degree of a polynomial
8. binomial
9. like terms
10. line of symmetry
11. vertex
12. axis
13. parabola

Objective 1

Now Try

1. 1, 7, –2; three terms

Practice Exercises

1. three terms; 3, –2, 1
3. three terms; 8, –1, –1

Objective 2

Now Try

2. $15m^3+29m^2$

Practice Exercises

5. $2.6z^8-0.9z^7$

Objective 3

Now Try

3a. $3x^3-7x^2+4x$; degree 3; trinomial

3b. $4w^5-3w^2$; degree 5; binomial

Practice Exercises

7. n^8-n^2; degree 8; binomial
9. $5c^5+3c^4-10c^2$; degree 5; trinomial

Objective 4

Now Try

4. 1285

Practice Exercises

11. a. 71; b. −19

Objective 5

Now Try

6a. $4x^3+x+12$

6b. $7x^3+8x^2-14x-1$

7. $-11x^3-7x+1$
9. $7+8x-4x^2$
10. x^2y+4xy

Practice Exercises

13. $3r^3+7r^2-5r-2$
15. $-7x^2y+3xy+5xy^2$

Objective 6
Now Try
11.

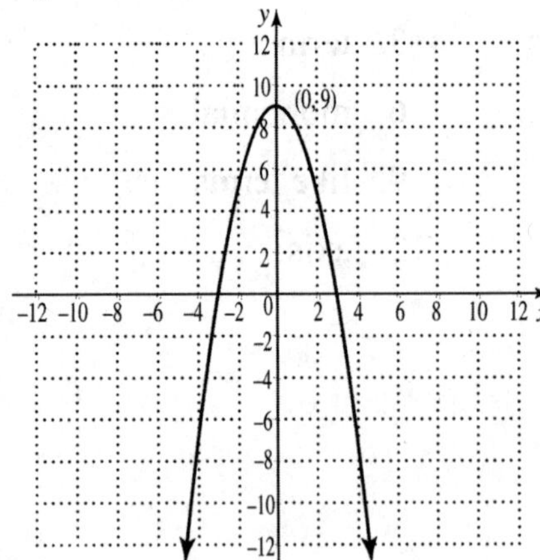

Practice Exercises
17.

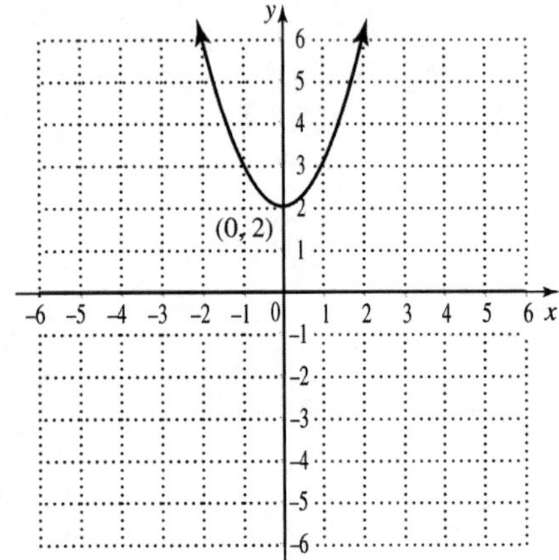

Vertex: (0, 2)

5.5 Multiplying Polynomials

Key Terms

1. inner product
2. FOIL
3. outer product

Objective 1

Now Try

1. $32x^4 + 64x^3$

Practice Exercises

1. $35z^4 + 14z$
3. $-6y^5 - 9y^4 + 12y^3 - 33y^2$

Objective 2

Now Try

2. $4x^7 - 2x^5 + 37x^4 - 18x^2 + 9x$

3. $28x^4 - 33x^3 + 51x^2 + 17x - 15$
4. $2x^5 - 6x^4 + 10x^3 - 29x^2 + 5$

Practice Exercises

5. $6m^5 + 4m^4 - 5m^3 + 2m^2 - 4m$

Objective 3

Now Try

5. $x^2 + 3x - 54$
6. $16xy + 72y - 14x - 63$
7. $15k^2 + 44kn + 32n^2$

Practice Exercises

7. $20a^2 + 11ab - 3b^2$
9. $-6m^2 - mn + 12n^2$

5.6 Special Products

Key Terms

1. binomial
2. conjugate

Objective 1

Now Try

2a. $4a^2 + 36ak + 81k^2$
2b. $9p^2 + p + \frac{1}{36}$

Practice Exercises

1. $49 + 14x + x^2$
3. $16y^2 - 5.6y + 0.49$

Objective 2

Now Try

3a. $x^2 - 81$
3b. $\frac{25}{36} - a^2$
4a. $121x^2 - y^2$
4b. $4p^5 - 144p$

Practice Exercises

5. $64k^2 - 25p^2$

Objective 3

Now Try

5a. $x^3 + 18x^2 + 108x + 216$
5b. $81x^4 - 540x^3 + 1350x^2 - 1500x + 625$

Practice Exercises

7. $a^3 - 9a^2 + 27a - 27$
9. $256s^4 + 768s^3t + 864s^2t^2 + 432st^3 + 81t^4$

5.7 Dividing Polynomials

Key Terms

1. dividend
2. quotient
3. divisor

Objective 1

Now Try

1. $10x^3 - 5x$
2. $3n^2 - 4n - \frac{2}{n}$
3. $\frac{7z^4}{2} - 4z^3 - \frac{5}{z} - \frac{3}{z^2}$
4. $-2a^3b^2 - 4a^2b + 3$

Practice Exercises

1. $2a^3 - 3a$
3. $-13m^2 + 4m - \frac{5}{m^2}$

Objective 2

Now Try

5. $4x + 3$
6. $2x^2 - 2x - 2 + \frac{-5}{5x - 1}$
7. $x^2 + 10x + 100$
8. $3x^2 + 5x + 5 + \frac{8x + 29}{x^2 - 4}$

Practice Exercises

5. $3x^2 - 6x + 2 + \frac{13x - 7}{2x^2 + 3}$

Objective 3

Now Try

10. $L = 2r^2 - r + 5$ units

Practice Exercises

7. $4y^2 + 24y + 100$ units

Chapter 6 FACTORING AND APPLICATIONS

6.1 The Greatest Common Factor; Factoring by Grouping

Key Terms

1. factoring
2. factored form
3. greatest common factor
4. factor

Objective 1

Now Try

1a. 6

1b. 8

1c. 1

2. $6x^4$

Practice Exercises

1. 28
3. $9xy^2$

Objective 2

Now Try

3. $4y^2(5y^2-3y+1)$
5a. $(y+8)(y+4)$
5b. $(z+5)(z^2-11)$

Practice Exercises

5. $(x-2y)(2a+9b)$

Objective 3

Now Try

6a. $(9+t)(4x+1)$
6b. $(x-7)(4x+5y)$
6c. $(x+7)(x^2-2)$
7. $(8x-3y)(7x+4)$

Practice Exercises

7. $(5-y)(3-x)$
9. $(r^2+s^2)(3r-2s)$

6.2 Factoring Trinomials

Key Terms

1. factoring
2. greatest common factor
3. prime polynomial

Objective 1

Now Try

1. $(x+3)(x+8)$
2. $(y-7)(y-5)$
3. $(p+9)(p-3)$
5. prime
6. $(p-7q)(p+2q)$

Practice Exercises

1. prime
3. $(x-11)(x+3)$

Objective 2

Now Try

7. $7x^4(x-5)(x-2)$

Practice Exercises

5. $2ab(a-3b)(a-2b)$

6.3 More on Factoring Trinomials

Key Terms

1. coefficient
2. trinomial
3. inner product
4. FOIL
5. outer product

Objective 1

Now Try

1. $(5x+2)(x+3)$

2a. $(7x-5)(2x+1)$

2b. $(3m-7)(m+2)$

2c. $(5x+3y)(2x-y)$

3. $3x^3(5x-3)(2x+7)$

Practice Exercises

1. $(4b+3)(2b+3)$

3. $(5c-7t)(2c-3t)$

Objective 2

Now Try

5. $(3x+1)(5x+7)$

6. $(4x-1)(5x-2)$

7. $(4x+7)(2x-3)$

8. $(6x-5y)(4x+3y)$

9. $-6a(3a-5)(a-2)$

Practice Exercises

5. $(a+2b)(3a+2b)$

6.4 Special Factoring Techniques

Key Terms

1. difference
2. perfect square trinomial

Objective 1

Now Try

1. $(z+6)(z-6)$

2a. $(2x+9)(2x-9)$

2b. $(5t+7)(5t-7)$

3a. $10(3x+7)(3x-7)$

3b. $(p^2+16)(p+4)(p-4)$

Practice Exercises

1. $(x-7)(x+7)$

3. prime

Objective 2

Now Try

4. $(p+8)^2$

5a. prime

5b. $5x(2x+5)^2$

5c. $(8m+3)^2$

Practice Exercises

5. $(3j+2)^2$

Objective 3

Now Try

6a. $(t-6)(t^2+6t+36)$

6b. $(3k-y)(9k^2+3ky+y^2$

6c. $(3x+7y^2)(9x^2-21xy^2+49y^4)$

Practice Exercises

7. $(2a-5b)(4a^2+10ab+25b^2)$

9. $2n(3m^2+n^2)$

Objective 4

Now Try

7a. $(6x+1)(36x^2-6x+1)$

7b. $6(x+2y)(x^2-2xy+4y^2)$

Practice Exercises

11. $8(a+2b)(a^2-2ab+4b^2)$

6.5 Solving Quadratic Equations Using the Zero-Factor Property

Key Terms

1. standard form
2. double solution
3. quadratic equation

Objective 1

Now Try

1a. $\left\{-12, \frac{7}{4}\right\}$

1b. $\left\{0, \frac{11}{6}\right\}$

3. $\left\{\frac{4}{5}, 3\right\}$

4. $\left\{-\frac{3}{10}, \frac{3}{10}\right\}$

Practice Exercises

1. $\left\{-\frac{5}{2}, 4\right\}$

3. $\left\{-4, \frac{3}{5}\right\}$

Objective 2

Now Try

6a. $\{-5, 0, 5\}$

6b. $\left\{\frac{2}{5}, 2, 9\right\}$

7. $\{-4, 3\}$

Practice Exercises

5. $\{-9, 0, 1\}$

6.6 Applications of Quadratic Equations

Key Terms

1. legs
2. hypotenuse

Objective 1

Now Try

1. width: 3 m, length: 5 m

Practice Exercises

1. width: 8 in., length: 24 in.
3. height: 4 ft, width: 6 ft

Objective 2

Now Try

2. 0, 1, 2, or 5, 6, 7

Practice Exercises

5. 6, 8

Objective 3

Now Try

3. 16 ft

Practice Exercises

7. 45 m, 60 m, 75 m
9. 20 mi

Objective 4

Now Try

4. 1 sec

Practice Exercises

11. 40 items or 110 items

Chapter 7 RATIONAL EXPRESSIONS AND APPLICATIONS

7.1 The Fundamental Property of Rational Expressions

Key Terms

1. rational expression
2. lowest terms

Objective 1

Now Try

1. 14

Practice Exercises

1. a. $-\frac{11}{9}$; b. -4
3. a. $-\frac{1}{6}$; b. $-\frac{7}{2}$

Objective 2

Now Try

2a. $y \neq \frac{1}{7}$

2b. $m \neq 5,\ m \neq -4$

2c. never undefined

Practice Exercises

5. none

Objective 3

Now Try

3. $\frac{3}{k^3}$

4a. $\frac{7}{9}$

4b. $\frac{m+6}{2m+3}$

5. -1

Practice Exercises

7. $\frac{-5b}{8c}$

9. $\frac{9(x+3)}{2}$

Objective 4

Now Try

7. $\frac{-(10x-7)}{4x-3}, \frac{-10x+7}{4x-3}, \frac{10x-7}{-(4x-3)}, \frac{10x-7}{-4x+3}$

Practice Exercises

11. $\frac{-(2p-1)}{-(1-4p)}; \frac{1-2p}{4p-1}; \frac{-(2p-1)}{4p-1}; \frac{1-2p}{-(1-4p)}$

7.2 Multiplying and Dividing Rational Expressions

Key Terms

1. reciprocal
2. rational expression
3. lowest terms

Objective 1

Now Try

1a. $\frac{4}{15}$

1b. $\frac{4}{3x}$

2. $\frac{s^2}{6(r-s)}$

3. $\frac{35}{x}$

Practice Exercises

1. $\frac{10m^3n}{3}$

3. $\frac{x+4}{2x-8}$

Objective 2

Now Try

4a. $\frac{15}{2}$

4b. $\frac{y-2}{6(y+2)}$

6. $\frac{8x}{(x-3)^2}$

7. $\dfrac{-(m-8)}{5m(m+9)}$

Practice Exercises

5. $\dfrac{(m-1)(m+n)}{m(m-n)}$

7.3 Least Common Denominators

Key Terms

1. equivalent expressions
2. least common denominator

Objective 1

Now Try

1. 72
2. $120a^4$
3a. $9w(w-2)$
3b. $(b+4)(b+1)(b-4)^2$
3c. $p-14$ or $14-p$

Practice Exercises

1. $108b^4$
3. $w(w+3)(w-3)(w-2)$

Objective 2

Now Try

4. $\dfrac{65}{30}$
5a. $\dfrac{76}{24c-20}$
5b. $\dfrac{3(z+2)}{z(z-7)(z+2)}$ or $\dfrac{3z+6}{z^3-5z-14z}$

Practice Exercises

5. $30r$

7.4 Adding and Subtracting Rational Expressions

Key Terms

1. greatest common factor
2. least common multiple

Objective 1

Now Try

1a. $\dfrac{4}{5}$
1b. $2x$

Practice Exercises

1. $\dfrac{4}{w^2}$
3. $\dfrac{1}{x-2}$

Objective 2

Now Try

2a. $\frac{137}{315}$

2b. $\frac{46}{63y}$

4. $\frac{7x^2+11x+8}{(x+2)(x+1)(x-4)}$

Practice Exercises

5. $\frac{7z^2-z-6}{(z+2)(z-2)^2}$

Objective 3

Now Try

9. $\frac{8x^2+37x+15}{(x+5)(x-5)^2}$

8. 7

Practice Exercises

7. $\frac{8z}{(z-2)(z+2)}$ or $\frac{8z}{z^2-4}$

9. $\frac{2m^2-m+2}{(m-2)(m+2)^2}$

7.5 Complex Fractions

Key Terms

1. complex fraction
2. LCD

Objective 1

Objective 2

Now Try

1. $\frac{90}{x}$

2. $\frac{bc^2}{a^2}$

3. $\frac{-9x+65}{2x-5}$

Practice Exercises

1. $\frac{7m^2}{2n^3}$

3. $\frac{9s+12}{6s^2+2s}$ or $\frac{3(3s+4)}{2s(3s+1)}$

Objective 3

Now Try

4. $\frac{12}{x}$

5. $\frac{5n-9}{3(6n+5)}$

Practice Exercises

5. $\frac{(x-2)^2}{x(x+2)}$

7.6 Solving Equations with Rational Expressions

Key Terms

1. extraneous solution
2. proposed solution

Objective 1

Now Try

1a. equation; {14}

1b. expression; $\frac{1}{2}x$

Practice Exercises

1. equation; {−2}

3. expression; $\frac{41x}{15}$

Objective 2

Now Try

4. ∅

7. {−4}

Practice Exercises

5. {−4, 16}

Objective 3

Now Try

9a. $b = aq + c$

9b. $y = \frac{x - wz}{w}$, or $\frac{x}{w} - z$

10. $x = \frac{yz}{y + z}$

Practice Exercises

7. $f = \frac{d_0 d_1}{d_0 + d_1}$

9. $q = \frac{2pf - Ab}{Ab}$ or $\frac{2pf}{Ab} - 1$

7.7 Applications of Rational Expressions

Key Terms

1. numerator
2. denominator
3. reciprocal

Objective 1

Now Try

1. 3

Practice Exercises

1. $-\frac{2}{3}$ or 1

3. $\frac{3}{5}$

Objective 2

Now Try

2. 3 miles per hour

Practice Exercises

5. 24 miles per hour

Objective 3

Now Try

3. $\frac{2}{5}$ hour

Practice Exercises

7. $2\frac{2}{5}$ hr

8. 3 hr

7.8 Variation

Key Terms

1. constant of variation
2. direct variation
3. inverse variation

Objective 1

Now Try

1. 168
2. 12.8 cm

Practice Exercises

1. 12
3. 275 mi

Objective 2

Now Try

3. 5
4. 7.2 pounds per square foot

Practice Exercises

5. 2.25

Chapter 8 ROOTS AND RADICALS

8.1 Evaluating Roots

Key Terms

1. radicand
2. perfect square
3. index (order)
4. square root
5. radical expression
6. principal square root
7. irrational number
8. radical
9. cube root

Objective 1

Now Try

1. 9, –9

2a. 13

2b. $-\frac{3}{7}$

3a. 19

3b. n^2+5

Practice Exercises

1. 25, –25

3. $\frac{30}{7}$

Objective 2

Now Try

4a. rational

4b. irrational

4c. not a real number

Practice Exercises

5. not a real number

Objective 3

Now Try

5. –26.038

Practice Exercises

7. 5.657

9. 14.491

Objective 4

Now Try

6. $c = 25$

7. 35 miles

Practice Exercises

11. 9

Objective 5

Now Try

8. $\sqrt{53}$

Practice Exercises

13. $\sqrt{34}$

15. $\sqrt{5}$

Objective 6

Now Try

9a. 7

9b. –5

10a. 6

10b. –5

Practice Exercises

17. 4

8.2 Multiplying, Dividing, and Simplifying Radicals

Key Terms

1. radical
2. perfect cube
3. radicand

Objective 1

Now Try

1a. $\sqrt{91}$ 1b. $\sqrt{42}$ 1c. $\sqrt{15c}$

Practice Exercises

1. $\sqrt{65}$ 3. $\sqrt{21x}$

Objective 2

Now Try

2a. $2\sqrt{3}$ 2b. $7\sqrt{2}$ 2c. $4\sqrt{5}$

3a. $15\sqrt{2}$ 3b. $6\sqrt{2}$

Practice Exercises

5. $11\sqrt{3}$

Objective 3

Now Try

4a. $\frac{13}{3}$ 4b. 9 4c. $\frac{\sqrt{7}}{8}$

5. $3\sqrt{10}$

Practice Exercises

7. $\frac{5}{9}$ 9. $\frac{2}{25}$

Objective 4

Now Try

7a. $7x^4$ 7b. $r^9\sqrt{r}$ 7c. $\frac{\sqrt{11}}{y^2}$

Practice Exercises

11. $4x^2y^2\sqrt{2y}$

Objective 5

Now Try

8a. $3\sqrt[3]{2}$ 8b. $3\sqrt[4]{2}$ 8c. $\frac{2}{7}$

9a. m^3 9b. $5x^2$ 9c. $2a\sqrt[4]{5a}$

9d. $\frac{x^7}{6}$

Practice Exercises

13. $-2\sqrt[5]{2}$

15. $\frac{5}{4}$

8.3 Adding and Subtracting Radicals

Key Terms

1. index
2. unlike radicals
3. like radicals

Objective 1

Now Try

1a. $10\sqrt{21}$

1b. $-6\sqrt{17}$

1c. cannot be combined

Practice Exercises

1. $5\sqrt{2}+\sqrt{3}$

3. $-5\sqrt{5}$

Objective 2

Now Try

2a. $8\sqrt{6}$

2b. $27\sqrt{6}$

2c. $23\sqrt[3]{2}$

Practice Exercises

5. $27\sqrt{2}$

Objective 3

Now Try

3a. $10\sqrt{5}$

3b. $5\sqrt{7k}$

3c. $32x\sqrt{3}$

3d. $11m\sqrt[3]{2}$

Practice Exercises

7. $4\sqrt{35}$

9. $39w\sqrt{6}$

8.4 Rationalizing the Denominator

Key Terms

1. product rule
2. rationalizing the denominator
3. quotient rule

Objective 1

Now Try

1a. $2\sqrt{7}$

1b. $\frac{\sqrt{3}}{3}$

Practice Exercises

1. $\frac{3\sqrt{10}}{2}$ 3. $\frac{3}{5}$

Objective 2

Now Try

2. $\frac{\sqrt{21}}{6}$ 3. $\frac{\sqrt{10}}{15}$ 4. $\frac{5\sqrt{ab}}{b}$

Practice Exercises

5. $\frac{m^2\sqrt{k}}{k^2}$

Objective 3

Now Try

5a. $\frac{\sqrt[3]{45}}{3}$ 5b. $\frac{\sqrt[3]{33}}{3}$ 5c. $\frac{\sqrt[3]{10x^2}}{2x}$

Practice Exercises

7. $\frac{\sqrt[3]{18}}{3}$ 9. $\frac{\sqrt[3]{35x^2}}{7x}$

8.5 More Simplifying and Operations with Radicals

Key Terms

1. rationalize the denominator 2. conjugate

Objective 1

Now Try

1a. $\sqrt{10}$ 1b. $-69-2\sqrt{30}$ 1c. $-2+2\sqrt{21}$

2a. $36-10\sqrt{11}$ 2b. $144-24\sqrt{m}+m$ 3a. 117

3b. $b-20$

Practice Exercises

1. $4\sqrt{14}-63$ 3. -38

Objective 2

Now Try

4a. $\frac{10(7-\sqrt{10})}{39}$ 4b. $\frac{-8\sqrt{6}-21}{3}$ 4c. $\frac{7(5+\sqrt{b})}{25-b}$

Practice Exercises

5. $\frac{\sqrt{3}+2\sqrt{6}+\sqrt{2}+4}{-7}$

Objective 3

Now Try

5. $\dfrac{\sqrt{5}+5}{9}$

Practice Exercises

7. $\dfrac{1+\sqrt{3}}{3}$

9. $27\sqrt{3}+5$

8.6 Solving Equations with Radicals

Key Terms

1. extraneous solution
2. radical equation

Objective 1

Now Try

1. $\{11\}$
2. $\{3\}$

Practice Exercises

1. $\left\{\dfrac{8}{3}\right\}$

3. $\{2\}$

Objective 2

Now Try

3. $\varnothing$
4. $\varnothing$

Practice Exercises

5. $\varnothing$

Objective 3

Now Try

5. $\{7\}$
7. $\{5\}$

Practice Exercises

7. $\{8\}$
9. $\{3\}$

Objective 4

Now Try

8. $\{2\}$

Practice Exercises

11. $\{-4\}$

Chapter 9 QUADRATIC EQUATIONS

9.1 Solving Quadratic Equations by the Square Root Property

Key Terms

1. quadratic equation
2. zero-factor property

Objective 1

Now Try

1a. $\{-7, -1\}$ 1b. $\{-10, 10\}$

Practice Exercises

1. $\{-4, -2\}$ 3. $\{-7, 5\}$

Objective 2

Now Try

2a. $\{-9, 9\}$ 2b. $\{-\sqrt{23}, \sqrt{23}\}$ 2c. $\varnothing$

2d. $\{3\sqrt{2}, -3\sqrt{2}\}$, or $\{\pm 3\sqrt{2}\}$

Practice Exercises

5. $\{-7\sqrt{2}, 7\sqrt{2}\}$

Objective 3

Now Try

3. $\{-6, 16\}$ 4. $\left\{\frac{3 \pm 4\sqrt{2}}{7}\right\}$ 5. $\varnothing$

Practice Exercises

7. $\{-6, 2\}$ 9. $\left\{\frac{1}{5}, \frac{4}{5}\right\}$

Objective 4

Now Try

6. About 16.9 in.

Practice Exercises

11. 2 in.

9.2 Solving Quadratic Equations by Completing the Square

Key Terms

1. perfect square trinomial
2. square root property
3. completing the square

Objective 1

Now Try

1a. 25; $(x+5)^2$

1b. 121; $(x-11)^2$

2. $-5\pm 4\sqrt{2}$

Practice Exercises

1. $\{-4-2\sqrt{3},\ -4+2\sqrt{3}\}$

3. $\{-9, 7\}$

Objective 2

Now Try

7. $\varnothing$

Practice Exercises

5. $\varnothing$

Objective 3

Now Try

8. $\{-3+\sqrt{30}, -3-\sqrt{30}\}$

Practice Exercises

7. $\varnothing$

9. $\{-2-\sqrt{2},\ -2+\sqrt{2}\}$

Objective 4

Now Try

9. 3 sec

Practice Exercises

11. 2 months, 4 months

9.3 Solving Quadratic Equations by the Quadratic Formula

Key Terms

1. quadratic formula
2. double solution
3. discriminant

Objective 1

Now Try

1a. $a = 12, b = -10, c = 7$

1b. $a = -4, b = -110, c = 8$

1c. $a = 15, b = 0, c = -8$

1d. $a = 30, b = 11, c = -39$

Practice Exercises

1. $a = 10,\ b = 4,\ c = 0$

3. $a = 1,\ b = 3,\ c = 9$

Objective 2

Now Try

2. $\left\{-2, \frac{7}{3}\right\}$

3. $\left\{3+\sqrt{5}, 3-\sqrt{5}\right\}$

Practice Exercises

5. $\varnothing$

Objective 3

Now Try

4. $\left\{\frac{11}{2}\right\}$

Practice Exercises

7. $\{2\}$

9. $\left\{\frac{9}{7}\right\}$

Objective 4

Now Try

5. $\left\{\frac{2\pm\sqrt{14}}{5}\right\}$

Practice Exercises

11. $\varnothing$

9.4 Graphing Quadratic Equations

Key Terms

1. line of symmetry
2. vertex
3. axis
4. parabola

Objective 1

Now Try

1.

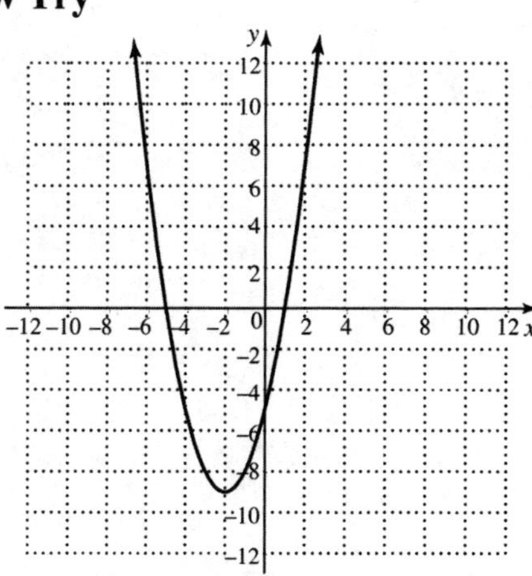

Vertex: (−2, −9)

2.

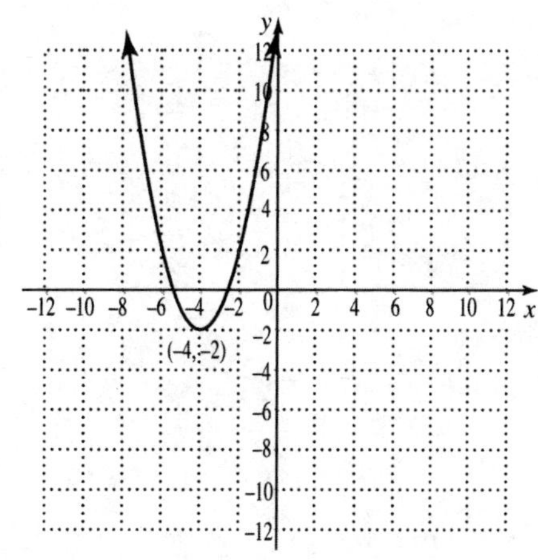

3.

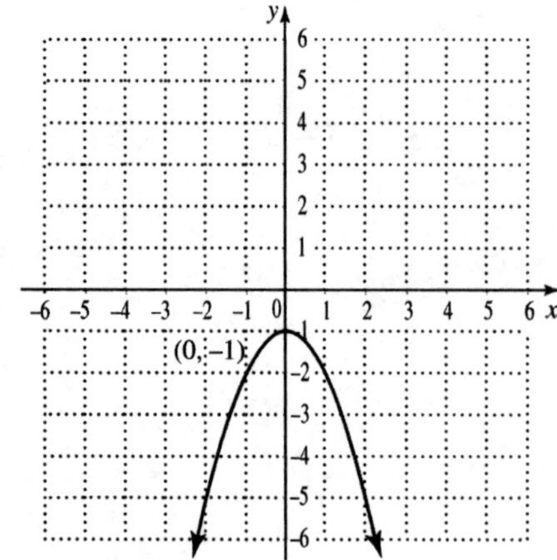

Vertex: (0, –1)
x-intercepts: none
y-intercept: (0, –1)

Practice Exercises

1.

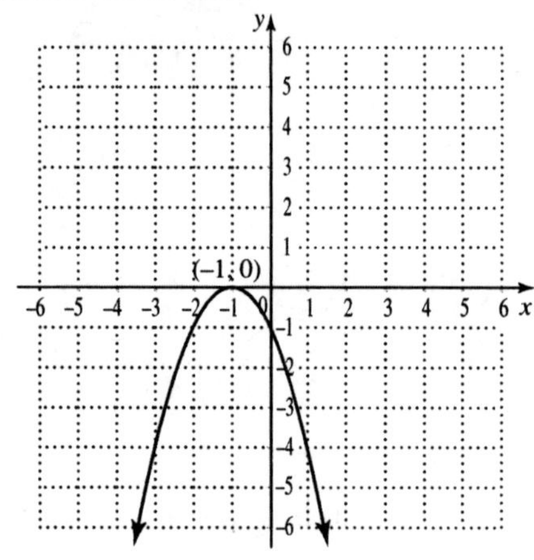

Vertex: (−1, 0)

3.

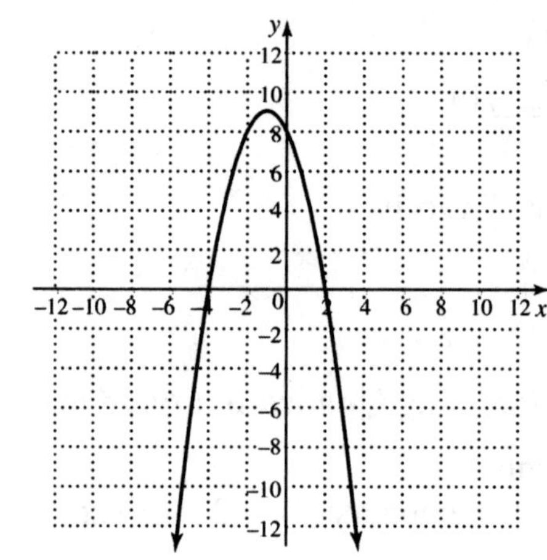

Vertex: (−1, 9)